CONTROL SYSTEM PROBLEMS

FORMULAS, SOLUTIONS, AND SIMULATION TOOLS

CONTROL SYSTEM PROBLEMS

FORMULAS, SOLUTIONS, AND SIMULATION TOOLS

ANASTASIA VELONI
ALEX PALAMIDES

CRC Press
Taylor & Francis Group
Boca Raton London New York

CRC Press is an imprint of the
Taylor & Francis Group, an **informa** business

CRC Press
Taylor & Francis Group
6000 Broken Sound Parkway NW, Suite 300
Boca Raton, FL 33487-2742

First issued in paperback 2019

© 2012 by Taylor & Francis Group, LLC
CRC Press is an imprint of Taylor & Francis Group, an Informa business

No claim to original U.S. Government works

ISBN-13: 978-1-4398-6850-8 (hbk)
ISBN-13: 978-0-367-38205-6 (pbk)

Library of Congress Cataloging-in-Publication Data

Veloni, Anastasia.
    Control system problems : formulas, solutions, and simulation tools / Anastasia Veloni, Alex Palamides.
      p. cm.
    Includes bibliographical references and index.
    ISBN 978-1-4398-6850-8 (hardback)
    1. Automatic control. 2. Automatic control--Simulation methods. I. Palamides, Alex. II. Title.

TJ213.V414 2011
629.8--dc23
                                               2011041488

Visit the Taylor & Francis Web site at
http://www.taylorandfrancis.com

and the CRC Press Web site at
http://www.crcpress.com

To my sister Mary

Anastasia Veloni

To my parents Panos and Nora

Alex Palamides

Contents

Preface

Instead of a dry introduction regarding the usefulness of control systems in our modern society, we prefer to start our preface with the words of two great men: The ancient Greek philosopher Aristotle in *Politics* in the part *On reason, the state, slavery and women* states: "For if every instrument could accomplish its own work, obeying or anticipating the will of others, like the statues of Daedalus, or the tripods of Hephaestus, which", says the poet, "of their own accord entered the assembly of the Gods; if, in like manner, the shuttle would weave and the plectrum touch the lyre without a hand to guide them, chief workmen would not want servants, nor masters slaves." The second is Norbert Wiener, founding thinker of cybernetics theory. In his book, *The Human Use of Human Beings: Cybernetics And Society*, he states: "The world of the future will be an even more demanding struggle against the limitations of our intelligence, not a comfortable hammock in which we can lie down to be waited upon by our robot slaves."

The authors believe that the "struggle against the limitations of our intelligence" demands highly educated scientists and engineers. Automatic control is a multidisciplinary subject covering topics of interest for electrical, mechanical, aerospace, chemical, and industrial engineers. The objective of this text is to provide a comprehensive but practical coverage of the concepts of control systems theory. The theory is written in a straightforward uncomplicated way in order to simplify as much as possible, and at the same time classify the problems met in classical automatic control. Each chapter includes an extensive section with formulas useful for dealing with the numerous solved problems that conclude the chapter. Finally, emphasis is given in the introduction of various simulation tools. The software packages covered are MATLAB®, Simulink®, Comprehensive Control (CC), Simapp, Scilab, and Xcos.

For MATLAB® and Simulink® product information, please contact:

The MathWorks, Inc.
3 Apple Hill Drive
Natick, MA, 01760-2098 USA
Tel: 508-647-7000
Fax: 508-647-7001
E-mail: info@mathworks.com
Web: www.mathworks.com

Acknowledgments

The authors would like to thank the acquiring editor of CRC Press, Nora Konopka, for her valuable help and support and Foteini Tsaglioti for reviewing the language of this text.

Authors

Professor Anastasia Veloni is with the Department of Electronic Computing Systems, Technological Educational Institute of Piraeus, Greece. She is currently head of the Signals and Systems lab at this institute. She has extensive educational experience in the field of automatic control and has authored six textbooks on the areas of signal processing and control systems.

Dr. Alex Palamides is with Otesat-Maritel, Piraeus, Greece. He was with the European Space Agency in Noordwijk, the Netherlands. His research interests lie in the areas of signal processing, dynamical systems, and differential equations. He has authored two other textbooks.

1

Laplace Transform

1.1 Introduction

Laplace transform is a mathematical tool particularly suitable for the study and design of linear time-invariant (LTI) systems. LTI systems are usually described by linear differential equations with constant coefficients. Laplace transform converts a linear differential equation with constant coefficients to an algebraic equation. The time response of a system is calculated by the use of Laplace transform as follows.

First, the system's mathematical models, that is, the differential equations that describe the system in the t-domain, are designated. Next, Laplace transform is applied to each differential equation of the model and the differential equations are converted to algebraic equations. Afterward one has to solve the algebraic equations in the s-domain and apply inverse Laplace transform in order to compute the time response.

Suppose that $f(t)$ is a function of time t, such as $f(t) = 0, t < 0$. The Laplace transform of $f(t)$ is defined as

$$F(s) = L\{f(t)\} = \int_0^\infty f(t)e^{-st}dt \tag{1.1}$$

where s is a complex variable of the form

$$s = \sigma + j\omega \tag{1.2}$$

The inverse Laplace transform of a function $F(s)$ is defined as

$$f(t) = ILT\{F(s)\} = L^{-1}\{F(s)\} = \frac{1}{2\pi j} \int_{c-j\infty}^{c+j\infty} F(s)e^{st}ds \tag{1.3}$$

where c is a real constant, greater than the real parts of the poles of $F(s)$.

1.1.1 Advantages of the Laplace Transform

Describing a system in the complex frequency domain s reveals information about the system, information that is not available when the system is described in the time domain. Moreover, by using graphical methods (Bode diagrams, polar diagrams, etc.), the behavior of a system is studied without having to solve the differential equations that describe it. Finally, as stated earlier, the use of Laplace transform converts a linear differential equation with constant coefficients into an algebraic equation, which can be easily solved.

1.2 Laplace Transform Properties and Theorems

The Laplace transform has many properties and theorems. Some of the most important for the study of automatic control systems are the following:

- **Linearity**: Laplace transform is a linear transform, which means that

$$L\{c_1 f_1(t) + c_2 f_2(t)\} = L\{c_1 f_1(t)\} + L\{c_2 f_2(t)\} = c_1 F_1(s) + c_2 F_2(s), \tag{1.4}$$

  where c_i are constant coefficients.

- **Laplace transform of derivatives**: The Laplace transform of a function's first derivative is given by

$$L\{f^{(1)}(t)\} = sF(s) - f(0^-), \tag{1.5}$$

  where $f^{(1)}(t) = df(t)/dt$. For the nth derivative the following relationship holds:

$$L\{f^{(n)}(t)\} = s^n F(s) - \sum_{k=0}^{n-1} s^k f^{(n-1-k)}(0^-) \tag{1.6}$$

  If all initial conditions are zero, (1.6) becomes

$$L\{f^{(n)}(t)\} = s^n F(s) \tag{1.7}$$

- **Time scaling**: If a function is scaled by a in the time domain, it is scaled by $1/a$ in the s-domain:

$$L\{f(at)\} = \frac{1}{a} F\left(\frac{s}{a}\right), \quad a > 0 \tag{1.8}$$

- **Frequency shifting**: This property is also known as modulation:

$$L\left[e^{-at} f(t)\right] = F(s+a) \tag{1.9}$$

- **Final value theorem**: This theorem is widely used in the study of automatic control systems as it allows the direct calculation of the system's steady-state response. The steady-state response of a system is the final value of a system's response.

$$\lim_{t \to \infty} f(t) = \lim_{s \to 0} sF(s) \tag{1.10}$$

  This relationship holds provided that the left limit exists and the roots of the denominator of $sF(s)$ have negative real parts.

1.3 Inverse Laplace Transform

The inverse Laplace transform is defined in (1.3). It is usually calculated directly by using tables, as solving the integral of (1.3) is a rather difficult task.

The procedure is first to convert a function $F(s)$ into a sum of partial fractions and then find the inverse Laplace transform of the fractions through the given tables.

The expansion into partial fractions is a very useful method for systems analysis and design, as the influence of every characteristic root or eigenvalue is visualized.

In the usual case the Laplace transform of a function is expressed as a rational function of s, that is, it is given as a ratio of two polynomials of s. Consider the complex function

$$F(s) = \frac{B(s)}{A(s)} = \frac{b_m s^m + b_{m-1} s^{m-1} + \cdots + b_1 s + b_0}{s^n + \alpha_{n-1} s^{n-1} + \cdots + \alpha_1 s + \alpha_0} \tag{1.11}$$

The roots of the numerator's polynomial are called **zeros**, while the roots of the of the denominator's polynomial are called **poles**.

In order to compute the inverse Laplace transform of $F(s)$, we must express $F(s)$ as a sum of partial fractions. Depending on the form of its poles, three cases can be distinguished

a. **The case of distinct real poles**
   In this case $F(s)$ is expanded into a sum of fractions of the form

$$F(s) = \frac{B(s)}{(s-p_1)\ldots(s-p_n)} = \frac{c_1}{(s-p_1)} + \cdots + \frac{c_n}{(s-p_n)} \tag{1.12}$$

The coefficients c_i are computed according to the Heaviside formula for distinct poles:

$$c_i = \lim_{s \to p_i} \left\{ (s-p_i) \frac{B(s)}{(s-p_1)(s-p_2)\ldots(s-p_n)} \right\} \tag{1.13}$$

Applying inverse Laplace transform in (1.12) we get

$$f(t) = L^{-1}\{F(s)\} = c_1 e^{p_1 t} + \cdots + c_k e^{p_k t} + \cdots + c_n e^{p_n t} \tag{1.14}$$

b. **The case of multiple real poles**
   Suppose that p_1 is a pole of multiplicity r. In this case $F(s)$ is written in the form

$$F(s) = \frac{B(s)}{(s-p_1)^r \ldots (s-p_n)} = \frac{c_{11}}{(s-p_1)} + \frac{c_{12}}{(s-p_1)^2} + \cdots + \frac{c_{1r}}{(s-p_1)^r} + \frac{c_{r+1}}{(s-p_{r+1})} + \cdots + \frac{c_n}{(s-p_n)} \tag{1.15}$$

The coefficients c_{1j} are calculated according to the Heaviside formula for multiple poles:

$$c_{1j} = \frac{1}{(r-j)!} \lim_{s \to p_i} \left\{ \frac{d^{(r-j)}}{ds^{(n-j)}} (s-p_i)^r F(s) \right\} \tag{1.16}$$

The rest of the coefficients $(c_{r+1} \ldots c_n)$ are calculated according to (1.13).

c. **The case of complex roots**

Suppose that p_1 is a complex root of the denominator and c_1 is the associated coefficient in the partial fraction expansion of $F(s)$. Coefficient c_1 is computed according to (1.13). Of course $\overline{p_1}$ (the complex conjugate of p_1) is also a pole. Moreover, the associated (in the partial fraction expansion) to $\overline{p_1}$ coefficient is $\overline{c_1}$. More specifically $F(s)$ is written as

$$F(s) = \frac{c_1}{s-p_1} + \frac{\overline{c_1}}{s-\overline{p_1}} + \cdots + \frac{c_k}{s-p_k} + \cdots + \frac{c_n}{s-p_n},\tag{1.17}$$

and consequently

$$f(t) = c_1 e^{p_1 t} + \overline{c_1} e^{\overline{p_1} t} + \cdots + c_k e^{p_k t} + \cdots + c_n e^{p_n t}\tag{1.18}$$

1.4 Solving Differential Equations with the Use of Laplace Transform

Consider the general nth-order differential equation with constant coefficients

$$y^{(n)}(t) + a_{n-1}y^{(n-1)}(t) + \cdots + a_0 y(t) = b_m x^{(m)}(t) + b_{m-1}x^{(m-1)} + \cdots + b_0 x(t)\tag{1.19}$$

and initial conditions

$$y(t_0) = y_0, \quad \dot{y}(t_0) = \dot{y}_0, \ldots, y^{(n-1)}(t_0) = y_0^{(n-1)}\tag{1.20}$$

By applying Laplace transform to both parts of the differential equation, we obtain

$$Y(s) = \frac{B(s)}{A(s)} X(s) + \frac{C(s)}{A(s)} = G(s)X(s) + \frac{C(s)}{A(s)}\tag{1.21}$$

where

$$G(s) = \frac{B(s)}{A(s)} = \frac{b_m s^m + \cdots + b_1 s + b_0}{s^n + a_{n-1}s^{n-1} + \cdots + a_1 s + a_0}\tag{1.22}$$

$G(s)$ is called the **transfer function** of the system. In order to solve the differential equation of (1.19) we apply at (1.21) inverse Laplace transform and get

$$y(t) = L^{-1}\left\{\frac{B(s)}{A(s)} X(s)\right\} + L^{-1}\left\{\frac{C(s)}{A(s)}\right\}\tag{1.23}$$

If the input $x(t)$ is zero, the system response $y(t)$ is called **zero-input response**, while if the initial conditions are zero, the system response is called **zero-state response**.

Formulas

TABLE F1.1

Laplace Transform Properties

$f(t)$	$F(s)$
1. $\alpha f_1(t) \pm \beta f_2(t)$	$\alpha F_1(s) \pm \beta F_2(s)$
2. $\dfrac{df(t)}{dt} = f'(t) = \dot{f}(t)$	$sF(s) - f(0)$
3. $\dfrac{d^2 f(t)}{dt^2} = f''(t) = \ddot{f}(t)$	$s^2 F(s) - sf(0) - f'(0)$
4. $\dfrac{d^n f(t)}{dt^n} = f^{(n)}(t)$	$s^n F(s) - s^{n-1} f(0) - \cdots - f^{(n-1)}(0)$
5. $\displaystyle\int_0^t f(t)\,dt$	$\dfrac{F(s)}{s} + \dfrac{f^{-1}(0)}{s}$
6. $\displaystyle\int_0^t \int_0^r f(t)\,dt$	$\dfrac{F(s)}{s^2} + \dfrac{f^{(-1)}(0)}{s^2} + \dfrac{f^{(-2)}(0)}{s}$
7. $\displaystyle\int \cdots \int f(t)(dt)^n$	$\dfrac{F(s)}{s^v} + \displaystyle\sum_{k=1}^{v}\left[\int\cdots\int f(t)(dt)^k\right]_{t=0}$
8. $f(t - t_0)u(t - t_0)$	$e^{-t_0 s} F(s)$
9. $t^n f(t)$	$(-1)^n \dfrac{d^n}{ds^n} F(s)$
10. $\dfrac{f(t)}{t}$	$\displaystyle\int_s^\infty F(\sigma)\,d\sigma$
11. $f_1(t) * f_2(t)$, where $*$ denotes convolution	$F_1(s) \cdot F_2(s)$
12. $f_1(t) f_2(t)$	$\dfrac{1}{2\pi j} \displaystyle\int_{c-j\infty}^{c+j\infty} F_1(p) F_2(s - p)\,dp$
13. $e^{-at} f(t)$	$F(s + a)$
14. $f\left(\dfrac{t}{a}\right)$	$aF(as)$
15. $f(t)\cos\omega t$	$\dfrac{1}{2} F(s - j\omega) + \dfrac{1}{2} F(s + j\omega)$
16. $f(at)$	$\dfrac{1}{a} F\left(\dfrac{s}{a}\right)$
17. $\dfrac{f(t)}{t^n}$	$\displaystyle\int_s^\infty \cdots \int_s^\infty F(s)(ds)^n$
18. $\displaystyle\lim_{t\to 0} f(t)$	$\displaystyle\lim_{s\to\infty} sF(s)$ (initial value theorem)
19. $\displaystyle\lim_{t\to\infty} f(t)$	$\displaystyle\lim_{s\to 0} sF(s)$ (final value theorem)

TABLE F1.2

Laplace Transform Pairs

$F(s) = LT\{f(t)\}$	$f(t) = ILT\{F(s)\}$
1. 1	$\delta(t)$
2. $\dfrac{1}{s}$	$u(t)$
3. $\dfrac{1}{s^2}$	t
4. $\dfrac{1}{s^n}$	$\dfrac{1}{(n-1)!}t^{n-1},$ n positive integer
5. s	$\dfrac{d\delta(t)}{dt} = \delta^{(1)}(t)$
6. s^n	$\delta^{(n)}(t),$ n positive integer
7. $\dfrac{1}{\sqrt{s}}$	$\dfrac{1}{\sqrt{\pi t}}$
8. $\dfrac{1}{s^{n+(1/2)}}$	$\dfrac{2^n t^{n-(1/2)}}{1 \cdot 3 \cdot 5 \cdots (2n-1)\pi^{(1/2)}}$
9. $\dfrac{e^{-as}}{s}$	$u(t - \alpha)$
10. $\dfrac{1}{s}(1 - e^{-as})$	$u(t) - u(t - \alpha)$
11. $\dfrac{1}{s+a}$	e^{-at}
12. $\dfrac{1}{(s+a)^n}$	$\dfrac{1}{(n-1)!}t^{n-1}e^{-at},$ n positive integer
13. $\dfrac{1}{s(s+a)}$	$\dfrac{1 - e^{-at}}{a}$
14. $\dfrac{1}{s(s+\alpha)(s+\beta)}$	$\dfrac{1 - (\beta/(\beta-\alpha))e^{-at} + (a/(\beta-\alpha))e^{-\beta t}}{\alpha\beta}$
15. $\dfrac{s+\gamma}{s(s+\alpha)(s+\beta)}$	$\dfrac{1}{\alpha\beta}\left(\gamma - \dfrac{\beta(\gamma-\alpha)}{\beta-\alpha}e^{-at} + \dfrac{a(\gamma-\beta)}{\beta-\alpha}e^{-\beta t}\right)$
16. $\dfrac{1}{(s+\alpha)(s+\beta)}$	$\dfrac{e^{-at} - e^{-\beta t}}{\beta-\alpha}$
17. $\dfrac{s}{(s+\alpha)(s+\beta)}$	$\dfrac{\alpha e^{-at} - \beta e^{-\beta t}}{\alpha-\beta}$
18. $\dfrac{s+\gamma}{(s+\alpha)(s+\beta)}$	$\dfrac{(\gamma-\alpha)e^{-at} - (\gamma-\beta)e^{-\beta t}}{\beta-\alpha}$
19. $\dfrac{1}{(s+\alpha)(s+\beta)(s+\gamma)}$	$\dfrac{e^{-at}}{(\beta-\alpha)(\gamma-\alpha)} + \dfrac{e^{-\beta t}}{(\gamma-\beta)(\alpha-\beta)} + \dfrac{e^{-\gamma t}}{(\alpha-\gamma)(\beta-\gamma)}$
20. $\dfrac{s+\delta}{(s+\alpha)(s+\beta)(s+\gamma)}$	$\dfrac{(\delta-\alpha)e^{-at}}{(\beta-\alpha)(\gamma-\alpha)} + \dfrac{(\delta-\beta)e^{-\beta t}}{(\gamma-\beta)(\alpha-\beta)} + \dfrac{(\delta-\gamma)e^{-\gamma t}}{(\alpha-\gamma)(\beta-\gamma)}$

TABLE F1.2 (continued)

Laplace Transform Pairs

$F(s) = LT\{f(t)\}$	$f(t) = ILT\{F(s)\}$
21. $\dfrac{\omega}{s^2 + \omega^2}$	$\sin \omega t$
22. $\dfrac{s}{s^2 + \omega^2}$	$\cos \omega t$
23. $\dfrac{s + a}{s^2 + \omega^2}$	$\dfrac{\sqrt{a^2 + \omega^2}}{\omega} \sin(\omega t + \phi)$ where $\phi = \tan^{-1} \dfrac{\omega}{\alpha}$
24. $\dfrac{s \cdot \sin \alpha + \omega \cdot \cos \alpha}{s^2 + \omega^2}$	$\sin(\omega t + \alpha)$
25. $\dfrac{1}{s(s^2 + \omega^2)}$	$\dfrac{1 - \cos \omega t}{\omega^2}$
26. $\dfrac{s + a}{s(s^2 + \omega^2)}$	$\dfrac{a}{\omega^2} - \dfrac{\sqrt{a^2 + \omega^2}}{\omega} \cos(\omega t + \phi)$ where $\phi = \tan^{-1} \dfrac{\omega}{\alpha}$
27. $\dfrac{1}{(s + \alpha)(s^2 + \omega^2)}$	$\dfrac{e^{-\alpha t}}{a^2 + \omega^2} + \dfrac{1}{\omega\sqrt{a^2 + \omega^2}} \sin(\omega t - \phi)$ $\phi = \tan^{-1} \dfrac{\omega}{\alpha}$
28. $\dfrac{1}{(s + \alpha)^2 + \beta^2}$	$\dfrac{e^{-\alpha t} \sin \beta t}{\beta}$
29. $\dfrac{s + a}{(s + \alpha)^2 + \beta^2}$	$e^{-\alpha t} \cos \beta t$
30. $\dfrac{s + \gamma}{(s + \alpha)^2 + \beta^2}$	$\dfrac{\sqrt{(\gamma - \alpha)^2 + \beta^2}}{\beta} e^{-\alpha t} \sin(\beta t + \phi)$ where $\phi = \tan^{-1} \dfrac{\beta}{\gamma - \alpha}$
31. $\dfrac{1}{s[(s + \alpha)^2 + \beta^2]}$	$\dfrac{1}{a^2 + \beta^2} + \dfrac{1}{\beta\sqrt{a^2 + \beta^2}} e^{-\alpha t} \sin(\beta t - \phi)$ where $\phi = \tan^{-1} \dfrac{\beta}{-\alpha}$
32. $\dfrac{s + \gamma}{s[(s + \alpha)^2 + \beta^2]}$	$\dfrac{\gamma}{a^2 + \beta^2} + \dfrac{1}{\beta} \sqrt{\dfrac{(\gamma - \alpha)^2 + \beta^2}{a^2 + \beta^2}} e^{-\alpha t} \sin(\beta t + \phi)$ where $\phi = \tan^{-1} \dfrac{\beta}{\gamma - \alpha} - \tan^{-1} \dfrac{\beta}{-\alpha}$
33. $\dfrac{1}{s^2(s + a)}$	$\dfrac{1}{a^2}(at - 1 + e^{-at})$

(continued)

TABLE F1.2 (continued)

Laplace Transform Pairs

$F(s) = LT\{f(t)\}$	$f(t) = ILT\{F(s)\}$
34. $\dfrac{1}{s(s+a)^2}$	$\dfrac{1}{a^2}(1 - e^{-at} - ate^{-at})$
35. $\dfrac{s+\beta}{s(s+a)^2}$	$\dfrac{1}{a^2}(\beta - \beta e^{-at} + a(\alpha - \beta)te^{-at})$
36. $\dfrac{s^2 + a_1 s + a_0}{s(s+a)(s+\beta)}$	$\dfrac{\alpha_0}{\alpha\beta} + \dfrac{\alpha^2 - \alpha_1\alpha + \alpha_0}{\alpha(\alpha - \beta)}e^{-at} - \dfrac{\beta^2 - \alpha_1\beta + \alpha_0}{\beta(\alpha - \beta)}e^{-\beta t}$
37. $\dfrac{1}{(s^2 + \omega^2)^2}$	$\dfrac{1}{2\omega^3}(\sin\omega t - \omega t\cos\omega t)$
38. $\dfrac{1}{s^2 - \omega^2}$	$\dfrac{1}{\omega}\sinh\omega t$
39. $\dfrac{1}{(s+\alpha)^2(s+\beta)}$	$\dfrac{1}{(\beta-\alpha)^2}[[(\beta-\alpha)t-1]e^{-at} + e^{-\beta t}]$
40. $\dfrac{1}{s^2(s^2 + \omega^2)}$	$\dfrac{1}{\omega^3}(\omega t - \sin\omega t)$
41. $\dfrac{s}{(s^2 + \omega^2)^2}$	$\dfrac{t}{2\omega}\sin\omega t$
42. $\dfrac{s^2}{(s^2 + \omega^2)^2}$	$\dfrac{1}{2\omega}(\omega t\cos\omega t + \sin\omega t)$
43. $\dfrac{s^n}{(s^2 + \omega^2)^{n+1}}$	$\dfrac{t^n \sin\omega t}{n!2^n\,\omega}$
44. $\dfrac{s^2 - \omega^2}{(s^2 + \omega^2)^2}$	$t\cos\omega t$
45. $\dfrac{s}{(s^2 + a^2)(s^2 + \beta^2)}$	$\dfrac{\cos\alpha t - \cos\beta t}{\beta^2 - \alpha^2}$
46. $\dfrac{s^2}{(s+a)(s+\beta)(s+\gamma)}$	$\dfrac{\alpha^2 e^{-at}}{(\beta-\alpha)(\gamma-\alpha)} + \dfrac{\beta^2 e^{-\beta t}}{(\alpha-\beta)(\gamma-\beta)} + \dfrac{\gamma^2 e^{-\gamma t}}{(\alpha-\gamma)(\beta-\gamma)}$
47. $\dfrac{s^2 + \alpha_1 s + a_0}{s^2(s+a)(s+\beta)}$	$\dfrac{a_1 + a_0 t}{a\beta} - \dfrac{a_0(\alpha+\beta)}{(a\beta)^2} - \dfrac{1}{a-\beta}\left(1 - \dfrac{a_1}{a} + \dfrac{a_0}{a^2}\right)e^{-at} - \dfrac{1}{a-\beta}\left(1 - \dfrac{a_1}{\beta} + \dfrac{a_0}{\beta^2}\right)e^{-\beta t}$
48. $\dfrac{s+\gamma}{s^2[(s+a)^2 + \beta^2]}$	$\dfrac{1}{a^2 + \beta^2}\left(\gamma t + 1 - \dfrac{2\gamma\alpha}{a^2 + \beta^2}\right) + \dfrac{\sqrt{\beta^2 + (\gamma-\alpha)^2}}{\beta(a^2 + \beta^2)}e^{-at}\sin(\beta t + \phi)$ where $\phi = 2\tan^{-1}\dfrac{\beta}{\alpha} + \tan^{-1}\dfrac{\beta}{\gamma-\alpha}$
49. $\dfrac{1}{(s+\gamma)[(s+a)^2 + \beta^2]}$	$\dfrac{e^{-\gamma t}}{(\gamma-\alpha)^2 + \beta^2} + \dfrac{e^{-at}\sin(\beta t - \phi)}{\beta\sqrt{(\gamma-\alpha)^2 + \beta^2}}$ where $\phi = \tan^{-1}\dfrac{\beta}{\gamma-\alpha}$
50. $\dfrac{1}{s^2(s+a)^2}$	$\dfrac{1}{a^2}t(1 + e^{-at}) - \dfrac{2}{a^3}(1 - e^{-at})$

TABLE F1.3

Expansion of $F(s)$ into Partial Fractions

1. $$F(s) = \frac{B(s)}{A(s)} = \frac{b_m s^m + b_{m-1} s^{m-1} + \cdots + b_1 s + b_0}{a_n s^n + a_{n-1} s^{n-1} + \cdots + a_1 s + a_0}$$

Case : $m < n$

Distinct poles $$F(s) = \frac{B(s)}{A(s)} = \frac{(s+z_1)(s+z_2)\cdots(s+z_m)}{(s+p_1)(s+p_2)\cdots(s+p_n)} = \frac{k_1}{s+p_1} + \frac{k_2}{s+p_2} + \cdots + \frac{k_n}{s+p_n}$$

where $k_i = \lim\limits_{s \to -p_i} F(s) \cdot (s+p_i)$: (Heaviside's formula for distinct poles)

$p_1, p_2, \ldots, p_n$ and $z_1, z_2, \ldots, z_n$ are either real or complex numbers

Multiple poles $$F(s) = \frac{B(s)}{A(s)} = \frac{(s+z_1)(s+z_2)\cdots(s+z_m)}{(s+p_1)^r (s+p_2)\cdots(s+p_n)}$$

$$= \frac{k_{11}}{s+p_1} + \frac{k_{12}}{(s+p_1)^2} + \cdots + \frac{k_{1r}}{(s+p_1)^r} + \frac{k_2}{s+p_2} + \cdots + \frac{k_n}{s+p_n}$$

$$= \frac{k_{11}}{s+p_1} + \frac{k_{12}}{(s+p_1)^2} + \cdots + \frac{k_{1r}}{(s+p_1)^r} + \frac{k_2}{s+p_2} + \cdots + \frac{k_n}{s+p_n}$$

where $k_{ij} = \dfrac{1}{(r-j)!} \lim\limits_{s \to -p_i} \left[\dfrac{d^{(r-j)}}{ds^{(r-j)}} \left(F(s) \cdot (s+p_i)^r \right) \right]$ (Heaviside's formula for multiple poles)

and $k_n = \lim\limits_{s \to -p_n} F(s) \cdot (s+p_n)$

2. $$F(s) = \frac{B(s)}{A(s)} = \frac{b_m s^m + b_{m-1} s^{m-1} + \cdots + b_1 s + b_0}{a_n s^n + a_{n-1} s^{n-1} + \cdots + a_1 s + a_0}$$

Case : $m > n$

If the degree of the polynomial of the numerator is greater than that of the denominator, then the division of $B(s)/A(s)$ is performed. This gives a polynomial of s, plus the fraction of the remainder divided by the denominator. The problem is then treated as described earlier.

TABLE F1.4

Passive Linear Elements of Electrical Circuits and Their Laplace Transform

Name	*t*-Domain	*s*-Domain
Electrical resistance	$i(t)$ R $A \bullet \longrightarrow \!\!\!\!\!\text{—WW—} \bullet B$ $u_R(t) = Ri(t)$	$I(s)$ R $A \bullet \longrightarrow \!\!\!\!\!\text{—WW—} \bullet B$ $V_R(s) = RI(s)$
Coil	$i(t)$ L $A \bullet \longrightarrow \!\!\!\!\text{—mm—} \bullet B$ $u_L(t) = L\dfrac{di(t)}{dt}$ $i(t) = \dfrac{1}{L}\displaystyle\int_0^t u_L(t)\,dt$	$I(s)$ sL $-\ +$ $A \bullet \longrightarrow \!\!\!\!\text{—mm—} \bigcirc \bullet B$ $Li(0)$ $V_L(s) = sLI(s) - Li(0)$ $<=>$ $\begin{array}{c} 1/sL \\ A \bullet \text{—} \begin{bmatrix} \longrightarrow \end{bmatrix} \text{—} \bullet B \\ i(0)/s \end{array}$ $I(s) = \dfrac{V_L(s)}{sL} + \dfrac{i(0)}{s}$
Capacitor	$i(t)$ C $A \bullet \longrightarrow \!\!\!\!\text{—}\|\|\text{—} \bullet B$ $i(t) = C\dfrac{du_C(t)}{dt}$ or $u_C(t) = \dfrac{1}{C}\displaystyle\int_0^t i(t)\,dt$	$I(s)$ $1/sC$ $A \bullet \longrightarrow \!\!\!\!\text{—}\|\|\text{—} \bigcirc \bullet B$ $-\ +$ $u_c(0)/s$ $I(s) = sCV_C(s) - Cu_C(0)$ $<=>$ $\begin{array}{c} sC \\ A \bullet \text{—} \begin{bmatrix} \longleftarrow \end{bmatrix} \text{—} \bullet B \\ Cu_c(0) \end{array}$ $V_C(s) = \dfrac{I(s)}{sC} + \dfrac{u_C(0)}{s}$

TABLE F1.5

Complex Numbers

$z = x + jy = \rho e^{j\theta} = \rho^{\angle\theta} = \rho(\cos\theta + j\sin\theta)$
Cartesian, polar, exponential, trigonometric form of a complex number

1. $\rho = \sqrt{x^2 + y^2}$ the modulus of a complex number

2.
$$
\left.
\begin{array}{ll}
\theta = \tan^{-1}\dfrac{y}{x} & x, y \rangle 0 \\[2mm]
\theta = -\tan^{-1}\dfrac{|y|}{x} & x \rangle 0, y \langle 0 \\[2mm]
\theta = 180° - \tan^{-1}\dfrac{y}{|x|} & x \langle 0, y \rangle 0 \\[2mm]
\theta = 180° + \tan^{-1}\dfrac{|y|}{|x|} & x, y \langle 0
\end{array}
\right\}
$$
the argument of a complex number

3. $e^{j\theta} + e^{-j\theta} = 2\cos\theta$
4. $e^{j\theta} - e^{-j\theta} = 2j\sin\theta$
5. $(x + jy) + (\alpha + j\beta) = (x + \alpha) + j(y + \beta)$, addition of complex numbers
6. $(x + jy) - (\alpha + j\beta) = (x - \alpha) + j(y - \beta)$, subtraction of complex numbers
7. $\rho_1 e^{j\theta_1} \cdot \rho_2 e^{j\theta_2} = \rho_1\rho_2 e^{j(\theta_1+\theta_2)}$, multiplication of complex numbers
8. $\dfrac{\rho_1 e^{j\theta_1}}{\rho_2 e^{j\theta_2}} = \dfrac{\rho_1}{\rho_2} e^{j(\theta_1-\theta_2)}$, division of complex numbers
9. $\bar{z} = x - jy$, complex-conjugate
   Properties of complex-conjugates
   $$\overline{z_1 + z_2} = \bar{z}_1 + \bar{z}_2$$
   $$\overline{z_1 - z_2} = \bar{z}_1 - \bar{z}_2$$
   $$\overline{\left(\dfrac{z_1}{z_2}\right)} = \dfrac{\bar{z}_1}{\bar{z}_2}$$

Problems

1.1 Compute the inverse Laplace transforms of the functions

a. $F(s) = \dfrac{2}{s(s+1)(s+2)}$

b. $F(s) = \dfrac{s+3}{(s+1)(s^2 + 4s + 5)}$

c. $F(s) = \dfrac{10}{(s+4)^3(s+3)^2}$

d. $F(s) = \dfrac{10(s+6)}{s(s+5)(s^2 + 3s + 1)(s^2 + 7s + 12)}$

e. $F(s) = \dfrac{3s+1}{(s-1)(s^2 + 1)}$

f. $F(s) = \dfrac{2s^2 - 4}{(s+1)(s-2)^2(s-3)}$

Solution

a. The first function is

$$F(s) = \frac{2}{s(s+1)(s+2)} = \frac{k_1}{s} + \frac{k_2}{s+1} + \frac{k_3}{s+2} \tag{P1.1.1}$$

Computation of k_1, k_2, k_3 according to case of distinct poles:

$$k_1 = \lim_{s \to 0} F(s)s = \lim_{s \to 0} \frac{2}{(s+1)(s+2)} = 1$$

$$k_2 = \lim_{s \to -1} F(s)(s+1) = \lim_{s \to -1} \frac{2}{s(s+2)} = -2$$

$$k_3 = \lim_{s \to -2} F(s)(s+2) = \lim_{s \to -2} \frac{2}{s(s+1)} = 1$$

By substituting in (P1.1.1), we get

$$F(s) = \frac{1}{s} - \frac{2}{s+1} + \frac{1}{s+2} \Rightarrow$$

$$f(t) = L^{-1}\left\{\frac{1}{s}\right\} - 2L^{-1}\left\{\frac{1}{s+1}\right\} + L^{-1}\left\{\frac{1}{s+2}\right\} = (1 - 2e^{-t} + e^{-2t})u(t)$$

b. The second function is

$$F(s) = \frac{s+3}{(s+1)(s^2+4s+5)} = \frac{s+3}{(s+1)(s+2-j)(s+2+j)} \Rightarrow$$

$$F(s) = \frac{k_1}{s+1} + \frac{k_2}{s+2-j} + \frac{k_3 = \overline{k_2}}{s+2+j} \tag{P1.1.2}$$

Computation of k_1, k_2, k_3 according to case of distinct and complex poles

$$k_1 = \lim_{s \to -1} F(s)(s+1) = \lim_{s \to -1} \frac{s+3}{s^2+4s+5} = 1$$

$$k_2 = \lim_{s \to -2+j} F(s)(s+2-j) = \lim_{s \to -2+j} \frac{s+3}{(s+1)(s+2+j)}$$

$$= \frac{-2+j+3}{(-2+j+1)(-2+j+2+j)} = \frac{1+j}{(-1+j)(2j)} = \frac{1+j}{-2-2j}$$

$$= \frac{(1+j)(-2+2j)}{(-2-2j)(-2+2j)} = \frac{-2+2j-2j+2j^2}{8} = -0.5$$

$$k_3 = \overline{k_2} = -0.5$$

By substitution in (P1.1.2) we get

$$F(s) = \frac{1}{s+1} - \frac{0.5}{s+2-j} - \frac{0.5}{s+2+j} \Rightarrow$$

$$f(t) = L^{-1}\left\{\frac{1}{s+1}\right\} - 0.5L^{-1}\left\{\frac{1}{s+2-j}\right\} - 0.5L^{-1}\left\{\frac{1}{s+2+j}\right\} \Rightarrow$$

$$f(t) = e^{-t} - 0.5e^{-(2-j)t} - 0.5e^{-(2+j)t} = e^{-t} - 0.5e^{-2t}(e^{jt} + e^{-jt}) \Rightarrow$$

$$f(t) = e^{-t} - 0.5e^{-2t} \cdot 2\cos t \Rightarrow f(t) = e^{-t}(1 - e^{-t}\cos t)$$

c. The third function is

$$F(s) = \frac{10}{(s+4)^3(s+3)^2} \Rightarrow F(s) = \frac{k_{11}}{s+4} + \frac{k_{12}}{(s+4)^2} + \frac{k_{13}}{(s+4)^3} + \frac{k_{21}}{s+3} + \frac{k_{22}}{(s+3)^2} \quad \text{(P1.1.3)}$$

First, we compute k_{11}, k_{12}, and k_{13} according to the case of multiple poles. The multiplicity of the pole $s_1 = -4$ is $r_1 = 3$. Thus,

$$k_{13} = \frac{1}{0!}\lim_{s\to-4}\left(\frac{d^{(0)}}{ds^{(0)}}\left(F(s)(s+4)^3\right)\right) = \lim_{s\to-4}\frac{10}{(s+3)^2} = 10$$

$$k_{12} = \frac{1}{1!}\lim_{s\to-4}\left(\frac{d^{(1)}}{ds^{(1)}}\left(F(s)(s+4)^3\right)\right) = \lim_{s\to-4}\left(\frac{d}{ds}\left(\frac{10}{(s+3)^2}\right)\right) \Rightarrow$$

$$k_{12} = \lim_{s\to-4}\left(-\frac{20}{(s+3)^3}\right) = 20$$

$$k_{11} = \frac{1}{2!}\lim_{s\to-4}\left(\frac{d^{(2)}}{ds^{(2)}}\left(F(s)(s+4)^3\right)\right) = \frac{1}{2}\lim_{s\to-4}\left(\frac{d^{(2)}}{ds^{(2)}}\left(\frac{10}{(s+3)^2}\right)\right) \Rightarrow$$

$$k_{11} = \frac{1}{2}\lim_{s\to-4}\left(\frac{60}{(s+3)^4}\right) = 30.$$

In the same way, we compute k_{21} and k_{22}. The multiplicity of the pole $s_2 = -3$ is $r_2 = 2$. Thus,

$$k_{22} = \frac{1}{0!}\lim_{s\to-3}\left(\frac{d^{(0)}}{ds^{(0)}}\left(F(s)(s+3)^2\right)\right) = \lim_{s\to-3}\frac{10}{(s+4)^3} = 10$$

$$k_{21} = \frac{1}{1!}\lim_{s\to-3}\left(\frac{d^{(1)}}{ds^{(1)}}\left(F(s)(s+3)^2\right)\right) = \lim_{s\to-3}\left(\frac{10}{(s+4)^3}\right)' = -30.$$

By substituting in (P1.1.3), we obtain

$$F(s) = \frac{30}{s+4} + \frac{20}{(s+4)^2} + \frac{10}{(s+4)^3} - \frac{30}{s+3} + \frac{10}{(s+3)^2} \Rightarrow$$

$$f(t) = 30L^{-1}\left\{\frac{1}{s+4}\right\} + 20L^{-1}\left\{\frac{1}{(s+4)^2}\right\} + 10L^{-1}\left\{\frac{1}{(s+4)^3}\right\}$$

$$-30L^{-1}\left\{\frac{1}{s+3}\right\} + 10L^{-1}\left\{\frac{1}{(s+3)^2}\right\}.$$

Thus,

$$f(t) == 30e^{-4t} + 20te^{-4t} + \frac{10}{2}t^2e^{-4t} - 30e^{-3t} + 10te^{-3t} = 10e^{-4t}(3+2t+0.5t^2) + 10e^{-3t}(t-3)$$

d. The fourth function is

$$F(s) = \frac{10(s+6)}{s(s+5)(s^2+3s+2)(s^2+7s+12)}$$

$$= \frac{10(s+6)}{s(s+1)(s+2)(s+3)(s+4)(s+5)} \Rightarrow$$

$$F(s) = \frac{k_1}{s} + \frac{k_2}{s+1} + \frac{k_3}{s+2} + \frac{k_4}{s+3} + \frac{k_5}{s+4} + \frac{k_6}{s+5} \qquad (P1.1.4)$$

Thus,

$$k_1 = \lim_{s \to 0} F(s)s = 0.5$$

$$k_2 = \lim_{s \to -1} F(s)(s+1) = -2.083$$

$$k_3 = \lim_{s \to -2} F(s)(s+2) = 3.333$$

$$k_4 = \lim_{s \to -3} F(s)(s+3) = -2.5$$

$$k_5 = \lim_{s \to -4} F(s)(s+4) = 0.833$$

$$k_6 = \lim_{s \to -5} F(s)(s+5) = -0.083$$

By substituting in (P1.1.4), we get

$$F(s) = \frac{0.5}{s} - \frac{2.083}{s+1} + \frac{3.333}{s+2} - \frac{2.5}{s+3} + \frac{0.833}{s+4} - \frac{0.083}{s+5} \Rightarrow$$

$$f(t) = (0.5 - 2.083e^{-t} + 3.333e^{-2t} - 2.5e^{-3t} + 0.833e^{-4t} - 0.083e^{-5t})u(t)$$

e. The fifth function is

$$F(s) = \frac{3s+1}{(s-1)(s^2+1)} = \frac{k_1}{s-1} + \frac{k_2}{s-j} + \frac{k_3}{s+j} \tag{P1.1.5}$$

The coefficients k_1, k_2, k_3 are computed as

$$k_1 = \lim_{s \to 1} F(s)(s-1) = \lim_{s \to 1} \frac{3s+1}{s^2+1} = 2$$

$$k_2 = \lim_{s \to j} F(s)(s-j) = \lim_{s \to j} \frac{3s+1}{(s-1)(s+j)} = \frac{1+3j}{(-1+j)2j} = -\left(1+\frac{1}{2}j\right)$$

$$k_3 = \overline{k_2} = -1 + \frac{1}{2}j$$

By substituting in (P1.1.5), we obtain

$$F(s) = \frac{2}{s-1} - \frac{(1+(1/2)j)}{s-j} + \frac{(-1+(1/2)j)}{s+j} \overset{I.L.T}{\Rightarrow}$$

$$f(t) = 2L^{-1}\left\{\frac{1}{s-1}\right\} - \left(1+\frac{1}{2}j\right)L^{-1}\left\{\frac{1}{s-j}\right\} + \left(-1+\frac{1}{2}j\right)L^{-1}\left\{\frac{1}{s+j}\right\}$$

$$= 2e^t - \left(1+\frac{1}{2}j\right)e^{jt} + \left(-1+\frac{1}{2}j\right)e^{-jt} = 2e^t - (e^{jt}+e^{-jt}) - \frac{1}{2}j(e^{jt}-e^{-jt}) \Rightarrow$$

$$\Rightarrow f(t) = 2e^t - 2\cos t - \frac{1}{2}j2j\sin t = 2e^t - 2\cos t + \sin t$$

f. The last function is

$$F(s) = \frac{2s^2-4}{(s+1)(s-2)^2(s-3)} \Rightarrow F(s) = \frac{k_1}{s+1} + \frac{k_{21}}{s-2} + \frac{k_{22}}{(s-2)^2} + \frac{k_3}{(s-3)} \tag{P1.1.6}$$

The coefficients k_1, k_3 are computed according to the case of distinct poles:

$$k_1 = \lim_{s \to -1} F(s)(s+1) = \lim_{s \to -1} \frac{2s^2-4}{(s-2)^2(s-3)} = \frac{1}{18}$$

$$k_3 = \lim_{s \to 3} F(s)(s-3) = \lim_{s \to 3} \frac{2s^2-4}{(s-1)(s-2)^2} = \frac{7}{2}$$

The coefficients k_{21}, k_{22} (corresponding to the root $s = 2$) are computed according to the case of multiple poles. The degree of multiplicity is $r = 2$, thus

$$k_{21} = \frac{1}{1!}\lim_{s \to 2}\left(\frac{d}{ds}\left(F(s)(s-2)^2\right)\right) = \lim_{s \to 2}\frac{d}{ds}\left(\frac{2s^2-4}{(s+1)(s-3)}\right) = -\frac{32}{9}$$

$$k_{22} = \frac{1}{0!}\lim_{s \to 2}\left(\frac{d^{(0)}}{ds^{(0)}}\left(F(s)(s-2)^2\right)\right) = \lim_{s \to 2}\left(\frac{2s^2-4}{(s+1)(s-3)}\right) = -\frac{4}{3}$$

By substituting in (P1.1.6), we have

$$F(s) = \frac{1/18}{s+1} - \frac{32/9}{s-2} - \frac{4/3}{(s-2)^2} + \frac{7/2}{s-3} \xrightarrow{I.L.T.}$$

$$f(t) = \frac{1}{18}L^{-1}\left\{\frac{1}{s+1}\right\} - \frac{32}{9}L^{-1}\left\{\frac{1}{s-2}\right\} - \frac{4}{3}L^{-1}\left\{\frac{1}{(s-2)^2}\right\} + \frac{7}{2}L^{-1}\left\{\frac{1}{s-3}\right\} \Rightarrow$$

$$\Rightarrow f(t) = \frac{1}{18}e^{-t} - \frac{32}{9}e^{2t} - \frac{4}{3}te^{2t} + \frac{7}{2}e^{3t}$$

1.2 Solve the differential equation

$$\frac{d^2y(t)}{dt^2} + 5\frac{dy(t)}{dt} + 4y(t) = 10, \quad y(0) = -1, \quad y'(0) = 1$$

Solution

The differential equation is

$$\frac{d^2y(t)}{dt^2} + 5\frac{dy(t)}{dt} + 4y(t) = 10, \quad y(0) = -1, \quad y'(0) = 1 \qquad \text{(P1.2.1)}$$

$$(P1.2.1) \xrightarrow{L.T.} L\left\{\frac{d^2y(t)}{dt^2} + 5\frac{dy(t)}{dt} + 4y(t)\right\} = L\{10\} \Rightarrow$$

$$s^2Y(s) - sy(0) - y'(0) + 5(sY(s) - y(0)) + 4Y(s) = \frac{10}{s} \Rightarrow \qquad \text{(P1.2.2)}$$

$$Y(s) = \frac{-s^2 - 4s + 10}{s(s^2 + 5s + 4)} = \frac{-s^2 - 4s + 10}{s(s+1)(s+4)} = \frac{k_1}{s} + \frac{k_2}{s+1} + \frac{k_3}{s+4}$$

Computation of the coefficients k_i:

$$k_1 = \lim_{s\to 0} Y(s)s = \frac{10}{4}$$

$$k_2 = \lim_{s\to -1} Y(s)(s+1) = -\frac{13}{3}$$

$$k_3 = \lim_{s\to -4} Y(s)(s+4) = \frac{10}{12}$$

$$(P1.2.2) \Rightarrow Y(s) = \frac{10/4}{s} - \frac{13/3}{s+1} + \frac{10/12}{s+4} \xrightarrow{I.L.T}$$

$$y(t) = \left(\frac{10}{4} - \frac{13}{3}e^{-t} + \frac{10}{12}e^{-4t}\right)u(t)$$

1.3 Solve the following system of differential equations:

$$\frac{dx(t)}{dt} = 2x(t) - 3y(t)$$

$$\frac{dy(t)}{dt} = y(t) - 2x(t)$$

The initial conditions are $x(0) = 8$ and $y(0) = 3$.

Solution

The system of differential equations is

$$\frac{dx(t)}{dt} = 2x(t) - 3y(t) \tag{P1.3.1}$$

$$\frac{dy(t)}{dt} = y(t) - 2x(t), \quad x(0) = 8, \quad y(0) = 3 \tag{P1.3.2}$$

$$(\text{P1.3.1}) \overset{L.T.}{\Rightarrow} sX(s) - x(0) = 2X(s) - 3Y(s) \Rightarrow X(s) = \frac{8 - 3Y(s)}{s - 2} \tag{P1.3.3}$$

$$(\text{P1.3.2}) \overset{L.T.}{\Rightarrow} sY(s) - y(0) = Y(s) - 2X(s) \Rightarrow Y(s) = \frac{3 - 2X(s)}{s - 1} \tag{P1.3.4}$$

$$(\text{P1.3.3}), (\text{P1.3.4}) \Rightarrow (s-1)Y(s) = 3 - 2\frac{8 - 3Y(s)}{s - 2} \Rightarrow (s-1)(s-2)Y(s) = 3s - 22 + 6Y(s) \Rightarrow$$

$$\tag{P1.3.5}$$

$$Y(s) = \frac{3s - 22}{s^2 - 3s - 4} = \frac{3s - 22}{(s - 4)(s + 1)} = \frac{k_1}{s - 4} + \frac{k_2}{s + 1}$$

$$k_1 = \lim_{s \to 4} Y(s)(s - 4) = -2$$

$$k_2 = \lim_{s \to -1} Y(s)(s + 1) = 5$$

$$(\text{P1.3.5}) \Rightarrow Y(s) = -\frac{2}{s - 4} + \frac{5}{s + 1} \overset{I.L.T}{\Rightarrow} y(t) = -2e^{4t} + 5e^{-t}$$

Moreover,

$$(\text{P1.3.3}), (\text{P1.3.4}) \Rightarrow (s - 2)X(s) = 8 - 3\frac{3 - 2X(s)}{s - 1} \Rightarrow$$

$$\tag{P1.3.6}$$

$$X(s) = \frac{8s - 17}{s^2 - 3s - 4} = \frac{8s - 17}{(s - 4)(s + 1)} = \frac{k_1'}{s - 4} + \frac{k_2'}{s + 1}$$

$$k_1' = \lim_{s \to 4} X(s)(s-4) = 3$$

$$k_2' = \lim_{s \to -1} X(s)(s+1) = 5$$

$$(P1.3.6) \Rightarrow X(s) = \frac{3}{s-4} + \frac{5}{s+1} \overset{I.L.T}{\Rightarrow} x(t) = 3e^{4t} + 5e^{-t}$$

1.4 Compute the voltage $u_2(t)$ in the circuit of the figure below, if

$$R = 1\Omega, \quad C = 1F, \quad u_1(t) = 2e^{-t}, \quad u_c(0) = 0.$$

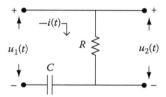

Solution

From the electrical circuit of the figure, we arrive at the equations:

$$u_1(t) = Ri(t) + \frac{1}{C}\int_0^t i(t)dt \tag{P1.4.1}$$

$$u_2(t) = Ri(t) \tag{P1.4.2}$$

Applying Laplace transform in (P1.4.1) and (P1.4.2), we get

$$(P1.4.1) \overset{L.T.}{\Rightarrow} U_1(s) = RI(s) + \frac{I(s)}{sC} = \frac{sRC+1}{sC}I(s) \tag{P1.4.3}$$

$$(P1.4.2) \overset{L.T.}{\Rightarrow} U_2(s) = RI(s) \tag{P1.4.4}$$

$$(P1.4.3), (P1.4.4) \Rightarrow \frac{U_2(s)}{U_1(s)} = \frac{RI(s)}{(sRC+1)/sC} = \frac{sRC}{sRC+1} \tag{P1.4.5}$$

Substituting the values of R, C we obtain

$$(P1.4.5) \Rightarrow \frac{U_2(s)}{U_1(s)} = \frac{s}{s+1} \Rightarrow U_2(s) = \frac{s}{s+1}U_1(s) \tag{P1.4.6}$$

but

$$U_1(s) = L\{u_i(t)\} = L\{2e^{-t}\} = \frac{2}{s+1}$$ (P1.4.7)

$$(P1.4.6), (P1.4.7) \Rightarrow U_2(s) = \frac{2s}{(s+1)^2} \overset{I.L.T.}{\Rightarrow} u_2(t) = L^{-1}\left\{\frac{2s}{(s+1)^2}\right\}$$ (P1.4.8)

and

$$\frac{s}{(s+1)^2} = \frac{k_{11}}{s+1} + \frac{k_{12}}{(s+1)^2}$$

Computation of k_{11}, k_{12}:
The multiplicity of the root $s = -1$ is $r = 2$. Thus,

$$k_{11} = \frac{1}{1!}\lim_{s\to-1}\left(\frac{d^{(1)}}{ds^{(1)}}\left(\frac{s}{(s+1)^2}(s+1)^2\right)\right) = \lim_{s\to-1}\left(\frac{d}{ds}s\right) = 1$$

$$k_{12} = \frac{1}{0!}\lim_{s\to-1}\left(\frac{d^{(0)}}{ds^{(0)}}\left(\frac{s}{(s+1)^2}(s+1)^2\right)\right) = \lim_{s\to-1}(s) = -1$$

Hence,

$$\frac{s}{(s+1)^2} = \frac{1}{s+1} - \frac{1}{(s+1)^2}.$$

$$(P1.4.8) \Rightarrow u_2(t) = 2L^{-1}\left\{\frac{1}{s+1}\right\} - 2L^{-1}\left\{\frac{1}{(s+1)^2}\right\} \Rightarrow u_2(t) = 2e^{-t} - 2te^{-t}$$

1.5 The differential equation of the circuit that is depicted in the following figure is

$$\frac{du(t)}{dt} + \frac{1}{RC}u(t) = \frac{1}{RC}u_\pi(t).$$

Compute and plot the voltage at the capacitor $u_c(t) = u(t)$ if the input voltage is (a) $u_\pi(t) = 20u(t)$ V and (b) $u_\pi(t) = 20\sin 5t$ V. The initial conditions are zero.

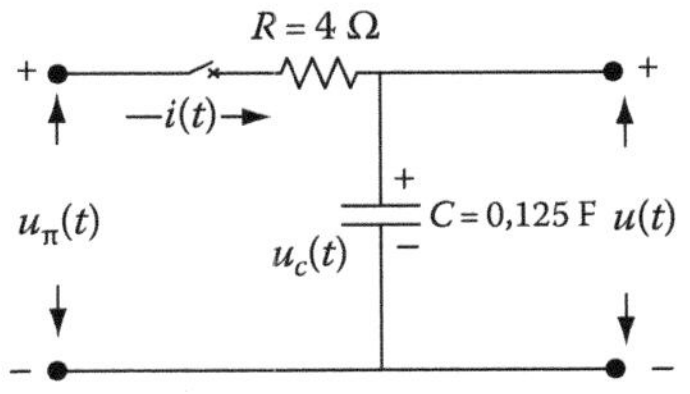

Solution

The differential equation that describes the circuit is

$$\frac{du(t)}{dt} + \frac{1}{RC}u(t) = \frac{1}{RC}u_\pi(t) \tag{P1.5.1}$$

$$(\text{P1.5.1}) \overset{L.T.}{\Rightarrow} SU(s) - u(0) + \frac{1}{RC}U(s) = \frac{1}{RC}U_\pi(s) \Rightarrow$$

$$U(s)\left(s + \frac{1}{RC}\right) = \frac{1}{RC}U_\pi(s) \Rightarrow U(s) = \frac{U_\pi(s)}{sRC+1} \tag{P1.5.2}$$

a. For $u_\pi(t) = 20u(t) \overset{L.T.}{\Rightarrow} U_\pi(s) = 20/s$. Thus,

$$(\text{P1.5.2}) \Rightarrow U(s) = \frac{20}{s(sRC+1)} = \frac{20}{s(0.5s+1)} = \frac{40}{s(s+2)} \Rightarrow \tag{P1.5.3}$$

$$U(s) = \frac{k_1}{s} + \frac{k_2}{s+2}$$

$$k_1 = \lim_{s \to 0} U(s)s = 20$$

$$k_2 = \lim_{s \to -2} U(s)(s+2) = -20$$

$$(\text{P1.5.3}) \Rightarrow U(s) = \frac{20}{s} - \frac{20}{s+2} \overset{I.L.T.}{\Rightarrow} u(t) = 20 - 20e^{-2t}$$

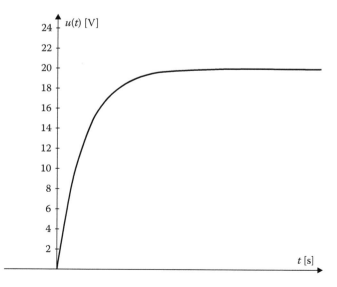

b. For $u_\pi(t) = 20 \sin 5t \overset{L.T.}{\Rightarrow} U_\pi(s) = 20\dfrac{5}{s^2+25} = \dfrac{100}{s^2+25}$

$$(\text{P1.5.2}) \Rightarrow U(s) = \dfrac{100}{(s^2+25)(0.5s+1)} = \dfrac{200}{(s^2+25)(s+2)} \Rightarrow$$

(P1.5.4)

$$U(s) = \dfrac{200}{(s+2)(s+5j)(s-5j)} = \dfrac{k_1}{s+2} + \dfrac{k_2}{s+j5} + \dfrac{k_3}{s-j5}$$

$$k_1 = \lim_{s \to -2} U(s)(s+2) = \dfrac{200}{29}$$

$$k_2 = \lim_{s \to -j5} U(s)(s+5j) = \lim_{s \to -j5} \dfrac{200}{(s+2)(s-j5)} = \dfrac{20}{-5-j2} = \dfrac{-100+j40}{29} \Rightarrow$$

$$k_2 = -\dfrac{100}{29} + j\dfrac{40}{29} = 3.717e^{-j201.8°}$$

$$k_3 = \overline{k_2} = -\dfrac{100}{29} - j\dfrac{40}{29} = 3.717e^{j201.8°}$$

$$(\text{P1.5.4}) \Rightarrow U(s) = \dfrac{200/29}{s+2} + \dfrac{(-100/29)+(j40/29)}{s+j5} + \dfrac{(-100/29)-(j40/29)}{s-j5} \overset{I.L.T}{\Rightarrow}$$

$$u(t) = \dfrac{200}{29}e^{-2t} + \left(-\dfrac{100}{29} + j\dfrac{40}{29}\right)e^{-j5t} + \left(-\dfrac{100}{29} - j\dfrac{40}{29}\right)e^{j5t} \Rightarrow$$

$$u(t) = \dfrac{200}{29}e^{-2t} - \dfrac{100}{29}(e^{j5t} + e^{-j5t}) - j\dfrac{40}{29}(e^{j5t} - e^{-j5t}) \Rightarrow$$

$$u(t) = \dfrac{200}{29}e^{-2t} - \dfrac{100}{29}2\cos 5t - j\dfrac{40}{29}2j\sin 5t \Rightarrow$$

$$u(t) = \dfrac{200}{29}e^{-2t} - \dfrac{200}{29}\cos 5t + \dfrac{80}{29}\sin 5t \Rightarrow$$

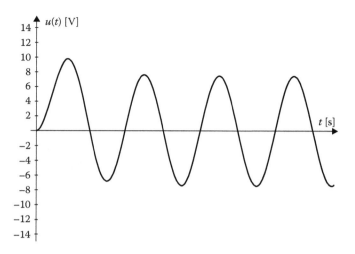

1.6 Given the electrical circuit of the figure below,

    a. Express the differential equation of the system and apply Laplace transform supposing that the initial conditions are zero.

    b. Verify that $U_0(s)/U_i(s) = (1/a)((ars + 1)/(rs + 1))$. Express a and r as functions of C_1, C_2, R_2.

    c. Find and plot the output of the circuit $u_0(t)$ for an input voltage

        i. $u_1(t) = 100\,\text{V}$

       ii. $u_1(t) = \sin t$ V

Suppose that $C_1 = C_2 = 1\,\mu\text{F}$ and $R_2 = 100\,\text{K}\Omega$.

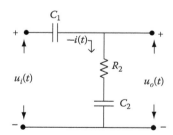

Solution

    a. The differential equation of the circuit is

$$u_i(t) = \frac{1}{C_1}\int_0^t i(t)dt + R_2 i(t) + \frac{1}{C_2}\int_0^t i(t)dt \qquad \text{(P1.6.1)}$$

$$\text{(P1.6.1)} \underset{I.C.=0}{\overset{L.T}{\Rightarrow}} U_i(s) = \frac{1}{C_1}\frac{I(s)}{s} + R_2 I(s) + \frac{1}{C_2}\frac{I(s)}{s} = I(s)\left(R_2 + \frac{1}{sC_1} + \frac{1}{sC_2} \right) \qquad \text{(P1.6.2)}$$

b. We have $u_0(t) = R_2 i(t) + \dfrac{1}{C_2} \displaystyle\int_0^t i(t)dt \overset{L.T}{\Rightarrow}$

$$U_0(s) = R_2 I(s) + \frac{1}{C_2}\frac{I(s)}{s} \Rightarrow U_0(s) = I(s)\left(R_2 + \frac{1}{sC_2}\right) \tag{P1.6.3}$$

$$(\text{P1.6.2}),(\text{P1.6.3}) \Rightarrow \frac{U_0(s)}{U_i(s)} = \frac{I(s)\left(R_2 + (1/sC_2)\right)}{I(s)\left(R_2 + (1/sC_1) + (1/sC_2)\right)} = \frac{(sR_2C_2 + 1)C_1}{sR_2C_1C_2 + C_1 + C_2} \Rightarrow \tag{P1.6.4}$$

$$\frac{U_0(s)}{U_i(s)} = \frac{C_1}{C_1 + C_2} \cdot \frac{sR_2C_2 + 1}{s(R_2C_1C_2/(C_1 + C_2)) + 1}$$

Hence, it is verified that

$$\frac{U_0(s)}{U_i(s)} = \frac{1}{a} \cdot \frac{ars + 1}{rs + 1}$$

where

$$\alpha = \frac{C_1 + C_2}{C_1} \quad \text{and} \quad r = \frac{R_2C_1C_2}{C_1 + C_2}$$

c. By substituting the values of R_2, C_1, C_2 in (4) we get

$$(\text{P1.6.4}) \Rightarrow \frac{U_0(s)}{U_i(s)} = \frac{10^{-6}}{10^{-6} + 10^{-6}} \cdot \frac{10^5 10^{-6} s + 1}{(10^5 10^{-6} 10^{-6}/(10^{-6} + 10^{-6}))s + 1} = \frac{1}{2}\frac{0.1s + 1}{0.05s + 1} \tag{P1.6.5}$$

i. For $u_i(t) = 100\,\text{V} \overset{L.T.}{\Rightarrow} U_i(s) = 100/s$

$$(\text{P1.6.5}) \Rightarrow U_o(s) = 0.5\frac{(0.1s + 1)100}{(0.05s + 1)s} = \frac{10^3(0.1s + 1)}{(s + 20)s} = \frac{k_1}{s} + \frac{k_2}{s + 20} \tag{P1.6.6}$$

$$k_1 = \lim_{s \to 0} U_o(s)s = 50$$

$$k_2 = \lim_{s \to -20} U_o(s)(s + 20) = 50$$

$$(\text{P1.6.6}) \Rightarrow U_o(s) = \frac{50}{s} + \frac{50}{s + 20} \overset{I.L.T.}{\Rightarrow} u_o(t) = (50 + 50e^{-20t})u(t)$$

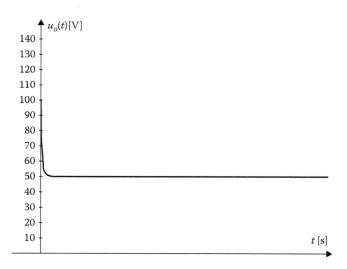

ii.   For $u_i(t) = \sin t \overset{L.T.}{\Longrightarrow} U_i(s) = \dfrac{1}{s^2+1}$

$$(P1.6.5) \Rightarrow U_o(s) = 0.5\frac{(0.1s+1)}{(0.05s+1)(s^2+1)} = \frac{10(0.1s+1)}{(s+20)(s^2+1)} \Rightarrow$$

$$(P1.6.7)$$

$$U_o(s) = \frac{k_1}{s+20} + \frac{k_2}{s+j} + \frac{k_3}{s-j}$$

where

$$k_1 = \lim_{s\to-20} U_o(s)(s+20) = -0.025$$

$$k_2 = \lim_{s\to-j} U_o(s)(s+j) = -0.0125 + j0.251 = 0.25e^{j92.8°}$$

$$k_3 = \overline{k_2} = -0.0125 - j0.251 = 0.25e^{-j92.8°}$$

$$(P1.6.7) \Rightarrow U_o(s) = -\frac{0.025}{s+20} + \frac{0.25e^{j92.8°}}{s+j} + \frac{0.25e^{-j92.8°}}{s-j} \overset{I.L.T.}{\Longrightarrow}$$

$$u_o(t) = -0.025e^{-20t} + 0.25e^{j92.8°}e^{-jt} + 0.25e^{-j92.8°}e^{jt} =$$

$$= -0.025e^{-20t} + 0.25(e^{j92.8°}e^{-jt} + e^{-j92.8°}e^{jt}) =$$

$$= -0.025e^{-20t} + 0.25(e^{j(92.8°-t)} + e^{-j(92.8°-t)}) \Rightarrow$$

$$u_o(t) = -0.025e^{-20t} + 0.125\cos(92.8° - t)$$

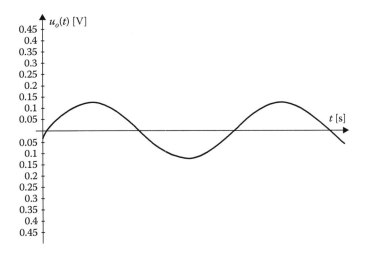

1.7 Given the electrical circuit of Figure (a)

a. Express the Laplace transformed mathematical model of the system (initial conditions are zero).

b. Find and plot the output voltage of the circuit.

c. Find and plot the currents in the loops.

d. If a Buffer is placed between the two partial circuits (see Figure (b)), compute and plot the new output voltage of the system.

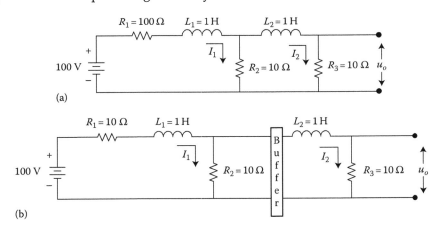

(a)

(b)

Solution

a. The Laplace transformed system equations are:

For the loop 12561

$$E_i(s) = (R_1 + sL_1)I_1(s) + R_2(I_1(s) - I_2(s))$$ (P1.7.1)

For the loop 23452

$$sL_2I_2(s) = R_2(I_1(s) - I_2(s)) - I_2(s)R_3$$ (P1.7.2)

For $E_o(s)$ it holds:

$$E_o(s) = R_3 I_2(s) \tag{P1.7.3}$$

b. By substituting in (P1.7.1) through (P1.7.3), we have

$$(P1.7.1) \Rightarrow \frac{100}{s} = (10+s)I_1(s) + 10(I_1(s) - I_2(s)) \Rightarrow$$

$$\frac{100}{s} = (20+s)I_1(s) - 10I_2(s) \tag{P1.7.4}$$

$$(P1.7.2) \Rightarrow sI_2(s) = 10(I_1(s) - I_2(s)) - 10I_2(s) \Rightarrow$$

$$I_1(s) = \frac{(s+20)I_2(s)}{10} \tag{P1.7.5}$$

$$(P1.7.3) \Rightarrow E_o(s) = 10I_2(s) \tag{P1.7.6}$$

$$(P1.7.4),(P1.7.5) \Rightarrow \frac{100}{s} = \frac{(20+s)^2 I_2(s)}{10} - 10I_2(s) = \frac{(20+s)^2 - 100}{10}I_2(s) \Rightarrow$$

$$I_2(s) = \frac{10^3}{s(s^2 + 40s + 300)} \tag{P1.7.7}$$

$$(P1.7.6),(P1.7.7) \Rightarrow E_o(s) = \frac{10^4}{s(s^2 + 40s + 300)} = \frac{10^4}{s(s+10)(s+30)} \Rightarrow$$

$$E_o(s) = \frac{k_1}{s} + \frac{k_2}{s+10} + \frac{k_3}{s+30} = \frac{100/3}{s} - \frac{50}{s+10} + \frac{100/6}{s+30} \stackrel{I.L.T}{\Rightarrow}$$

$$E_o(t) = \left(\frac{100}{3} - 50e^{-10t} + \frac{100}{6}e^{-30t} \right)u(t)$$

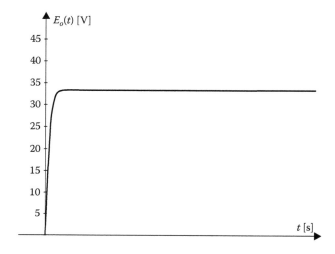

c. Computation of the currents

$$(P1.7.6) \Rightarrow I_2(s) = \frac{E_o(s)}{10} \overset{I.L.T}{\Rightarrow} i_2(t) = \frac{E_o(t)}{10} \Rightarrow$$

$$i_2(t) = \left(\frac{10}{3} - 5e^{-10t} + \frac{10}{6} e^{-30t} \right) u(t)$$

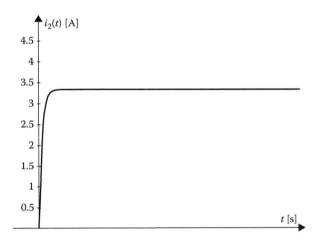

$$(P1.7.5) \Rightarrow I_1(s) = \frac{(s+20)I_2(s)}{10} = \frac{(s+20)E_o(s)}{100} = \frac{100(s+20)}{s(s+10)(s+30)} \Rightarrow$$

$$I_1(s) = \frac{20/3}{s} + \frac{-5}{s+10} + \frac{-5/3}{s+30} \overset{I.L.T}{\Rightarrow}$$

$$i_1(t) = \left(\frac{20}{3} - 5e^{-10t} - \frac{5}{3} e^{-30t} \right) u(t)$$

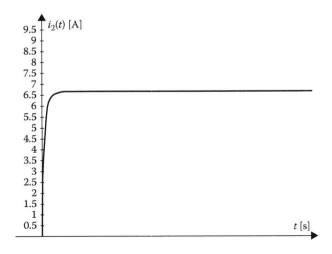

d. The Buffer isolates the two circuits, hence

$$\frac{E_o(s)}{E_i(s)} = \underbrace{\frac{E_{o1}(s)}{E_i(s)}}_{\substack{\text{Input-output}\\\text{relationship of}\\\text{the first circuit}}} \cdot \underbrace{\frac{E_o(s)}{E_{o1}(s)}}_{\substack{\text{Input-output}\\\text{relationship of}\\\text{the second circuit}}} \tag{P1.7.8}$$

Thus,

$$\frac{E_{o1}(s)}{E_i(s)} = \frac{R_2 I(s)}{(R_1 + R_2 + sL_1)I(s)} = \frac{R_2}{R_1 + R_2 + sL_1} = \frac{10}{s+20}$$

Moreover,

$$\frac{E_o(s)}{E_{o1}(s)} = \frac{R_3 I(s)}{(R_3 + sL_2)I(s)} = \frac{R_3}{R_3 + sL_2} = \frac{10}{s+10}$$

$$(\text{P1.7.8}) \Rightarrow \frac{E_o(s)}{E_i(s)} = \frac{10}{s+20} \cdot \frac{10}{s+10} = \frac{100}{(s+10)(s+20)} \Rightarrow$$

$$E_o(s) = \frac{100}{(s+10)(s+20)} E_i(s) \Rightarrow E_o(s) = \frac{10^4}{s(s+10)(s+20)} \Rightarrow$$

$$E_o(s) = \frac{50}{s} - \frac{100}{s+10} + \frac{50}{s+20} \overset{I.L.T}{\Rightarrow}$$

$$E_o(t) = (50 - 100e^{-10t} + 50e^{-20t})u(t)$$

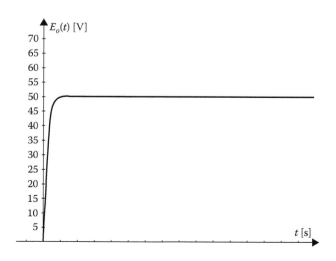

1.8 A mass (M) moves on the x-axis (see figure below). The force exercised on the mass is eight times the distance from the position $t_s = 0$. During this time ($t_s = 0$) the mass has covered space $x(0) = 10$ m. Find and plot the displacement $x(t)$ of the mobile without taking into account the friction.

$$f(t) \quad M = 2$$

$$x(0) \qquad \longleftarrow \bullet \qquad\qquad\qquad x(t)$$

$$\vdash\! x(t) \longrightarrow\!\!\dashv$$

$$0$$

Solution

The differential equation of the system is:

$$M\frac{d^2x(t)}{dt^2} = -f(t) \tag{P1.8.1}$$

where

$$x(0) = 10 \tag{P1.8.2}$$

$$f(t) = 8x(t) \tag{P1.8.3}$$

$$x'(0) = 0 \tag{P1.8.4}$$

$$(\text{P1.8.1}),(\text{P1.8.3}) \Rightarrow M\frac{d^2x(t)}{dt^2} = -8x(t) \overset{\text{L.T.}}{\Rightarrow} M(s^2X(s) - sx(0) - x'(0)) = -8X(s)$$

$$\underset{(\text{P1.8.4})}{\overset{(\text{P1.8.2})}{\Rightarrow}} M(s^2X(s) - 10s) = -8X(s) \Rightarrow X(s)(Ms^2 + 8) = 10Ms \Rightarrow$$

$$X(s) = \frac{20s}{2s^2 + 8} = \frac{10s}{s^2 + 4} \overset{\text{I.L.T.}}{\Rightarrow} x(t) = 10\cos 2t$$

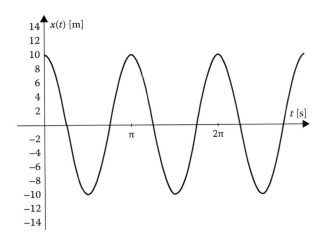

1.9 The differential equation which relates the movement of a mass M with the applied force $f(t)$ for the mechanical system shown in the following figure is

$$f(t) = M\frac{d^2x(t)}{dt^2} + f\frac{dx(t)}{dt} + kx(t),$$

where
  k is the spring constant
  f is the friction constant

Find the rate of change of the displacement $x(t)$ if $f(t) = 20\,\text{N}$, $f = 80\,\text{N}\cdot\text{s/m}$, $k = 160\,\text{N/m}$, and (a) $M = 10\,\text{kg}$ and (b) $M = 20\,\text{kg}$.

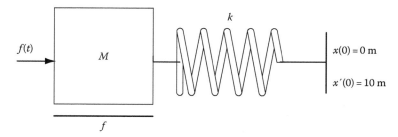

Solution

The differential equation of the systems is

$$f(t) = M\frac{d^2x(t)}{dt^2} + f\frac{dx(t)}{dt} + kx(t) \tag{P1.9.1}$$

a. Thus for $M = 10\,\text{kg}$

$$(\text{P1.9.1}) \Rightarrow 20u(t) = 10x''(t) + 80x'(t) + 160x(t) \overset{L.T.}{\Rightarrow}$$

$$\frac{20}{s} = 10(s^2X(s) - sx(0) - x'(0)) + 80(sX(s) - x(0)) + 160X(s) \Rightarrow$$

$$X(s) = \frac{10(s+0.2)}{s(s+4)^2} \overset{I.L.T.}{\Rightarrow} x(t) = \frac{10}{16}(0.2 - 0.2e^{-4t} + 4(4 - 0.2)te^{-4t}) \Rightarrow$$

$$x(t) = 0.125 + 9.5te^{-4t} - 0.125e^{-4t}$$

b. For $M = 20\,\text{kg}$

$$(\text{P1.9.1}) \Rightarrow 20u(t) = 20x''(t) + 80x'(t) + 160x(t) \overset{L.T.}{\Rightarrow}$$

$$\frac{20}{s} = 20(s^2X(s) - sx(0) - x'(0)) + 80(sX(s) - x(0)) + 160X(s) \Rightarrow$$

$$X(s) = \frac{10s+1}{s(s^2 + 4s + 8)} = \frac{10(s+0.1)}{s(s+2+j2)(s+2-j2)} \overset{I.L.T.}{\Rightarrow}$$

$$x(t) = \frac{10}{(2+j2)(2-j2)}\left[0.1 - \frac{(2-j2)(0.1-2-j2)}{2-j2-2-j2}e^{-(2+j2)t} + + \frac{(2+j2)(0.1-2+j2)}{2-j2-2-j2}e^{-(2-j2)t}\right] \Rightarrow$$

$$x(t) = 10\left[\frac{0.1}{8} + \frac{-1.9-j2}{(2+j2)4j}e^{-(2+j2)t} + \frac{-1.9+j2}{(2-j2)(-j4)}e^{-(2-j2)t}\right] =$$

$$= 10\left(\frac{(-1.9-j2)(-8-j8)}{128}e^{-(2+j2)t} + \frac{(-1.9+j2)(-8+j8)}{128}e^{-(2-j2)t} + 0.0125\right) \Rightarrow$$

$$x(t) = 10\left(\frac{-0.8+j31.2}{128}e^{-(2+j2)t} + \frac{-0.8-j31.2}{128}e^{-(2-j2)t} + 0.0125\right) =$$

$$\frac{10}{128}(-0.8e^{-2t}(e^{j2t}+e^{-j2t}) - 31.2je^{-2t}(e^{j2t}-e^{-j2t}) + 1.6) \Rightarrow$$

$$x(t) = 0.125e^{-2t}\cos 2t + 4.875e^{-2t}\sin 2t + 0.125$$

1.10 For the mechanical system that is depicted in the following figure, compute and plot the time response $y_2(t)$ if

a. $f_1(t) = 10\delta(t)$
b. $f_1(t) = 10u(t)$
c. $f_1(t) = 10r(t) = 10t$.

The values of the system's parameters are $B_1 = B_2 = 2\,\text{N} \cdot \text{s/m}$, $k_1 = k_2 = 1\,\text{N/m}$, $m_2 = 1\,\text{kg}$, $m_1 \cong 0\,\text{kg}$.

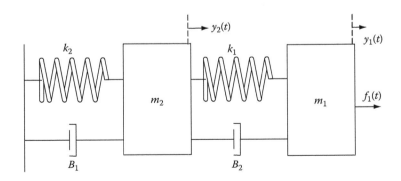

Solution

The mathematical model of the mechanical system is

$$f_1(t) - B_1\left(\frac{dy_1(t)}{dt} - \frac{dy_2(t)}{dt}\right) - k_1(y_1(t) - y_2(t)) = m_1\frac{d^2y_1(t)}{dt^2} \qquad (\text{P1.10.1})$$

$$B_1\left(\frac{dy_1(t)}{dt}-\frac{dy_2(t)}{dt}\right)+k_1(y_1(t)-y_2(t))-B_2\frac{dy_2(t)}{dt}-k_2y_2(t)=m_2\frac{d^2y_2(t)}{dt^2} \qquad \text{(P1.10.2)}$$

$$\text{(P1.10.1)}\overset{\text{L.T.}}{\underset{\text{I.C.=0}}{\Rightarrow}}L\left\{f_1(t)-B_1\left(\frac{dy_1(t)}{dt}-\frac{dy_2(t)}{dt}\right)-k_1(y_1(t)-y_2(t))\right\}=L\left\{m_1\frac{d^2y_1(t)}{dt^2}\right\}\Rightarrow$$

$$\text{(P1.10.3)}$$

$$F_1(s)=Y_1(s)(m_1s^2+B_1s+k_1)-Y_2(s)(sB_1+k_1)$$

$$\text{(P1.10.2)}\overset{\text{L.T.}}{\underset{\text{I.C.=0}}{\Rightarrow}}L\left\{B_1\left(\frac{dy_1(t)}{dt}-\frac{dy_2(t)}{dt}\right)+k_1(y_1(t)-y_2(t))-B_2\frac{dy_2(t)}{dt}-k_2y_2(t)\right\}=$$

$$=L\left\{m_2\frac{d^2y_2(t)}{dt^2}\right\}\Rightarrow$$

$$\text{(P1.10.4)}$$

$$-Y_1(s)(B_1s+k_1)+Y_2(s)(m_2s^2+(B_2+B_1)s+(k_2+k_1))=0\Rightarrow$$

$$Y_1(s)=Y_2(s)\frac{m_2s^2+(B_2+B_1)s+k_1+k_2}{B_1s+k_1}$$

$$\text{(P1.10.3)},\text{(P1.10.4)}\Rightarrow Y_2(s)=\frac{(B_1s+k_1)F_1(s)}{(m_1s^2+B_1s+k_1)(m_2s^2+(B_2+B_1)s+k_1+k_2)-(sB_1+k_1)^2} \qquad \text{(P1.10.5)}$$

By substituting $B_1=B_2=2\,\text{N}\cdot\text{s/m}$, $k_1=k_2=1\,\text{N/m}$, $m_2=1\,\text{kg}$, $m_1\cong0\,\text{kg}$ we arrive at

$$Y_2(s)=\frac{(2s+1)F_1(s)}{(2s+1)(s^2+4s+2)-(2s+1)^2}=\frac{F_1(s)}{(s+1)^2} \qquad \text{(P1.10.6)}$$

a. For $f_1(t)=10\,\delta(t)$ we have:

$$f_1(t)=10\delta(t)\overset{\text{L.T.}}{\Rightarrow}F_1(s)=10$$

$$\text{(P1.10.6)}\Rightarrow Y_2(s)=\frac{10}{(s+1)^2}\overset{\text{I.L.T.}}{\Rightarrow}y_2(t)=10te^{-t}$$

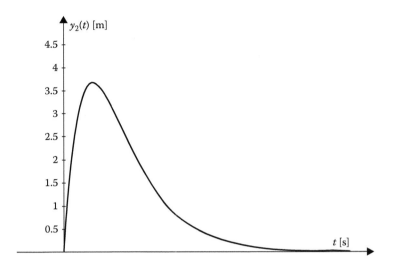

$$\lim_{t \to 0} y_2(t) = \lim_{s \to \infty} sY_2(s) = 0, \text{ initial value theorem}$$

$$\lim_{t \to \infty} y_2(t) = \lim_{s \to 0} sY_2(s) = 0, \text{ final value theorem}$$

b. For $f_1(t) = 10\,u(t)$ we have:

$$f_1(t) = 10u(t) \overset{L.T.}{\Rightarrow} F_1(s) = \frac{10}{s}$$

$$(\text{P1.10.6}) \Rightarrow Y_2(s) = \frac{10}{s(s+1)^2} \overset{I.L.T.}{\Rightarrow} y_2(t) = 10(1-(t+1)e^{-t})$$

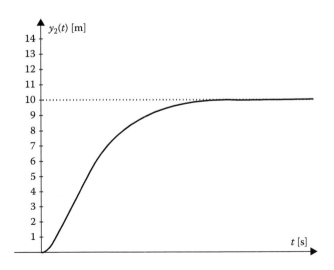

$$\lim_{t \to 0} y_2(t) = \lim_{s \to \infty} sY_2(s) = 0$$

$$\lim_{t \to \infty} y_2(t) = \lim_{s \to 0} sY_2(s) = 10$$

c. For $f_1(t) = 10\,r(t)$ we have:

$$f_1(t) = 10r(t) = 10t \overset{L.T.}{\Rightarrow} F_1(s) = \frac{10}{s^2}$$

$$(\text{P1.10.6}) \Rightarrow Y_2(s) = \frac{10}{s^2(s+1)^2} \overset{I.L.T.}{\Rightarrow} y_2(t) = t(1+e^{-t}) - 2(1-e^{-t})$$

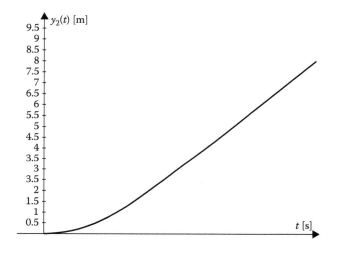

$$\lim_{t \to 0} y_2(t) = \lim_{s \to \infty} sY_2(s) = 0$$

$$\lim_{t \to \infty} y_2(t) = \lim_{s \to 0} sY_2(s) = \infty$$

2

Transfer Functions, Block Diagrams, and Signal Flow Graphs

2.1 Transfer Function

A single-input, single-output (SISO) linear time invariant (LTI) system is completely described by a linear differential equation with constant coefficients, that is, by an nth order differential equation of the form

$$a_n \frac{d^n y(t)}{dt^n} + \cdots + a_1 \frac{dy(t)}{dt} + a_0 y(t) = b_m \frac{d^m x(t)}{dt^m} + \cdots + b_1 \frac{dx(t)}{dt} + b_0 x(t) \tag{2.1}$$

The coefficients a_i, b_j are constants. Moreover, suppose that the initial conditions are zero, that is,

$$y^{(k)}(0) = 0, \quad k = 0, 1, 2, \ldots, n-1$$
$$x^{(k)}(0) = 0, \quad k = 0, 1, 2, \ldots, m-1 \tag{2.2}$$

Applying Laplace transform to both sides of (2.1) yields

$$a_n s^n Y(s) + a_{n-1} s^{n-1} Y(s) + \cdots + a_0 Y(s) = b_m s^m X(s) + b_{m-1} s^{m-1} X(s) + \cdots + b_0 X(s) \tag{2.3}$$

Solving for $Y(s)/X(s)$ yields

$$G(s) = \frac{Y(s)}{X(s)}\bigg|_{I.C.=0} = \frac{b_m s^m + b_{m-1} s^{m-1} + \cdots + b_0}{a_n s^n + a_{n-1} s^{n-1} + \cdots + a_0} \tag{2.4}$$

The function $G(s)$ is called **transfer function** and is a complete description of a LTI system. Thus the transfer function of a LTI system is defined as the ratio of the Laplace transform $Y(s)$ of the output signal $y(t)$ to the Laplace transform $X(s)$ of the input signal $x(t)$, which is applied to the system, supposing zero initial conditions.

2.1.1 Easy Calculation of the Transfer Function

If the differential equation of a system is provided, the system's transfer function can be found directly by substituting $D = d/dt$ with variable s and by substituting the time functions with their corresponding Laplace transforms. More specifically, assume that a SISO system is described by the differential equation

$$\frac{d^n y(t)}{dt^n} + a_{n-1}\frac{d^{n-1}y(t)}{dt^{n-1}} + \cdots + a_0 y(t) = \frac{d^m x(t)}{dt^m} + b_{m-1}\frac{d^{m-1}x(t)}{dt^{m-1}} + \cdots + b_0 x(t) \tag{2.5}$$

or equivalently by

$$(D^n + a_{n-1}D^{n-1} + \cdots + a_0)y(t) = (D^m + b_{m-1}D^{m-1} + \cdots + b_0)x(t) \tag{2.6}$$

Assuming zero initial conditions, the system's transfer function is given directly by the relationship

$$G(s) = \frac{Y(s)}{X(s)} = \frac{s^m + b_{m-1}s^{m-1} + \cdots + b_0}{s^n + a_{n-1}s^{n-1} + \cdots + a_0} \tag{2.7}$$

The polynomial of the transfer function's denominator is called **characteristic polynomial**, while the **characteristic equation** of the system is formed when we suppose that the **characteristic polynomial** equals zero. Certain properties of the system, like the stability or the behavior of the system response, can be examined by the study of the characteristic polynomial. A system's transfer function is sometimes expressed in the zero-pole gain form as shown in the following equation:

$$G(s) = \frac{k(s+z_1)(s+z_2)\ldots(s+z_m)}{(s+p_1)(s+p_2)\ldots(s+p_n)} \tag{2.8}$$

The roots of the numerator's polynomial $z_1, \ldots, z_m$ are called **zeros**, the roots of the denominator's polynomial $p_1, \ldots, p_n$ are called **poles**, while k is called the **gain constant**. The gain constant k is always a real number, in contrast to the zeros and poles which can be complex numbers.

The transfer function $G(s)$ becomes infinite for $s = p_1, \ldots, p_n$ and zero for $s = z_1, \ldots, z_m$. A zero/pole plot in the s-plane reveals the type of a system's transient response.

Finally we mention that the differential equation of a system can be found from the system's transfer function by substituting s with D and the Laplace transforms with the relevant time functions.

2.2 Block Diagrams

A **block diagram** is a graphical representation of the interconnection of subsystems that form a system. It represents both the overall model of the system and the operation of the system's components. Each block of the diagram is symbolized by a rectangle. The name of the block (e.g., an amplifier) or the relevant transfer function of the block is indicated on the rectangle. The following figure represents a block diagram of one element.

The use of block diagrams provides simplicity in the modeling of a system and reveals information for the dynamic behavior. The block diagram of a system can be simplified to fewer blocks.

The next figure illustrates the block diagram of a closed-loop control system with input $u(t)$ and output $y(t)$.

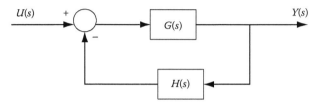

In the depicted block diagram, $U(s)$ is the Laplace transform of the input signal, $G(s)$ is the forward (-path) transfer function, $Y(s)$ is the Laplace transform of the output, and $H(s)$ is the feedback transfer function.

The **closed-loop transfer function** is given by

$$G_{cl}(s) = \frac{Y(s)}{U(s)} = \frac{G(s)}{1 + G(s) \cdot H(s)} \tag{2.9}$$

while the **open-loop transfer function** is defined as

$$F(s) = G(s)H(s) \tag{2.10}$$

The formulas in the next section provide various rules for the simplification and reduction of block diagrams. The analysis of control systems with the use of block diagram reduction is more helpful for understanding how each element contributes to the system, in comparison to the study of the equations of the mathematical model.

2.3 Signal Flow Graphs

Signal flow graphs (SFGs), like block diagrams, provide an alternative method for describing graphically a system. SFG theory was introduced by **Samuel J. Mason** and it can be applied to any system without having to simplify the block diagram, which sometimes can be a quite difficult process.

An **SFG** is actually a simplified version of a block diagram. It consists of nodes, branches, and loops. Each node represents a variable (or signal). There are three categories of nodes:

a. **An input node (source)** is a node which has only outgoing branches (Figure (a)).

b. **An output node (sink)** is a node which has only incoming branches (Figure (b)).

c. **A mixed node** is a node which has both incoming and outgoing branches (Figure (c)).

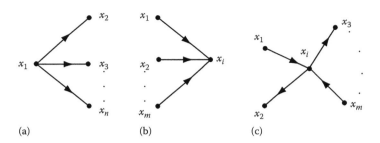

(a)   (b)   (c)

A **branch** connects two nodes and has two characteristics, the direction and the gain. The direction shows the signal flow, while the gain is a coefficient a (corresponding to a transfer function), which relates the variables x_1 and x_2. For the figure below the relationship $x_2 = ax_1$ holds.

Path is a succession of branches with the same direction.

Forward path is the path that starts at an input node and ends at an output node. Every node is traversed only once.

Loop is the closed path that starts and ends at the same node. Two branches of a SFG are called **nontouching** when they do not have any common node.

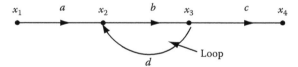

2.4 Mason's Gain Formula

Mason's gain formula or **Mason's rule** is a method for finding the transfer function of a system through its SFG. By Mason's rule there is no need to use block diagram reduction. The mathematical equation is

$$G(s) = \frac{\sum_{n=1}^{k} T_n \Delta_n}{\Delta} \tag{2.11}$$

where

T_n is the gain of the n-order input-output forward path

Δ is the determinant of the diagram, and is given by the following relationship

$$\Delta = 1 - \sum L_1 + \sum L_2 - \sum L_3 + \cdots \tag{2.12}$$

where

L_1 is the gain of each closed loop in the system

L_2 is the product of the gains of any two non-touching loops

L_3 is the product of the gains of any three pairwise non-touching loops and so on

Finally, Δ_n is the determinant of the forward path T_n, and it can be calculated from (2.12) without considering that part of the SFG that is non-touching with the nth forward path.

In case we have a complex block diagram and we need to compute the system transfer function we first convert the block diagram into an SFG and then we apply Mason's gain formula. Finally note that two nodes that are connected in series must be separated by branches of unity transfer functions.

2.5 Response of a Multiple Input System

The linear system depicted in the figure below is excited by three inputs: $X(s)$, $V(s)$, and $R(s)$.

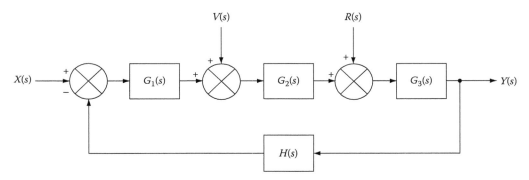

The system's response to each input is independent from the other responses, as the system is linear. Hence, the total output is equal to the sum of the individual outputs (**superposition principle**), that is,

$$Y(s) = Y_X(s) + Y_V(s) + Y_R(s) \tag{2.13}$$

where
  $Y_X(s)$ is the output when $V(s) = R(s) = 0$
  $Y_V(s)$ is the output when $X(s) = R(s) = 0$
  $Y_R(s)$ is the output when $V(s) = X(s) = 0$

In order to find $Y_X(s)$, we plot the following block diagram, assuming that $V(s) = R(s) = 0$.

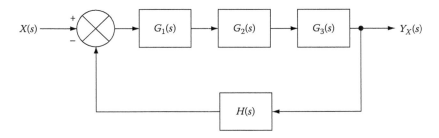

Thus,

$$Y_X(s) = \frac{G_1(s)G_2(s)G_3(s)}{1+G_1(s)G_2(s)G_3(s)H(s)} \cdot X(s) \qquad (2.14)$$

In order to find $Y_V(s)$, we plot the following block diagram, assuming that $X(s) = R(s) = 0$:

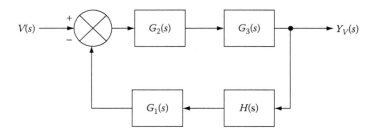

Thus,

$$Y_V(s) = \frac{G_2(s)G_3(s)}{1+G_1(s)G_2(s)G_3(s)H(s)} \cdot V(s) \qquad (2.15)$$

Finally in order to find $Y_R(s)$, we plot the following block diagram, assuming that $V(s) = X(s) = 0$:

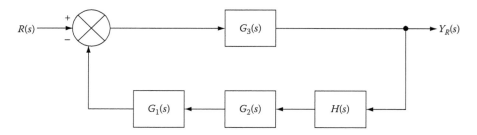

Thus,

$$Y_R(s) = \frac{G_3(s)}{1+G_1(s)G_2(s)G_3(s)H(s)} \cdot R(s) \qquad (2.16)$$

By substituting (2.14) through (2.16) to (2.13), we find the total output of the system:

$$Y(s) = \frac{G_1(s)G_2(s)G_3(s)X(s)+G_2(s)G_3(s)V(s)+G_3(s)R(s)}{1+G_1(s)G_2(s)G_3(s)H(s)} \qquad (2.17)$$

Formulas

TABLE F2.1
Block Diagram Transformations

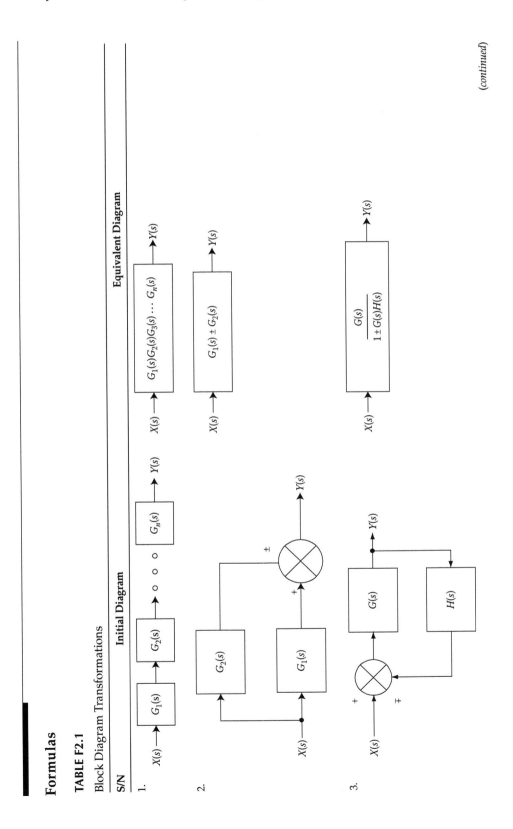

(continued)

TABLE F2.1 (continued)

Block Diagram Transformations

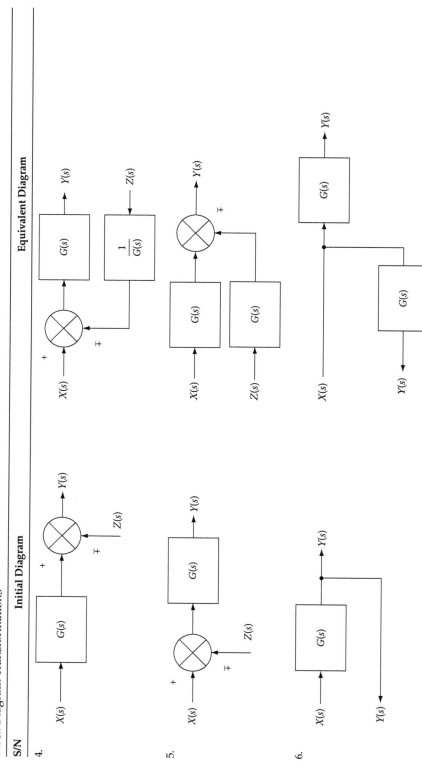

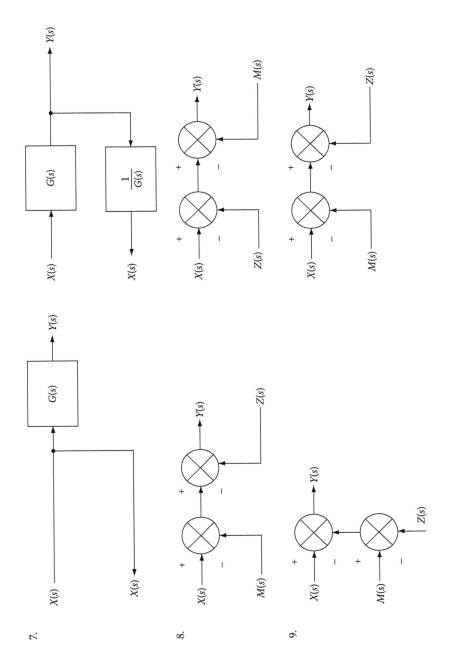

(continued)

TABLE F2.1 (continued)

Block Diagram Transformations

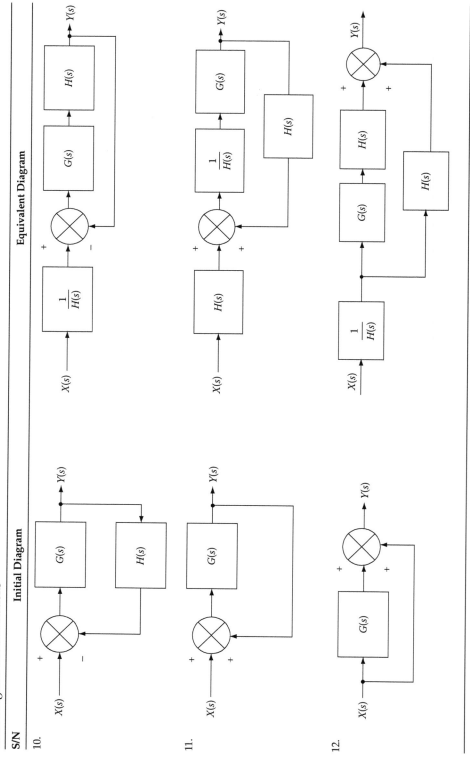

Problems

2.1 Compute the transfer function of the depicted block diagram
   a. By reduction
   b. By plotting the relevant SFG and applying Mason's gain formula

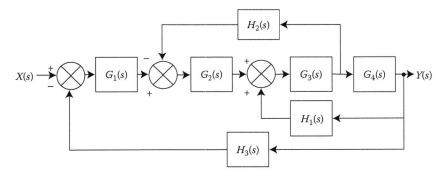

Solution

   a. By applying transformation 7 (Table F2.1), the branch point at the left of the block with transfer function $G_4(s)$ is moved at the right of $G_4(s)$. The equivalent block diagram is:

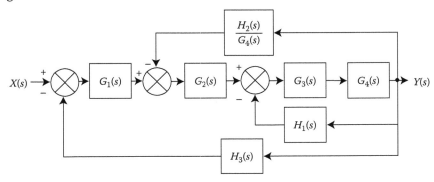

The next block diagram emerges when transformations 1 and 3 are applied to the loop that contains the blocks with transfer functions $G_3(s)$, $G_4(s)$, and $H_1(s)$.

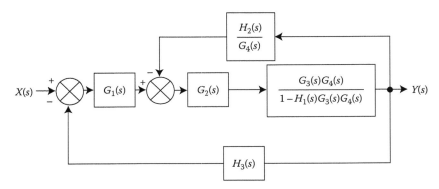

Next we apply transformations 1 and 3 to the loop that contains the transfer function $H_2(s)/G_4(s)$ as feedback and get the following block diagram:

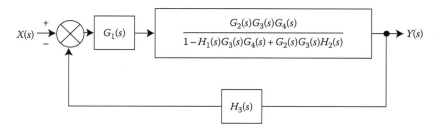

Similarly, by applying transforms 1 and 3 we obtain the simplified block diagram that represents the system's transfer function.

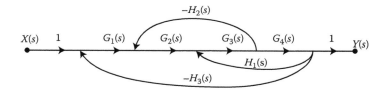

b. The corresponding SFG of the given block diagram is

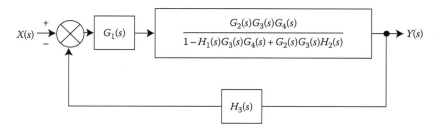

From Mason's gain formula, the transfer function is

$$G(s) = \frac{Y(s)}{X(s)} = \frac{T_1 \Delta_1}{\Delta} \tag{P2.1.1}$$

where

$$T_1 = 1 \cdot G_1(s) \cdot G_2(s) \cdot G_3(s) \cdot G_4(s) \cdot 1 = G_1(s) \cdot G_2(s) \cdot G_3(s) \cdot G_4(s)$$

$$\Delta = 1 - \sum L_1 \tag{P2.1.2}$$

where

$$\sum L_1 = G_1(s)G_2(s)G_3(s)G_4(s)(-H_3(s)) + G_3(s)G_4(s)H_1(s) + G_2(s)G_3(s)(-H_2(s))$$

Thus,

$$\Delta = 1 + G_1(s)G_2(s)G_3(s)G_4(s)H_3(s) + G_2(s)G_3(s)H_2(s) - G_3(s)G_4(s)H_1(s) \tag{P2.1.3}$$

Finally,

$$\Delta_1 = 1 - (0) = 1 \qquad \text{(P2.1.4)}$$

Substituting (P2.1.2), (P2.1.3), and (P2.1.4) to (P2.1.1), we get

$$G(s) = \frac{Y(s)}{X(s)} = \frac{G_1(s)G_2(s)G_3(s)G_4(s)}{1 - G_3(s)G_4(s)H_1(s) + G_2(s)G_3(s)H_2(s) + G_1(s)G_2(s)G_3(s)G_4(s)H_2(s)}$$

As expected the derived transfer function is the same. However it is obvious that the computation of a system's transfer function is much simpler when using Mason's gain formula than using block diagram reduction.

2.2 For the system depicted in the figure below suppose that $a = 1$, $k_1 = 2$, $k_2 = 20$, $F_1 = 0.1$, $F_0 = 1.125$

  a. Express the differential equation of the two input system.

  b. For $N(s) = 0$, find the system's response to $r(t) = 10u(t)$.

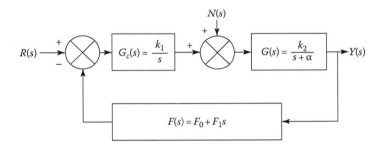

Solution

  a. The total output $Y(s)$ of the system is computed by applying the superposition principle

$$Y(s) = Y_1(s) + Y_2(s) \qquad \text{(P2.2.1)}$$

where
  $Y_1(s)$ is the output for $N(s) = 0$
  $Y_2(s)$ is the output for $R(s) = 0$

For $N(s) = 0$ and from the block diagram of the system, we have

$$Y_1(s) = \frac{G_c(s)G(s)}{1 + G_c(s)G(s)F(s)} R(s) \qquad \text{(P2.2.2)}$$

Similarly, for $R(s) = 0$ and from the block diagram of the system, we get

$$Y_2(s) = \frac{G(s)}{1 + G_c(s)G(s)F(s)} N(s) \qquad \text{(P2.2.3)}$$

From (P2.2.1) and due to (P2.2.2) and (P2.2.3), it follows that

$$Y(s) = \frac{G_c(s)G(s)R(s) + G(s)N(s)}{1 + G_c(s)G(s)F(s)} \qquad \text{(P2.2.4)}$$

With substitution to (P2.2.4)

$$Y(s) = \frac{k_1 k_2 R(s) + k_2 s N(s)}{s(s+a) + k_1 k_2 (F_0 + F_1 s)} \qquad \text{(P2.2.5)}$$

$$(\text{P2.2.5}) \Rightarrow (s^2 + s(a + k_1 k_2 F_1) + k_1 k_2 F_0)Y(s) = k_1 k_2 R(s) + k_2 s N(s)$$

Thus, the differential equation is

$$\frac{d^2 y(t)}{dt^2} + (a + k_1 k_2 F_1)\frac{dy(t)}{dt} + k_1 k_2 F_0 y(t) = k_1 k_2 r(t) + k_2 \frac{dN(t)}{dt} \qquad \text{(P2.2.6)}$$

b. By substituting the parameter values to (P2.2.5) for $N(s) = 0$ and $R(s) = 10/s$, we have

$$Y(s) = \frac{20}{s^2 + 3s + 22.5}R(s) = \frac{200}{(s^2 + 3s + 22.5)s} \qquad \text{(P2.2.7)}$$

Hence

$$Y(s) = \frac{200}{s(s + 1.5 - j4.5)(s + 1.5 + j4.5)}$$

$$(\text{P2.2.8}) \Rightarrow y(t) = 200 \left[\frac{-(1.5 + j4.5)e^{-(1.5 - j4.5)t} + (1.5 - j4.5)e^{-(1.5 + j4.5)t}}{22.5 \cdot j9} + \frac{1}{22.5} \right]$$

$$\Rightarrow y(t) = \frac{200}{22.5} - j\frac{200}{22.5}e^{-1.5t}\left[-1.5\underbrace{(e^{j4.5t} - e^{-j4.5t})}_{2j\sin 4.5t} + j4.5\underbrace{(e^{j4.5t} + e^{-j4.5t})}_{2\cos 4.5t} \right]$$

$$\Rightarrow y(t) = \frac{200}{22.5}\left[1 - e^{-1.5t}\left(\frac{1}{3}\sin 4.5t + \cos 4.5t \right) \right] \qquad \text{(P2.2.8)}$$

2.3 The equations that describe an automatic control system are

$$x_2 = x_1 t_{12} - x_2 t_{22} - x_4 t_{42} - x_3 t_{32} \qquad \text{(P2.1)}$$

$$x_3 = x_2 t_{23} - x_3 t_{33} - x_4 t_{43} \qquad \text{(P2.2)}$$

$$x_4 = x_2 t_{24} + x_3 t_{34} - x_4 t_{44} \qquad \text{(P2.3)}$$

$$x_5 = x_4 t_{45} \qquad \text{(P2.4)}$$

a. Plot the SFG of the system.

b. Compute the transfer function $G(s) = X_5(s)/X_1(s)$ of the system by applying Mason's gain formula.

Solution

a. The equations that describe the system are

$$x_2 = x_1 t_{12} - x_2 t_{22} - x_4 t_{42} - x_3 t_{32} \tag{P2.3.1}$$

$$x_3 = x_2 t_{23} - x_3 t_{33} - x_4 t_{43} \tag{P2.3.2}$$

$$x_4 = x_2 t_{24} + x_3 t_{34} - x_4 t_{44} \tag{P2.3.3}$$

$$x_5 = x_4 t_{45} \tag{P2.3.4}$$

We plot the SFG based on the aforementioned equations.

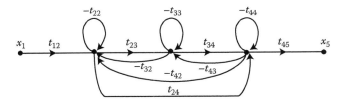

b. From Mason's rule, the transfer function is

$$G(s) = \frac{X_5(s)}{X_1(s)} = \frac{\sum_{n=1}^{2} T_n \Delta_n}{\Delta} = \frac{T_1 \Delta_1 + T_2 \Delta_2}{\Delta} \tag{P2.3.5}$$

where

$$T_1 = t_{12} t_{23} t_{34} t_{45} \tag{P2.3.6}$$

$$T_2 = t_{12} t_{24} t_{45} \tag{P2.3.7}$$

$$\Delta = 1 - \sum L_1 + \sum L_2 - \sum L_3 \tag{P2.3.8}$$

and

$$\sum L_1 = -t_{22} - t_{33} - t_{44} - t_{23} t_{32} - t_{34} t_{43} - t_{23} t_{34} t_{42} - t_{24} t_{42} + t_{24} t_{43} t_{32}$$

$$\sum L_2 = t_{22} t_{33} + t_{22} t_{44} + t_{33} t_{44} + t_{22} t_{34} t_{43} + t_{33} t_{24} t_{42} + t_{44} t_{23} t_{32} \tag{P2.3.9}$$

$$\sum L_3 = -t_{22} t_{33} t_{44}$$

$$\Delta_1 = 1-(0) = 1 \tag{P2.3.10}$$

$$\Delta_2 = 1-(-t_{33}) = 1+t_{33} \tag{P2.3.11}$$

Substituting (P2.3.6) through (P2.3.11) to (P2.3.5) we come up with the transfer function of the system.

2.4 Given the block diagram of the system of next figure, show that if $1 + Ak_mk_T$ is much greater than the time constant T_m, then the system behaves like an integrator.

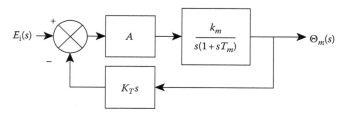

Solution

The transfer function of the system is

$$G(s) = \frac{\Theta_m(s)}{E_i(s)} = \frac{A(k_m/s(1+sT_m))}{1+k_Ts(Ak_m/s(1+sT_m))} = \frac{Ak_m}{s\left[1+Ak_mk_T+sT_m\right]} \Rightarrow$$

$$G(s) = \frac{Ak_m/(1+Ak_mk_T)}{s\left[1+(sT_m/(1+Ak_mk_T))\right]} \tag{P2.4.1}$$

but

$$1+Ak_mk_T \gg T_m \Rightarrow \frac{T_m}{1+Ak_mk_T} \ll 1$$

Thus,

$$1+\frac{sT_m}{1+Ak_mk_T} \simeq 1 \tag{P2.4.2}$$

$$(\text{P2.4.1}),(\text{P2.4.2}) \Rightarrow G(s) = \frac{Ak_m}{1+Ak_mk_T}\cdot\frac{1}{s} = \frac{\Theta_m(s)}{E_i(s)}$$
$$\tag{P2.4.3}$$

$$(\text{P2.4.3}) \Rightarrow \Theta_m(s) = \frac{Ak_m}{1+Ak_mk_T}\frac{E_i(s)}{s} \overset{I.L.T.}{\Rightarrow}$$

$$\theta_m(t) = \frac{Ak_m}{1+Ak_mk_T}\int_0^t e_i(t)\,dt \tag{P2.4.4}$$

Form (P2.4.4) it is obvious that the system behaves like an integrator.

2.5 Find the transfer function of the block diagram depicted in the figure:

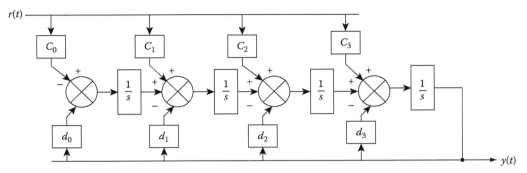

Solution

There are two ways to proceed with the solution:

i. The system equations are

$$C_3 r - d_3 + z_1 = sy \tag{P2.5.1}$$

$$C_2 r - d_1 y + z_2 = s z_1 \tag{P2.5.2}$$

$$C_1 r - d_1 y + z_3 = s z_2 \tag{P2.5.3}$$

$$C_0 r - d_0 y = s z_3 \tag{P2.5.4}$$

Starting from (P2.5.4) and substituting consecutively to the following relationships, we arrive at the transfer function:

$$G(s) = \frac{Y(s)}{R(s)} = \frac{C_0 + C_1 s + C_2 s^2 + C_3 s^3}{d_0 + d_1 s + d_2 s^2 + d_3 s^3 + s^4} \tag{P2.5.5}$$

ii. We plot the SFG

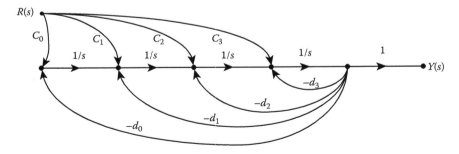

The transfer function is

$$G(s) = \frac{Y(s)}{R(s)} = \frac{T_1 \Delta_1 + T_2 \Delta_2 + T_3 \Delta_3 + T_4 \Delta_4}{\Delta} \tag{P2.5.6}$$

where

$$T_1 = \frac{C_0}{s^4}, \quad T_2 = \frac{C_1}{s^3}, \quad T_3 = \frac{C_2}{s^2}, \quad T_4 = \frac{C_3}{s} \qquad \text{(P2.5.7)}$$

$$\Delta = 1 - \sum L_1 \qquad \text{(P2.5.8)}$$

and

$$\sum L_1 = -\frac{d_3}{s} - \frac{d_2}{s^2} - \frac{d_1}{s^3} - \frac{d_0}{s^4} \qquad \text{(P2.5.9)}$$

$$\Delta_1 = \Delta_2 = \Delta_3 = \Delta_4 = 1 - (0) = 1 \qquad \text{(P2.5.10)}$$

By substituting (P2.5.7) through (P2.5.10) to (P2.5.6) we compute the transfer function of the system:

$$G(s) = \frac{Y(s)}{R(s)} = \frac{(C_0/s^4) + (C_1/s^3) + (C_2/s^2) + (C_3/s)}{1 + (d_3/s) + (d_2/s^2) + (d_1/s^3) + (d_0/s^4)} = \frac{C_0 + C_1 s + C_2 s^2 + C_3 s^3}{d_0 + d_1 s + d_2 s^2 + d_3 s^3 + s^4} \qquad \text{(P2.5.11)}$$

2.6 Find the transfer function of the depicted circuit.

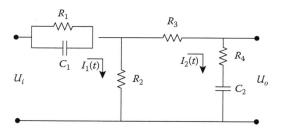

Solution

We will compute the transfer function of the circuit in two ways (i and ii):

i. The circuit is transformed as follows:

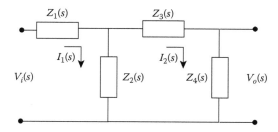

The following Laplace transforms emerge from the circuit:

$$V_i(s) = Z_1(s)I_1(s) + Z_2(s)(I_1(s) - I_2(s)) \qquad \text{(P2.6.1)}$$

$$V_o(s) = Z_4(s)I_2(s) \qquad\qquad (P2.6.2)$$

$$(I_1(s) - I_2(s))Z_2(s) = I_2(s)(Z_3(s) + Z_4(s)) \qquad\qquad (P2.6.3)$$

where

$$Z_1(s) = \frac{R_1 \cdot (1/sC_1)}{R_1 + (1/sC_1)} = \frac{R_1}{sR_1C_1 + 1}$$

$$Z_2(s) = R_2$$

$$Z_3(s) = R_3$$

$$Z_4(s) = R_4 + \frac{1}{sC_2} = \frac{sR_4C_2 + 1}{sC_2}$$

Therefore,

$$G(s) = \frac{V_o(s)}{V_i(s)} \overset{(P2.6.2)}{\underset{(P2.6.1)}{=}} \frac{((sR_4C_2+1)/sC_2)I_2(s)}{(R_1/(sR_1C_1+1))I_1(s) + R_2(I_1(s)-I_2(s))} \qquad (P2.6.4)$$

$$(P2.6.3) \Rightarrow I_1(s) = I_2(s)\frac{(R_2 + R_3 + R_4)C_2s + 1}{R_2C_2s} \qquad\qquad (P2.6.5)$$

$$(P2.6.4),(P2.6.5) \Rightarrow G(s) = \frac{sR_4C_2 + 1}{\left(R_1\left[(R_2+R_3+R_4)C_2s+1\right]/R_2\left(R_1C_1s+1\right)\right) + (R_2+R_3+R_4)C_2s + 1 - R_2C_2s}$$

$$= \frac{R_2(sR_4C_2+1)(R_1C_1s+1)}{R_1R_2(R_3+R_4)C_1C_2s^2 + \left[R_1(R_2+R_3+R_4)C_2 + (R_3+R_4)R_2C_2 + R_1R_2C_1\right]s + R_1 + R_2}$$

The denominator is of the form $as^2 + bs + c = 0 \Rightarrow a(s + s_1)(s + s_2) = 0$ where

$$a = R_1R_2(R_3 + R_4)C_1C_2$$

We now compute the roots s_1, s_2. The discriminant is

$$\Delta = b^2 - 4ac = R_1^2(R_2+R_3+R_4)^2C_2^2 + R_2^2(R_3+R_4)^2C_2^2 + R_1^2R_2^2C_1^2 +$$

$$+ 2R_1(R_2+R_3+R_4)C_2R_2(R_3+R_4)C_2 + + 2R_1(R_2+R_3+R_4)C_2R_1C_1R_2 +$$

$$+ 2R_2(R_3+R_4)C_2R_1C_1R_2 - 4R_1R_2(R_3+R_4)C_1C_2(R_1+R_2) \Rightarrow$$

$$\Delta = \left[R_1(R_2+R_3+R_4)C_2 + R_2(R_3+R_4)C_2 - R_1R_2C_1\right]^2$$

The roots s_1, s_2 are

$$s_{1,2} =$$

$$\frac{-\left[R_1(R_2+R_3+R_4)C_2 + (R_3+R_4)R_2C_2 + R_1R_2C_1\right] \pm \left[R_1(R_2+R_3+R_4)C_2 + R_2(R_3+R_4)C_2 - R_1R_2C_1\right]}{2R_1R_2(R_3+R_4)C_1C_2}$$

Thus

$$s_1 = -\frac{1}{(R_3 + R_4)C_2}$$

$$s_2 = -\frac{R_1R_2 + R_1R_3 + R_1R_4 + R_2R_3 + R_2R_4}{R_1R_2C_1(R_3 + R_4)}$$

Thus, the transfer function is

$$G(s) = \frac{V_o(s)}{V_i(s)}$$

$$= \frac{R_2(R_4C_2s + 1)(R_1C_1s + 1)}{R_1R_2(R_3 + R_4)C_1C_2\left[s + (1/(R_3 + R_4)C_2)\right]\left[s + ((R_1R_2 + R_1R_3 + R_1R_4 + R_2R_3 + R_2R_4)/(R_1R_2(R_3 + R_4)C_1))\right]} \Rightarrow$$

$$G(s) = \frac{V_o(s)}{V_i(s)} = k\frac{(\tau_1s + 1)(\tau_2s + 1)}{(\tau_3s + 1)(\tau_4s + 1)}$$

where

$$\left.\begin{aligned}
k &= \frac{R_2(R_3 + R_4)}{R_1R_2 + R_1R_3 + R_1R_4 + R_2R_3 + R_2R_4} \\[2mm]
\tau_1 &= R_4C_2 \\[2mm]
\tau_2 &= R_1C_1 \\[2mm]
\tau_3 &= (R_3 + R_4)C_2 \\[2mm]
\tau_4 &= \frac{R_1R_2C_1(R_3 + R_4)}{R_1R_2 + R_1R_3 + R_1R_4 + R_2R_3 + R_2R_4}
\end{aligned}\right\} \qquad \text{(P2.6.6)}$$

ii. The SFG is plotted according to the circuit equations.

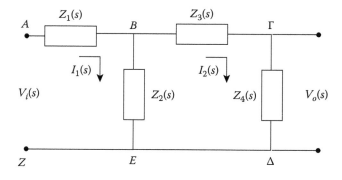

$$I_1(s) = \frac{V_i(s) - V_{BE}(s)}{Z_1(s)} \qquad\qquad \text{(P2.6.7)}$$

$$V_{BE}(s) = (I_1(s) - I_2(s))Z_2(s) \qquad\qquad \text{(P2.6.8)}$$

$$I_2(s) = \frac{V_{BE}(s) - V_o(s)}{Z_3(s)} \qquad \text{(P2.6.9)}$$

$$V_o(s) = I_2(s)Z_4(s) \qquad \text{(P2.6.10)}$$

We plot the SFG of the circuit based on Equations P2.6.7 through P2.6.10.

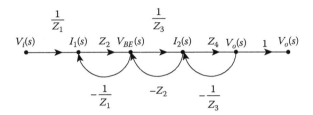

From Mason's gain formula, we have

$$G(s) = \frac{V_o(s)}{V_i(s)} = \frac{T_1 \Delta_1}{\Delta} \qquad \text{(P2.6.11)}$$

where

$$T_1 = \frac{1}{Z_1} \cdot Z_2 \cdot \frac{1}{Z_3} \cdot Z_4 \cdot 1 = \frac{Z_2 Z_4}{Z_1 Z_3} = \frac{(R_2(sR_4C_2 + 1)/sC_2)}{R_1 R_3/(sR_1C_1 + 1)} \Rightarrow$$

$$T_1 = \frac{R_2(sR_4C_2 + 1)(sR_1C_1 + 1)}{sC_2 R_1 R_3} \qquad \text{(P2.6.12)}$$

and

$$\Delta = 1 - \sum L_1 + \sum L_2$$

where

$$\sum L_1 = Z_2 \cdot \left(-\frac{1}{Z_1}\right) + \frac{1}{Z_3} \cdot (-Z_2) + Z_4 \cdot \left(-\frac{1}{Z_3}\right) = -\frac{Z_2}{Z_1} - \frac{Z_2}{Z_3} - \frac{Z_4}{Z_3} \Rightarrow$$

$$\sum L_1 = -\frac{R_2(sR_1C_1 + 1)}{R_1} - \frac{R_2}{R_3} - \frac{sR_4C_2 + 1}{sC_2 R_3}$$

$$\sum L_2 = Z_2 \cdot \left(-\frac{1}{Z_1}\right) Z_4 \cdot \left(-\frac{1}{Z_3}\right) = \frac{Z_2 Z_4}{Z_1 Z_3} \Rightarrow$$

$$\sum L_2 = \frac{R_2(sR_4C_2 + 1)/sC_2}{(R_1/(sR_1C_1 + 1))R_3} = \frac{R_2(sR_4C_2 + 1)(sR_1C_1 + 1)}{sC_2 R_1 R_3}$$

Consequently,

$$\Delta = 1 + \frac{R_2(sR_1C_1 + 1)}{R_1} + \frac{R_2}{R_3} + \frac{sR_4C_2 + 1}{sC_2R_3} + \frac{R_2(sR_4C_2 + 1)(sR_1C_1 + 1)}{sC_2R_1R_3} \quad \text{(P2.6.13)}$$

Finally,

$$\Delta_1 = 1 - (0) = 1 \quad \text{(P2.6.14)}$$

By substituting (P2.6.12) through (P2.6.14) to (P2.6.11), we obtain the transfer function of the circuit.

2.7 Compute the transfer functions $G(s) = V_o(s)/V_i(s)$ of the electrical circuits of Figure (a) and (b):

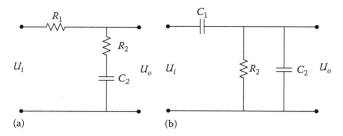

(a)   (b)

Solution

Both circuits have the form shown below:

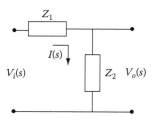

Thus,

$$G(s) = \frac{V_o(s)}{V_i(s)} = \frac{Z_2 I(s)}{(Z_1 + Z_2) I(s)} \Rightarrow \frac{V_o(s)}{V_i(s)} = \frac{Z_2}{Z_1 + Z_2} \quad \text{(P2.7.1)}$$

i. For the electrical circuit in Figure (a), we have

$$\left. \begin{array}{l} Z_1 = R_1 \\[2mm] Z_2 = R_2 + \dfrac{1}{sC_2} = \dfrac{sR_2C_2 + 1}{sC_2} \end{array} \right\} \quad \text{(P2.7.2)}$$

$$(\text{P2.7.1}),(\text{P2.7.2}) \Rightarrow G(s) = \frac{V_o(s)}{V_i(s)} = \frac{(sR_2C_2+1)/sC_2}{R_1+((sR_2C_2+1)/sC_2)} = \frac{sR_2C_2+1}{sR_1C_2+sR_2C_2+1} \Rightarrow$$

$$G(s) = \frac{V_o(s)}{V_i(s)} = \frac{sR_2C_2+1}{sC_2(R_1+R_2)+1}$$

ii. For the electrical circuit of Figure (b), we have

$$\left. \begin{aligned} Z_1 &= \frac{1}{sC_1} \\[2mm] Z_2 &= \frac{R_2(1/sC_2)}{R_2+(1/sC_2)} = \frac{R_2}{sR_2C_2+1} \end{aligned} \right\} \tag{P2.7.3}$$

$$(\text{P2.7.1}),(\text{P2.7.3}) \Rightarrow G(s) = \frac{V_o(s)}{V_i(s)} = \frac{R_2/(sR_2C_2+1)}{(1/sC_1)+(R_2/(sR_2C_2+1))} = \frac{sR_2C_1}{sR_2C_2+1+sR_2C_1} \Rightarrow$$

$$G(s) = \frac{V_o(s)}{V_i(s)} = \frac{sR_2C_1}{sR_2(C_1+C_2)+1}$$

2.8 Figure (b) illustrates the SFG of the electrical circuit shown in Figure (a).

a. Indicate the gains of the branches on the SFG.

b. Find the transfer function $G(s) = V_2(s)/V_1(s)$ of the circuit.

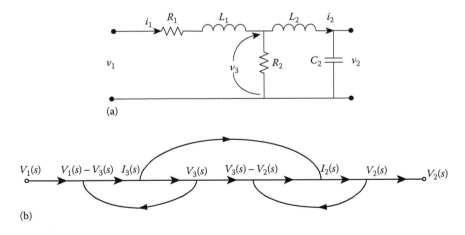

(a)

(b)

Solution

a. The system equations of the electrical circuit are

$$V_1(s) - V_3(s) = (R_1 + sL_1)I_1(s) \tag{P2.8.1}$$

$$I_1(s) = I_2(s) + I_3(s) \tag{P2.8.2}$$

$$(P2.8.1),(P2.8.2) \Rightarrow I_3(s) = \frac{V_1(s) - V_3(s)}{R_1 + sL_1} - I_2(s) \tag{P2.8.3}$$

$$V_3(s) = R_2 I_3(s) \tag{P2.8.4}$$

$$I_2(s) = \frac{V_3(s) - V_2(s)}{sL_2} \tag{P2.8.5}$$

$$V_2(s) = I_2(s)\frac{1}{sC_1} \tag{P2.8.6}$$

The gains of the SFG's circuit branches are derived directly from Equations P2.8.3 through P2.8.6.

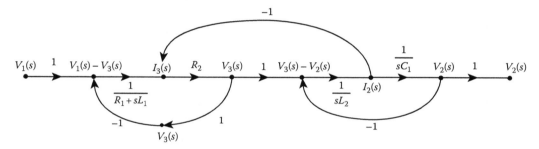

b. From Mason's gain formula, we have

$$G(s) = \frac{V_2(s)}{V_1(s)} = \frac{T_1 \Delta_1}{\Delta} \tag{P2.8.7}$$

where

$$T_1 = 1 \cdot \frac{1}{R_1 + sL_1} \cdot R_2 \cdot 1 \cdot \frac{1}{sL_2} \cdot \frac{1}{sC_1} \cdot 1 = \frac{R_2}{s^2 C_1 L_2 (R_1 + sL_1)} \tag{P2.8.8}$$

$$\Delta = 1 - \sum L_1 + \sum L_2 \tag{P2.8.9}$$

and

$$\left. \begin{aligned} \sum L_1 &= -\frac{R_2}{R_1 + sL_1} - \frac{R_2}{sL_2} - \frac{1}{s^2 L_1 L_2} \\ \sum L_2 &= \frac{R_2}{(R_1 + sL_1)} \cdot \frac{1}{s^2 C_1 L_2} \end{aligned} \right\} \tag{P2.8.10}$$

$$\Delta_1 = 1 - (0) = 1 \tag{P2.8.11}$$

By substitution at (P2.8.7), we get

$$G(s) = \frac{V_2(s)}{V_1(s)}$$

$$= \frac{R_2}{s^3 C_1 L_1 L_2 + s^2 \left(R_1 C_1 L_2 + R_2 C_1 L_2 + R_2 C_1 L_1\right) + s\left(R_1 C_1 R_2 + L_1\right) + R_1 + R_2}$$

2.9 Compute the transfer functions of the electrical circuits shown in Figure (a) and (b).

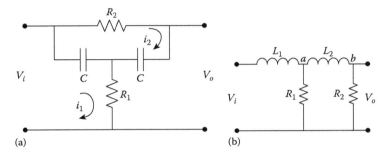

(a)   (b)

Solution

a. It holds that

$$V_i(s) = \left(R_1 + \frac{1}{Cs}\right) I_1(s) - \frac{1}{Cs} I_2(s) \tag{P2.9.1}$$

$$0 = -\frac{1}{Cs} I_1(s) + \left(R_2 + \frac{2}{Cs}\right) I_2(s) \tag{P2.9.2}$$

Solving for $I_2(s)$, we get

$$I_2(s) = \frac{\begin{vmatrix} R_1 + \dfrac{1}{Cs} & V_i(s) \\[2mm] -\dfrac{1}{Cs} & 0 \end{vmatrix}}{\begin{vmatrix} R_1 + \dfrac{1}{Cs} & -\dfrac{1}{Cs} \\[2mm] -\dfrac{1}{Cs} & R_2 + \dfrac{2}{Cs} \end{vmatrix}} = \frac{V_i(s)Cs}{1 + (2R_1 + R_2)Cs + R_1 R_2 C^2 s^2} \tag{P2.9.3}$$

as

$$V_o(s) = V_i(s) - I_2(s)R_2 \tag{P2.9.4}$$

Substituting $I_2(s)$ to (P2.9.4) from (P2.9.3), we get

$$\frac{V_o(s)}{V_i(s)} = \frac{1 + 2R_1 Cs + R_1 R_2 C^2 s^2}{1 + (2R_1 + R_2)Cs + R_1 R_2 C^2 s^2}$$

b. The equations for the nodes a and b, after applying the node method, are

$$\frac{V_a(s)-V_i(s)}{sL_1} + \frac{V_a(s)}{R_1} + \frac{V_a(s)-V_\beta(s)}{sL_2} = 0 \tag{P2.9.5}$$

$$\frac{V_\beta(s)-V_\alpha(s)}{sL_2} + \frac{V_\beta(s)}{R_2} = 0 \tag{P2.9.6}$$

but $V_0(s) = V_\beta(s)$. Thus,

$$(\text{P2.9.1}),(\text{P2.9.2}),(\text{P2.9.3}) \Rightarrow \frac{V_o(s)}{V_i(s)} = \frac{R_1R_2}{s^2L_1L_2 + s(L_1R_1 + L_1R_2 + L_2R_1) + R_1R_2}$$

2.10 Compute the transfer function $G(s) = V_o(s)/V_i(s)$ of the circuit depicted in the following figure. Assume that $RC_1 \gg RC_2$. This system is called PID controller.

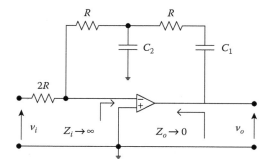

Solution
The system equations are

$$V_i(s) = 2RI_1(s) \tag{P2.10.1}$$

$$V_o(s) = I_2(s)\left(R + \frac{1}{sC_1}\right) - I_1(s)R \tag{P2.10.2}$$

$$I_1(s)R + (I_1(s) + I_2(s))\frac{1}{sC_2} = 0 \tag{P2.10.3}$$

$$(\text{P2.10.1}),(\text{P2.10.2}) \Rightarrow \frac{V_o(s)}{V_i(s)} = G(s) = \frac{I_2(s)((sC_1R+1)/sC_1) - I_1(s)R}{2RI_1(s)} \tag{P2.10.4}$$

$$(\text{P2.10.3}) \Rightarrow I_2(s) = -I_1(s)(sC_2R + 1) \tag{P2.10.5}$$

$$(P2.10.4), (P2.10.5) \Rightarrow G(s) = -\frac{(sC_1R+1)(sC_2R+1)-sC_1R}{2sC_1R} \Rightarrow$$

$$G(s) = -\frac{sC_1R(1+(1/sC_1R))((sC_2R/sC_1R)+(1/sC_1R))-(sC_1R/sC_1R)}{2} \Rightarrow$$

$$G(s) = -\frac{sC_1R+1-1}{2} \Rightarrow G(s) = -\frac{sC_1R}{2}$$

2.11 The next system depicts how voltage feedback can attain a practically constant output voltage, despite the variations in the load current.

a. Describe the operation of the system.

b. Plot the block diagram of the system supposing a voltage drop of $-I_a(s)R_a(s)$ as a disturbance.

c. Find the output $V_L(s)$ of the system.

d. Suppose the parameters of the system are

$$k_a = 2000 \text{ V/V}, \quad R_f = 200\,\Omega, \quad k_1 = 300 \text{ V/A}, \quad f = R'/R_g = 0.1$$

Calculate the error voltage $V_e(t)$ when the desired output voltage of the generator, with its load disconnected, is 200 V.

e. Find the demanded reference voltage $V_r(t)$, given the parameters of question (d).

f. Explain how voltage is practically stabilized after connecting the load (given that $R_a = 1\,\Omega$, $i_a = 20\,\text{A}$).

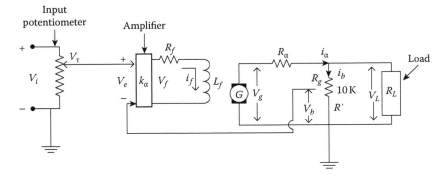

Solution

a. We assume that the load current is practically equal to the current $i_a(t)$. This is so because the resistance of the feedback potentiometer has been conveniently chosen ($R_g = 10\,\text{K}$), so that the passing current to be insignificant.

The operation of the system can be described in the following way. The reference voltage $V_r(t)$ and the feedback voltage $V_b(t)$ are adjusted so that when $R_L \to \infty$, the error voltage $V_e(t) = V_r(t) - V_b(t)$ is such that the generator provides the needed voltage $V_L(t)$.

However, when the load's resistance R_L is reduced, then the load current becomes $i_L(t) \neq 0$, and as $i_b(t) = i_a(t) - i_L(t)$ the current is $i_b(t) \neq i_a(t)$. Hence the voltage $V_b(t)$ is reduced, which leads to the increase of the error voltage $V_e(t)$ ($V_e(t) = V_r(t) - V_b(t)$).

This increase of $V_e(t)$ results into an increase of the amplifier's output voltage, that is, of $V_f(t)$ $(V_f(t) = k_a V_e(t))$, which in turn increases the field current $i_f(t)$. The increase of $i_f(t)$ provokes an increase in the electromotive force of the generator $V_g(t)$. Finally, an increase of $V_g(t)$ results in an increase of $V_L(t)$, which leads $V_L(t)$ to take its initial value, that is, the value it had when $R_L \to \infty$.

b. The equations that describe the voltage-control system are

$$V_e(t) = V_r(t) - V_b(t) \tag{P2.11.1}$$

$$V_f(t) = k_a V_e(t) \tag{P2.11.2}$$

$$V_f(t) = R_f i_f(t) + L_f \frac{di_f(t)}{dt} \tag{P2.11.3}$$

$$V_g(t) = k_1 i_f(t) \tag{P2.11.4}$$

$$V_g(t) = R_a i_a(t) + V_L(t) \tag{P2.11.5}$$

$$V_b(t) = \frac{R'}{R_g} V_L(t) = f V_L(t) \tag{P2.11.6}$$

Supposing zero initial conditions we apply Laplace transform to the system equations and get

$$(\text{P2.11.1}) \overset{L.T.}{\Rightarrow} V_e(s) = V_r(s) - V_b(s) \tag{P2.11.1'}$$

$$(\text{P2.11.2}) \overset{L.T.}{\Rightarrow} V_f(s) = k_a V_e(s) \tag{P2.11.2'}$$

$$(\text{P2.11.3}) \overset{L.T.}{\Rightarrow} V_f(s) = (R_f + sL_f)I_f(s) \tag{P2.11.3'}$$

$$(\text{P2.11.4}) \overset{L.T.}{\Rightarrow} V_g(s) = k_1 I_f(s) \tag{P2.11.4'}$$

$$(\text{P2.11.5}) \overset{L.T.}{\Rightarrow} V_g(s) = R_a I_a(s) + V_L(s) \tag{P2.11.5'}$$

$$(\text{P2.11.6}) \overset{L.T.}{\Rightarrow} V_b(s) = f V_L(s) \tag{P2.11.6'}$$

From Equations P2.11.1 through P2.11.6, we plot the block diagram of the system. The voltage drop $-R_a I_a(s)$ is considered disturbance.

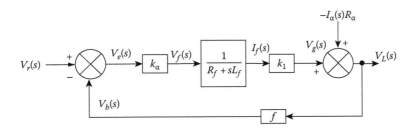

c. According to the superposition principle, the output $V_L(s)$ is

$$V_L(s) = V_L'(s)\big|_{-R_aI_a(s)=0} + V_L''(s)\big|_{V_r(s)=0} \tag{P2.11.7}$$

- Computation of $V_L'(s)$
  We have

$$V_L'(s) = G_1(s)V_r(s) \tag{P2.11.8}$$

We now compute $G_1(s)$ from the block diagram of the system, supposing that $R_aI_a(s) = 0$. Hence,

$$G_1(s) = \frac{k_a k_1}{sL_f + R_f + fk_a k_1} \tag{P2.11.9}$$

$$(P2.11.8),(P2.11.9) \Rightarrow V_L'(s) = \frac{k_a k_1}{sL_f + R_f + fk_a k_1}V_r(s) \tag{P2.11.10}$$

- Computation of $V_L''(s)$
  We have

$$V_L''(s) = G_2(s)(-I_a(s)R_a) \tag{P2.11.11}$$

Similarly, for the computation of $G_2(s)$ we suppose that $V_r(s) = 0$. Hence,

$$G_2(s) = \frac{sL_f + R_f}{sL_f + R_f + fk_a k_1} \tag{P2.11.12}$$

$$(P2.11.11),(P2.11.12) \Rightarrow V_L''(s) = \frac{sL_f + R_f}{sL_f + R_f + fk_a k_1}(-I_a(s)R_a) \tag{P2.11.13}$$

Substitution (P2.11.10) and (P2.11.13) to (P2.11.7), we get

$$V_L(s) = \frac{k_a k_1 V_r(s) - (sL_f + R_f)I_a(s)R_a}{sL_f + R_f + fk_a k_1} \tag{P2.11.14}$$

d. Let the desired output voltage of the generator be 200 V, with its load disconnected. Then the demanded input voltage $V_e(t)$ at the amplifier is

$$V_e = \frac{R_f}{k_a k_1} E_g = \frac{200}{200 \cdot 300} \cdot 200 = 0.067 \text{ V} \tag{P2.11.15}$$

e. The demanded reference voltage is

$$V_e = V_r - V_b = V_r - fV_L \Rightarrow V_r = V_e + fV_L = 0.067 + 0.1 \cdot 200 \Rightarrow$$

$$V_r = 20.067 \text{ V} \tag{P2.11.16}$$

f. If we connect the load, then the voltage at its ends is

$$V_L = V_g - R_a I_a \tag{P2.11.17}$$

Hence, for $R_a = 1 \, \Omega$ and $I_a = 20 \, \text{A}$ we have

$$(\text{P2.11.17}) \Rightarrow V_L = 200 - 20 \cdot 1 \Rightarrow V_L = 180 \text{ V} \quad \text{(voltage decrease)} \tag{P2.11.18}$$

However, when V_L becomes 180 V, the feedback voltage becomes

$$V_b = fV_L = 0.1 \cdot 180 = 18 \text{ V} \quad \text{(voltage decrease)} \tag{P2.11.19}$$

Thus, the error voltage V_e is

$$V_e = V_r - V_b = 20.067 - 18 = 2.067 \text{ V} \quad \text{(voltage increase)} \tag{P2.11.20}$$

The increase of V_e corresponds to an increase in the excitation current i_f which results an increase in the electromotive force at the generator that compensates the induced decrease.

Suppose that initially $i_a = 0 \, \text{A}$. Then from the relationships (P2.11.8) and (P2.11.9) we have

$$V_L'(s) = \frac{2000 \cdot 300}{(sL_f + 200) + 2000 \cdot 300 \cdot 0.1} \cdot \frac{20.067}{s} \Rightarrow \tag{P2.11.21}$$

$$\lim_{t \to \infty} V_L'(t) = \lim_{s \to 0} V_L'(s) \cdot s = 200 \text{ V}$$

Assuming that $i_a = 20 \, \text{A}$, then from (P2.11.7) we get

$$\lim_{t \to \infty} V_L(t) = \lim_{s \to 0} V_L(s) \cdot s = 200 + \lim_{s \to 0} s \cdot G_2(s)(-I_a(s)R_a)$$

$$= 200 - \lim_{s \to 0} s \cdot \frac{sL_f + R_f}{sL_f + R_f + fk_a k_1} \cdot 1 \cdot \frac{20}{s} = 200 - \frac{200 \cdot 20}{200 + 2000 \cdot 300 \cdot 0.1} \Rightarrow$$

$$\lim_{t \to \infty} V_L(t) = 199.93 \text{ V} \tag{P2.11.22}$$

We notice that the closed system's voltage drop is insignificant (199.93 V instead of 200 V), and that the output voltage practically remains at the desired value of 200 V.

2.12 The depicted linear mechanical system is described by the equations

$$f(t) = k_1 x_1(t) + M_1 \frac{d^2 x_1(t)}{dt^2} + k_2(x_1(t) - x_2(t)) \tag{P2.5}$$

$$0 = M_2 \frac{d^2 x_2(t)}{dt^2} + x_2(t)(k_2 + k_3) - k_2 x_1(t) \tag{P2.6}$$

a. Find the transfer functions

$$G_1(s) = \frac{X_1(s)}{F(s)} \quad \text{and} \quad G_2(s) = \frac{X_2(s)}{F(s)}$$

b. Compute the displacement $x_2(t)$ of the mass M_2 given that the force is $f(t) = 10e^{-t}$, and the system's parameters are: $M_1 = 0.25\,\text{Kgr}$, $k_1 = 2\,\text{N/m}$, $k_2 = k_3 = 4\,\text{N/m}$, $M_2 \simeq 0$. All initial conditions are zero.

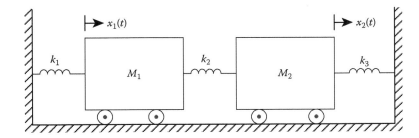

Solution

a. We apply Laplace transform to the system equations that describe the mechanical system and have

$$(P2.5) \Rightarrow f(t) = k_1 x_1(t) + M_1 \frac{d^2 x_1(t)}{dt^2} + k_2(x_1(t) - x_2(t)) \overset{L.T.}{\Rightarrow} \tag{P2.12.1}$$

$$F(s) = X_1(s)(M_1 s^2 + k_1 + k_2) - k_2 X_2(s)$$

$$(P2.6) \Rightarrow 0 = M_2 \frac{d^2 x_2(t)}{dt^2} + x_2(t)(k_2 + k_3) - k_2 x_1(t) \overset{L.T.}{\Rightarrow} \tag{P2.12.2}$$

$$k_2 X_1(s) = X_2(s)(M_2 s^2 + k_2 + k_3)$$

$$(P2.12.1), (P2.12.2) \Rightarrow F(s) = \left(\frac{(M_1 s^2 + k_1 + k_2)(M_2 s^2 + k_2 + k_3)}{k_2} - k_2 \right) X_2(s) \Rightarrow$$

$$G_2(s) = \frac{X_2(s)}{F(s)} = \frac{k_2}{(M_1 s^2 + k_1 + k_2)(M_2 s^2 + k_2 + k_3) - k_2^2} \tag{P2.12.3}$$

$$(P2.12.2), (P2.12.3) \Rightarrow G_1(s) = \frac{X_1(s)}{F(s)} = \frac{M_2 s^2 + k_2 + k_3}{(M_1 s^2 + k_1 + k_2)(M_2 s^2 + k_2 + k_3) - k_2^2} \quad \text{(P2.12.4)}$$

b.
$$(P2.12.3) \Rightarrow \frac{X_2(s)}{F(s)} = \frac{4}{(0.25 s^2 + 6) \cdot 8 - 16} \quad \text{(P2.12.5)}$$

$$x_2(t) = ILT\{X_2(s)\} \overset{(P2.12.5)}{=} ILT\left\{ \frac{4}{(0.25 s^2 + 6) \cdot 8 - 16} F(s) \right\} \quad \text{(P2.12.6)}$$

but

$$F(s) = LT\{f(t)\} = LT\{10 e^{-t}\} = \frac{10}{s+1} \quad \text{(P2.12.7)}$$

$$(P2.12.6), (P2.12.7) \Rightarrow x_2(t) = ILT\left\{ \frac{4}{(0.25 s^2 + 6) \cdot 8 - 16} \cdot \frac{10}{s+1} \right\} \Rightarrow$$

$$x_2(t) = L^{-1}\left\{ \frac{20}{(s^2 + 16)(s+1)} \right\} \quad \text{(P2.12.8)}$$

We have

$$\frac{1}{(s^2 + 16)(s+1)} = \frac{As + B}{s^2 + 16} + \frac{C}{s+1}$$

where

$$C = \lim_{s \to -1} \frac{1}{(s^2 + 16)(s+1)}(s+1) = \frac{1}{17}$$

$$(As + B)(s+1) + C(s^2 + 16) \equiv 1 \Rightarrow$$

$$(A + \Gamma)s^2 + (A + B)s + B + 16C \equiv 1 \Rightarrow$$

$$A = -\frac{1}{17}, \quad B = \frac{1}{17}$$

Hence,

$$\frac{1}{(s^2 + 16)(s+1)} = \frac{1}{17}\left(\frac{-s+1}{s^2 + 16} + \frac{1}{s+1} \right) \quad \text{(P2.12.9)}$$

$$(P2.12.8), (P2.12.9) \Rightarrow x_2(t) = \frac{20}{17} L^{-1}\left\{ \frac{-s+1}{s^2 + 16} + \frac{1}{s+1} \right\} \Rightarrow$$

$$x_2(t) = \frac{20}{17}(e^{-t} - \sin 4t + \cos 4t) \quad \text{(P2.12.10)}$$

2.13 Find the transfer functions of the systems depicted in Figure (a) and (b).

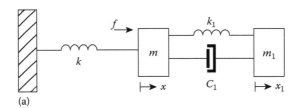

(a)

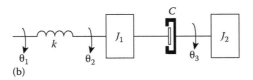

(b)

Solution

a. The differential equations that describe the movement of the system are

$$m\ddot{x} = -kx - k_1(x - x_1) - C_1(\dot{x} - \dot{x}_1) + f \Rightarrow$$
$$m\ddot{x} + C_1\dot{x} + (k + k_1)x = C_1\dot{x}_1 + k_1x_1 + f$$

(P2.13.1)

and

$$m_1\ddot{x}_1 = k_1(x - x_1) + C_1(\dot{x} - \dot{x}_1) \Rightarrow$$
$$m_1\ddot{x}_1 + C_1\dot{x}_1 + k_1x_1 = C_1\dot{x} + k_1x$$

(P2.13.2)

Assuming zero initial conditions, we apply Laplace transform to (P2.13.1) and (P2.13.2) and get

$$(ms^2 + C_1s + k + k_1)X(s) = (C_1s + k_1)X_1(s) + F(s)$$

(P2.13.1')

$$(m_1s^2 + C_1s + k_1)X_1(s) = (C_1s + k_1)X(s)$$

(P2.13.2')

$$(\text{P2.13.2}') \Rightarrow X_1(s) = \frac{C_1s + k_1}{m_1s^2 + C_1s + k_1}X(s)$$

(P2.13.3)

$$(\text{P2.13.1}'), (\text{P2.13.3}) \Rightarrow \left[(ms^2 + C_1s + k + k_1) - \frac{C_1s + k_1}{m_1s^2 + C_1s + k_1}\right]X(s) = F(s)$$

(P2.13.4)

The system transfer function $G(s) = X(s)/F(s)$ is given by

$$G(s) = \frac{X(s)}{F(s)} \overset{(\text{P2.13.4})}{=} \frac{(ms^2 + C_1s + k + k_1)(m_1s^2 + C_1s + k_1) - (C_1s + k_1)}{m_1s^2 + C_1s + k_1}$$

(P2.13.5)

b. Let us suppose that $\theta_1 > \theta_2$, $\dot{\theta}_2 > \dot{\theta}_3$. The system equations are

$$k(\theta_1 - \theta_2) = J_1\ddot{\theta}_2 + C(\dot{\theta}_2 - \dot{\theta}_3) \Rightarrow$$

$$J_1\ddot{\theta}_2 + C\dot{\theta}_2 + k\theta_2 = C\dot{\theta}_3 + k\theta_1 \tag{P2.13.6}$$

and

$$C(\dot{\theta}_2 - \dot{\theta}_3) = J_2\ddot{\theta}_3 \Rightarrow$$

$$J_2\ddot{\theta}_3 + C\dot{\theta}_3 = C\dot{\theta}_2 \tag{P2.13.7}$$

We apply Laplace transform to the system equations and have

$$(J_1s^2 + Cs + k) = Cs\Theta_3(s) + k\Theta_1(s) \tag{P2.13.6'}$$

$$(J_2s^2 + Cs)\Theta_3(s) = Cs\Theta_2(s) \tag{P2.13.7'}$$

$$(\text{P2.13.7'}) \Rightarrow \Theta_2(s) = \left(\frac{J_2}{C}s + 1\right)\Theta_3(s) \tag{P2.13.8}$$

By substituting (P2.13.8) to (P2.13.6'), we compute the transfer function

$$G(s) = \frac{\Theta_3(s)}{\Theta_1(s)} = \frac{k}{(J_1J_2/C)s^3 + (J_1 + J_2)s^2 + (J_2k/C)s + k} \tag{P2.13.9}$$

2.14 Compute the transfer function $G(s) = \Theta_3(s)/T(s)$ for the system depicted in the following figure. Assume zero initial conditions.

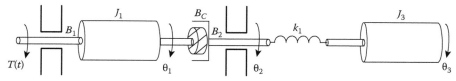

Solution

We have

$$J_1\frac{d^2\theta_1}{dt^2} = T(t) - B_1\frac{d\theta_1}{dt} - B_C\frac{d(\theta_1 - \theta_2)}{dt} \tag{P2.14.1}$$

$$0 = -B_C\frac{d(\theta_2 - \theta_1)}{dt} - k_1(\theta_2 - \theta_3) - B_2\frac{d\theta_2}{dt} \tag{P2.14.2}$$

$$J_3\frac{d^2\theta_3}{dt^2} = -k_1(\theta_3 - \theta_2) \tag{P2.14.3}$$

Assuming zero initial conditions and applying Laplace transform to the aforementioned equations, we obtain

$$(J_1s^2 + B_1s + B_Cs)\Theta_1(s) - B_Cs\Theta_2(s) = T(s) \tag{P2.14.1'}$$

$$-B_Cs\Theta_1(s) + (B_Cs + k_1 + B_2s)\Theta_2(s) - k_1\Theta_3(s) = 0 \tag{P2.14.2'}$$

$$-k_1\Theta_2(s) + (J_3s^2 + k_1)\Theta_3(s) = 0 \tag{P2.14.3'}$$

Solving for $\Theta_3(s)$, we have

$$\Theta_3(s) = \frac{\begin{vmatrix} J_1s^2 + B_1s + B_Cs & -B_Cs & T(s) \\ -B_Cs & B_Cs + k_1 + B_2s & 0 \\ 0 & -k_1 & 0 \end{vmatrix}}{\begin{vmatrix} J_1s^2 + B_1s + B_Cs & -B_Cs & 0 \\ -B_Cs & B_Cs + k_1 + B_2s & -k_1 \\ 0 & -k_1 & J_3s^2 + k_1 \end{vmatrix}} \qquad \text{(P2.14.4)}$$

By solving (4) we compute the transfer function

$$G(s) = \frac{\Theta_3(s)}{T(s)}$$

$$= \frac{B_Ck_1/(s(B_C + B_2))}{J_3[J_1s^3 + B_1s^2 + k_1((B_C + B_1)/(B_C + B_2))s + k_1B_1] + J_3((J_1k_1 + BB_C)/(B_C + B_2))s^2 + J_1k_1s + B_CB_2}$$

$$\text{(P2.14.5)}$$

2.15 Compute the transfer function $G(s) = X(s)/E_1(s)$ for the position control system that is depicted below. Assume that all initial conditions are zero.

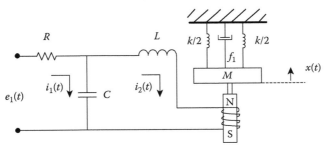

Solution

The mathematical model of the system shown in the above figure is

$$e_1(t) = Ri_1(t) + \frac{1}{C}\int_0^t i_1(t) - \frac{1}{C}\int_0^t i_2(t) \qquad \text{(P2.15.1)}$$

$$0 = -\frac{1}{C}\int_0^t i_1(t) + L\frac{di_2(t)}{dt} + \frac{1}{C}\int_0^t i_2(t)d(t) + e_b(t) \qquad \text{(P2.15.2)}$$

$$e_b(t) = k_1\frac{dx(t)}{dt} \qquad \text{(P2.15.3)}$$

$$0 = f(t) + M\frac{d^2x(t)}{dt^2} + f_1\frac{dx(t)}{dt} + kx(t) \qquad \text{(P2.15.4)}$$

$$f(t) = k_2i_2(t) \qquad \text{(P2.15.5)}$$

We apply Laplace transform to the equations of the model, and supposing zero initial conditions we get

$$(P2.15.1) \overset{\text{L.T.}}{\Rightarrow} E_1(s) = \left(R + \frac{1}{sC}\right)I_1(s) - \frac{I_2(s)}{sC} \tag{P2.15.1'}$$

$$(P2.15.2) \overset{\text{L.T.}}{\Rightarrow} 0 = -\frac{1}{C}\frac{I_1(s)}{s} + LsI_2(s) + \frac{1}{C}\frac{I_2(s)}{s} + E_b(s) \Rightarrow$$

$$\frac{I_1(s)}{sC} = I_2(s)\left(Ls + \frac{1}{sC}\right) + E_b(s) \tag{P2.15.2'}$$

$$(P2.15.3) \overset{\text{L.T.}}{\Rightarrow} E_b(s) = k_1 s X(s) \tag{P2.15.3'}$$

$$(P2.15.4) \overset{\text{L.T.}}{\Rightarrow} F(s) = -X(s)(Ms^2 + sf_1 + k) \tag{P2.15.4'}$$

$$(P2.15.5) \overset{\text{L.T.}}{\Rightarrow} F(s) = k_2 I_2(s) \tag{P2.15.5'}$$

The transfer function is

$$G(s) = \frac{X(s)}{E_1(s)} \overset{(P2.15.1')}{=} \frac{X(s)}{I_1(s)((RsC+1)/sC) - I_2(s)(1/sC)} = \frac{sCX(s)}{I_1(s)(RsC+1) - I_2(s)} =$$

$$\overset{(P2.15.2')}{=} \frac{sCX(s)}{\left[I_2(s)((s^2LC+1)/sC) + E_b(s)\right]sC(RsC+1) - I_2(s)} =$$

$$\overset{(P2.15.3')}{=} \frac{sCX(s)}{I_2(s)(s^2LC+1)(RsC+1) + sC(RsC+1)k_1 s X(s) - I_2(s)} =$$

$$= \frac{sCX(s)}{I_2(s)[(s^2LC+1)(RsC+1) - 1] + k_1 s^2 C(RsC+1)X(s)} =$$

$$\overset{(P2.15.5')}{=} \frac{sCX(s)}{(F(s)/k_2)[(s^2LC+1)(RsC+1) - 1] + k_1 s^2 C(RsC+1)X(s)} =$$

$$\overset{(P2.15.4')}{=} \frac{sCX(s)}{-(X(s)(Ms^2 + sf_1 + k)/k_2)[(s^2LC+1)(RsC+1) - 1] + k_1 s^2 C(RsC+1)X(s)} =$$

$$= \frac{sCk_2}{-(Ms^2 + sf_1 + k)[(s^2LC+1)(RsC+1) - 1] + k_1 k_2 s^2 C(RsC+1)} \Rightarrow$$

$$G(s) = \frac{X(s)}{E_1(s)} = \frac{sCk_2}{(Ms^2 + sf_1 + k)[1 - (s^2LC+1)(RsC+1)] + k_1 k_2 s^2 C(RsC+1)}$$

Hence

$$G(s) = \frac{sk_2}{k_1 k_2 s^2 (RsC+1) - (Ms^2 + sf_1 + k)(s^3 RLC + s^2 L + sR)} \tag{P2.15.6}$$

2.16 The motor of the figure below, is functioning under a constant field current. The parameters of the motor are

$$R_a = 0.5\,\Omega, \quad L_a = 0.05\,\text{H}, \quad k_\omega = 2\,\text{V/r/s}, \quad k_t = 5\,\text{N}\cdot\text{m/A}, \quad B = 1\,\text{N}\cdot\text{m/r/s}$$

and $J = 45\,\text{N}\cdot\text{m/r/s}^2$

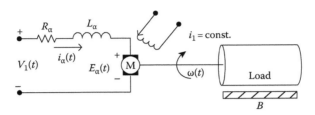

a. Plot the block diagram and the SFG of the system.
b. Find the transfer function $G(s) = \Omega(s)/V_i(s)$.
c. If $V_i(t) = 200u(t)$ V, compute the steady-state angular velocity of the motor.
d. Find the new steady-state velocity of the motor if a load torque $T'(t) = 40\,\text{N}\cdot\text{m}$ is applied on the axis of the motor.

Solution

a. The equations that describe the system are

$$v_i(t) = R_a i_a(t) + L_a \frac{di_a(t)}{dt} + E_a(t) \tag{P2.16.1}$$

$$T(t) = k_t i_a(t) \tag{P2.16.2}$$

$$T(t) = J \frac{d\omega(t)}{dt} + B\omega(t) \tag{P2.16.3}$$

$$E_a(t) = k_\omega \omega(t) \tag{P2.16.4}$$

Assuming zero initial conditions, we apply Laplace transform to the system equations and get

$$(\text{P2.16.1}) \overset{\text{L.T.}}{\Rightarrow} V_i(s) = (R_a + sL_a)I_a(s) + E_a(s) \tag{P2.16.1'}$$

or

$$V_i(s) - E_a(s) = (R_a + sL_a)I_a(s) \tag{P2.16.1''}$$

$$(\text{P2.16.2}) \overset{\text{L.T.}}{\Rightarrow} T(s) = k_t I_a(s) \tag{P2.16.2'}$$

$$(P2.16.3) \overset{\text{L.T.}}{\Rightarrow} T(s) = (Js+B)\Omega(s) \tag{P2.16.3'}$$

$$(P2.16.4) \overset{\text{L.T.}}{\Rightarrow} E_a(s) = k_\omega \Omega(s) \tag{P2.16.4'}$$

Based on Equations P2.16.1″ and P2.16.2′ through P2.16.4′ we plot the block diagram of the system.

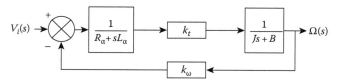

The relevant SFG is depicted below:

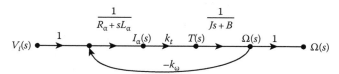

b. The system transfer function, after Mason's gain formula, is

$$G(s) = \frac{\Omega(s)}{V_i(s)} = \frac{T_1 \Delta_1}{\Delta} \tag{P2.16.5}$$

where

$$T_1 = \frac{1}{R_a + sL_a} \cdot k_t \cdot \frac{1}{Js+B} = \frac{k_t}{(R_a + sL_a)(Js+B)} \tag{P2.16.6}$$

$$\Delta = 1 - \sum L_1 \tag{P2.16.7}$$

and

$$\sum L_1 = \frac{1}{R_a + sL_a} \cdot k_t \cdot \frac{1}{Js+B} \cdot (-k_\omega) = -\frac{k_t k_\omega}{(R_a + sL_a)(Js+B)} \tag{P2.16.8}$$

$$(P2.16.7),(P2.16.8) \Rightarrow \Delta = 1 + \frac{k_t k_\omega}{(R_a + sL_a)(Js+B)} \tag{P2.16.9}$$

$$\Delta_1 = 1 - (0) = 1 \tag{P2.16.10}$$

$$(P2.16.5),(P2.16.6),(P2.16.9),(P2.16.10) \Rightarrow$$

$$G(s) = \frac{\Omega(s)}{V_i(s)} = \frac{k_t}{(R_a + sL_a)(Js+B) + k_t k_\omega} \tag{P2.16.11}$$

c. The steady-state response is

$$\omega_{ss}(t) = \lim_{t\to\infty}\omega(t) = \lim_{s\to 0} s\Omega(s) = \lim_{s\to 0} sG(s)V_i(s) \Rightarrow$$

$$\omega_{ss}(t) = \lim_{s\to 0} s \cdot \frac{k_t}{(R_a + sL_a)(Js + B) + k_t k_\omega} \cdot \frac{200}{s} = \frac{200 k_t}{R_a B + k_t k_\omega} \Rightarrow \qquad \text{(P2.16.12)}$$

$$\omega_{ss}(t) = 95.24\,\text{rad/s}$$

d. The following figure depicts the SFG of the system under the effect of a disturbance torque $T'(s)$ on the motor's axis.

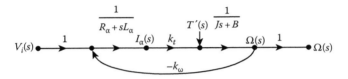

The system that emerges has two inputs ($V_i(s)$, $T'(s)$) and one output ($\Omega(s)$). From
the superposition principle the total output $\Omega(s)$ is given by

$$\Omega(s) = \Omega_{V_i}(s) + \Omega_{T'}(s) \qquad \text{(P2.16.13)}$$

However,

$$\Omega_{V_i}(s) = V_i(s)\cdot G(s) \qquad \text{(P2.16.14)}$$

and

$$\Omega_{T'}(s) = T'(s)\cdot \frac{(1/(Js+B))}{1 + (1/(R_a + sL_a))\cdot(1/(Js+B))\cdot k_t k_\omega} \Rightarrow$$

$$\qquad \text{(P2.16.15)}$$

$$\Omega_{T'}(s) = T'(s)\cdot \frac{R_a + sL_a}{(R_a + sL_a)(Js+B) + k_t k_\omega}$$

$$(\text{P2.16.13}),(\text{P2.16.14}),(\text{P2.16.15}) \Rightarrow \Omega(s) = \frac{k_t V_i(s) + T'(s)(R_a + sL_a)}{(R_a + sL_a)(Js+B) + k_t k_\omega} \qquad \text{(P2.16.16)}$$

The new steady-state response (i.e., the steady state of the angular velocity) of the
motor is

$$\omega_{ss}(t) = \lim_{s\to 0} s\Omega(s) = \lim_{s\to 0} s \cdot \frac{k_t(200/s) + (40/s)(R_a + sL_a)}{(R_a + sL_a)(Js+B) + k_t k_\omega} \Rightarrow$$

$$\qquad \text{(P2.16.17)}$$

$$\omega_{ss}(t) = 95.24 + \frac{40 R_a}{R_a B + k_t k_\omega}$$

Thus,

$$\omega_{ss}(t) = 95.24 + \underbrace{1.9}_{\substack{\text{effect of} \\ \text{disturbance} \\ \text{torque}}} = 97.14\,\text{rad/s}$$

We notice that the steady-state angular velocity fluctuates in a percentage of $(97.14 - 95.24)/95.24 \cdot 100\% \cong 2\%$ under the effect of the disturbance.

2.17 The reference and feedback tachometers of the angular velocity control system shown in the following are the same. The parameters of the system are k_a [V/V] (amplifier), R_f [Ω], L_f [H], k_f [V/A] field (generator), R [Ω], k_t [N · m/A] (torque constant of the motor), a [V/r/s] (velocity constant of the tachometers), and B [N·m·s/rad] (friction constant).

a. Write down the Laplace transform of the system equations, assuming zero initial conditions.

b. Plot the block diagram and the SFG of the system, and find the transfer function $G(s) = \Omega(s)/E_i(s)$ by applying Mason's gain formula.

c. Compute the steady-state step response of the system

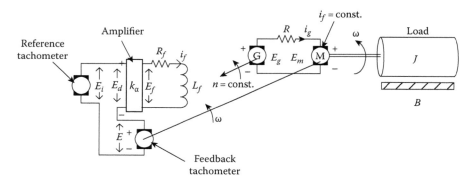

Solution

a. Assuming zero initial conditions the equations that describe the system are

$$E_i(t) - E(t) = E_d(t) \tag{P2.17.1}$$

$$E_f(t) = k_a E_d(t) \tag{P2.17.2}$$

$$E_f(t) = R_f i_f(t) + L_f \frac{di_f(t)}{dt} \tag{P2.17.3}$$

$$E_g(t) = k_f i_f(t) \tag{P2.17.4}$$

$$E_g(t) = R i_g(t) + E_m(t) \tag{P2.17.5}$$

$$T(t) = k_t i_g(t) \qquad\qquad \text{(P2.17.6)}$$

$$T(t) = J\frac{d\omega(t)}{dt} + B\omega(t) \qquad\qquad \text{(P2.17.7)}$$

$$E_m(t) = k_\omega \omega(t) \qquad\qquad \text{(P2.17.8)}$$

$$E(t) = a\omega(t) \qquad\qquad \text{(P2.17.9)}$$

The Laplace transform of the mathematical model is

$$E_i(s) - E(s) = E_d(s) \qquad\qquad \text{(P2.17.1')}$$

$$E_f(s) = k_a E_d(s) \qquad\qquad \text{(P2.17.2')}$$

$$E_f(s) = (R_f + sL_f)I_f(s) \qquad\qquad \text{(P2.17.3')}$$

$$E_g(s) = k_f I_f(s) \qquad\qquad \text{(P2.17.4')}$$

$$E_g(s) - E_m(s) = RI_g(s) \qquad\qquad \text{(P2.17.5')}$$

$$T(s) = k_t I_g(s) \qquad\qquad \text{(P2.17.6')}$$

$$T(s) = (Js + B)\Omega(s) \qquad\qquad \text{(P2.17.7')}$$

$$E_m(s) = k_\omega \Omega(s) \qquad\qquad \text{(P2.17.8')}$$

$$E(s) = a\Omega(s) \qquad\qquad \text{(P2.17.9)'}$$

Based on Equations P2.17.1 through P2.17.9, we plot the block diagram of the angular velocity control system.

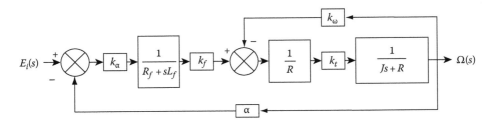

Then we plot the SFG of the system

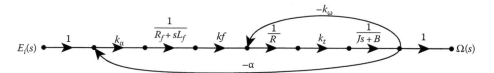

b. From Mason's gain formula, we have

$$G(s) = \frac{\Omega(s)}{E_i(s)} = \frac{T_1 \Delta_1}{\Delta} \tag{P2.17.10}$$

where

$$T_1 = 1 \cdot k_a \cdot \frac{1}{R_f + sL_f} \cdot k_f \cdot \frac{1}{R} \cdot k_t \cdot \frac{1}{Js + B} \cdot 1 = \frac{k_a k_f k_t}{R(R_f + sL_f)(Js + B)} \tag{P2.17.11}$$

$$\Delta = 1 - \sum L_1 \tag{P2.17.12}$$

and

$$\sum L_1 = -\frac{k_t k_\omega}{R(Js + B)} - \frac{a k_a k_f k_t}{R(R_f + sL_f)(Js + B)} \tag{P2.17.13}$$

$$\Delta_1 = 1 - (0) = 1 \tag{P2.17.14}$$

We substitute (P2.17.11) through (P2.17.14) to (P2.17.10) and get

$$G(s) = \frac{\Omega(s)}{E_i(s)} = \frac{k_a k_f k_t}{R(R_f + sL_f)(Js + B) + \alpha k_a k_f k_t + k_\omega k_t (R_f + sL_f)} \Rightarrow$$

$$G(s) = \frac{\Omega(s)}{E_i(s)} = \frac{k_a k_f k_t}{RL_f Js^2 + (RR_f J + RL_f B + k_\omega k_t L_f)s + RR_f B + \alpha k_a k_f k_t + k_\omega k_t R_f} \tag{P2.17.15}$$

c. The output voltage is a constant voltage of amplitude A (Volts), that is,

$$E_i(s) = \frac{A}{s} \tag{P2.17.16}$$

Thus, the steady-state step response of the system is

$$\omega_{ss}(t) = \lim_{t \to \infty} \omega(t) = \lim_{s \to 0} s\Omega(s) \Rightarrow$$

$$\omega_{ss}(t) \overset{\text{(P2.17.15),(P2.17.16)}}{=} \frac{A k_a k_f k_t}{RR_f B + \alpha k_a k_f k_t + k_\omega k_t R_f} \tag{P2.17.17}$$

2.18 Consider the position-control system depicted in the following figure.

   a. Plot its block diagram.

   b. Find the transfer function $G(s) = \Theta_o(s)/E_f(s)$ by using block diagram reduction.

   c. Given that $R_f = 5\,\Omega$, $L_f \simeq 0\,H$, $k_G = 100\,V/A$, $R_a = 5\,\Omega$, $k_m = 100\,V/r/s$, $k_M = 10\,N \cdot m/A$, $J = 200\,Kgr^2$, $f = 100\,N \cdot m/r/s$ express the differential equation that describes the system.

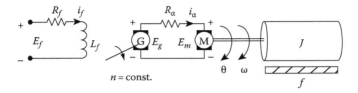

$$n = \text{const.}$$

Solution

   a. The equations that describe the control system are

$$E_f(t) = R_f i_f(t) + L_f \frac{di_f(t)}{dt} \tag{P2.18.1}$$

$$E_g(t) = k_G i_f(t) \tag{P2.18.2}$$

$$E_g(t) = R_a i_a(t) + E_M(t) \tag{P2.18.3}$$

$$E_M(t) = k_m \omega(t) \tag{P2.18.4}$$

$$M(t) = k_M i_a(t) \tag{P2.18.5}$$

$$M(t) = J \frac{d\omega(t)}{dt} + f\omega(t) \tag{P2.18.6}$$

$$\omega(t) = \frac{d\theta(t)}{dt} \tag{P2.18.7}$$

Assuming zero initial conditions, we apply Laplace transform to the equations of the model.

$$E_f(s) = (R_f + sL_f)I_f(s) \tag{P2.18.1'}$$

$$E_g(s) = k_G I_f(s) \tag{P2.18.2'}$$

$$E_g(s) = R_a I_a(s) + E_M(s) \tag{P2.18.3'}$$

$$E_M(s) = k_m \Omega(s) \tag{P2.18.4'}$$

$$M(s) = k_M I_a(s) \tag{P2.18.5'}$$

$$M(s) = (Js + f)\Omega(s) \qquad \text{(P2.18.6')}$$

$$\Omega(s) = s\Theta(s) \qquad \text{(P2.18.7')}$$

The block diagram of the system is plotted from Equations P2.18.1 through P2.18.7:

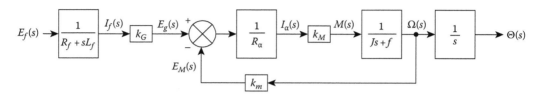

b. A first reduction of the block diagram is:

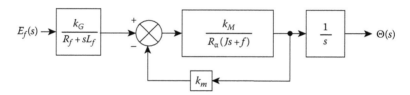

Reducing further the block diagram, we get:

$$E_f(s) \longrightarrow \boxed{\dfrac{k_G}{R_f + sLf}} \longrightarrow \boxed{\dfrac{k_M}{R_\alpha(Js+f)+k_Mk_m}} \longrightarrow \boxed{\dfrac{1}{s}} \longrightarrow \Theta(s)$$

Therefore, the transfer function is

$$G(s) = \frac{\Theta(s)}{E_f(s)} = \frac{k_G k_M}{s(R_f + sL_f)[R_\alpha(Js+f)+k_Mk_m]} \qquad \text{(P2.18.8)}$$

c. Substituting the parameter values in (P2.18.8) we get

$$\frac{\Theta(s)}{E_f(s)} = \frac{10}{50s+75} \Rightarrow (50s+75)\Theta(s) = 10E_f(s) \overset{I.L.T.}{\Rightarrow}$$

and the differential equation is

$$50\frac{d\Theta(t)}{dt} + 75\Theta(t) = 10E_f(t) \qquad \text{(P2.18.9)}$$

2.19 A Ward Leonard position control system is described by the depicted SFG. Compute the response $\theta(t)$ of the system to the input voltage $u_f(t) = 0.4e^{-5t}$. Also compute the steady-state response if a disturbance-torque $T(s) = 20/s$ is applied on the load.

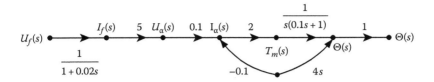

Solution

From the SFG and by using Mason's formula, we compute the transfer function $G(s) = \Theta(s)/V_f(s)$ as follows:

$$G(s) = \frac{\Theta(s)}{V_f(s)} = \frac{T_1 \Delta_1}{\Delta} \tag{P2.19.1}$$

where

$$T_1 = \frac{1}{s(0.1s+1)(1+0.02s)} \tag{P2.19.2}$$

$$\Delta = 1 - \sum L_1 \tag{P2.19.3}$$

and

$$L_1 = -\frac{8s}{10s(0.1s+1)} \tag{P2.19.4}$$

$$\Delta_1 = 1 - (0) = 1 \tag{P2.19.5}$$

Substituting (P2.19.2) through (P2.19.5) to (P2.19.1), we get

$$G(s) = \frac{\Theta(s)}{V_f(s)} = \frac{10}{(1+0.02s)\left[10s(0.1s+1)+8s\right]} = \frac{500}{s(s+50)(s+18)} \tag{P2.19.6}$$

It holds that

$$u_f(t) = 0.4e^{-5t} \Rightarrow V_f(s) = LT\{0.4e^{-5t}\} \Rightarrow V_f(s) = \frac{0.4}{s+5} \tag{P2.19.7}$$

Hence, from (P2.19.6) we have

$$\Theta(s) = G(s)V_f(s) \overset{(P2.19.6),(P2.19.7)}{=} \frac{200}{s(s+5)(s+18)(s+50)} \tag{P2.19.8}$$

or

$$\Theta(s) = \frac{k_1}{s} + \frac{k_2}{s+5} + \frac{k_3}{s+18} + \frac{k_4}{s+50} \tag{P2.19.9}$$

- Computation of k_1, k_2, k_3, k_4:

$$\left.\begin{array}{l} k_1 = \lim_{s \to 0} s\Theta(s) = 0.044 \\[2mm] k_2 = \lim_{s \to -5}(s+5)\Theta(s) = -0.068 \\[2mm] k_3 = \lim_{s \to -18}(s+18)\Theta(s) = -0.0267 \\[2mm] k_4 = \lim_{s \to -50}(s+50)\Theta(s) = -2.77 \cdot 10^{-3} \end{array}\right\}$$ (P2.19.10)

Thus,

$$(P2.19.9) \overset{(P2.19.10)}{\Rightarrow} \Theta(s) = \frac{0.044}{s} - \frac{0.068}{s+5} + \frac{0.0267}{s+18} - \frac{2.77 \cdot 10^{-3}}{s+50}$$ (P2.19.11)

So,

$$\theta(t) = L^{-1}\{\Theta(s)\} \overset{(P2.19.10)}{=}$$

$$= 0.044 - 0.068 \cdot e^{-5t} + 0.0267 \cdot e^{-18t} - 2.77 \cdot 10^{-3} \cdot e^{-50t}$$ (P2.19.12)

Consequently if $u_f(t)$ is the only input, the steady-state response is

$$\theta_{ss}(t) = \lim_{t \to \infty} \theta(t) = \lim_{s \to 0} s\Theta(s) \overset{(P2.19.11)}{=} 0.044\,\text{rad}$$ (P2.19.13)

If a disturbance torque $T(s) = 20/s$ is applied to the load, in order to compute the system response we have to use the superposition principle. In this case, the system response is given by

$$\Theta(s) = \Theta'(s)\big|_{T(s)=0} + \Theta''(s)\big|_{V_f(s)=0}$$ (P2.19.14)

For $T(s) = 0$ we have already found that $\Theta'(s)$ is given by relationship (P2.19.11).
  For $V_f(s) = 0 \Rightarrow$

$$\Theta''(s) = G'(s)T(s)$$ (P2.19.15)

where $G'(s)$ is the transfer function of the system which accepts as its input only the disturbance. We will compute $G'(s)$ after plotting the new SFG.

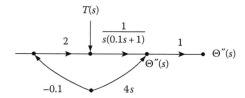

From Mason's rule, we have

$$G'(s) = \frac{T_1' \Delta_1'}{\Delta'}$$ (P2.19.16)

where

$$T_1' = \frac{1}{s(0.1s+1)}$$ (P2.19.17)

$$\Delta' = 1 - \sum L_1'$$ (P2.19.18)

and

$$\sum L_1' = -\frac{8s}{10s(0.1s+1)}$$ (P2.19.19)

$$\Delta_1' = 1 - (0) = 1$$ (P2.19.20)

From (P2.19.16) and on the basis of (P2.19.17) through (P2.19.20) we arrive at

$$G'(s) = \frac{10}{s(s+18)}$$ (P2.19.21)

$$\text{(P2.19.14)} \overset{\text{(P2.19.20)}}{\Rightarrow} \Theta''(s) = \frac{10}{s(s+18)} T(s)$$ (P2.19.22)

Therefore, from (P2.19.14) the total response of the system is

$$\Theta(s) \overset{\text{(P2.19.11)}}{\underset{\text{(P2.19.21)}}{=}} \frac{200}{s(s+5)(s+18)(s+50)} + \frac{200}{s^2(s+18)}$$ (P2.19.23)

The new steady-state system response by taking into account the effect of the disturbance is

$$\theta_{ss}(t) = \lim_{t \to \infty} \theta(t) = \lim_{s \to 0} s\Theta(s) = 0.044 + \lim_{s \to 0} \frac{200}{s(s+18)} \to \infty$$ (P2.19.24)

In conclusion for an input $u_f(t) = 0.4e^{-5t}$ the system is led to instability under the effect of the disturbance $T(t) = 20\,\text{Nm}$.

2.20 Find the transfer function $G(s) = E_o(s)/E_{in}(s)$ of the electromechanical system depicted in the figure below. Assume that $Eo(s) = E_2\Theta_3(s)/10 \cdot 2\pi$ and $T_d(s) = k_m E_{in}(s)$.

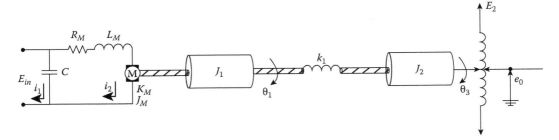

Solution

We have

$$e_{in}(t) = \frac{1}{C} \int_0^t (i_1(t) - i_2(t)) \, dt \qquad \text{(P2.20.1)}$$

and

$$\frac{1}{C} \int_0^t (i_2(t) - i_1(t)) \, dt + R_m i_2(t) + L_m \frac{d i_2(t)}{dt} + k_n \frac{d \theta_1(t)}{dt} = 0 \qquad \text{(P2.20.2)}$$

Assuming zero initial conditions and applying Laplace transform to (P2.20.1) and (P2.20.2), we get

$$(\text{P2.20.1}) \overset{\text{L.T.}}{\Rightarrow} E_{in}(s) = \frac{I_1(s) - I_2(s)}{sC} \qquad \text{(P2.20.1')}$$

$$(\text{P2.20.2}) \overset{\text{L.T.}}{\Rightarrow} \frac{1}{sC}\left(I_1(s) - I_2(s)\right) + R_m I_2(s) + s L_m I_2(s) + k_n s \Theta_1(s) = 0 \qquad \text{(P2.20.2')}$$

By combining (P2.20.1') and (P2.20.2'), we get

$$I_2(s) = \frac{E_{in}(s) - k_n s \Theta_1(s)}{R_m + s L_m} \qquad \text{(P2.20.3)}$$

We know that

$$T_d(s) = k_T I_2(s) \qquad \text{(P2.20.4)}$$

$$(\text{P2.20.3}), (\text{P2.20.4}) \Rightarrow T_d(s) = k_T \frac{E_{in}(s) - k_n s \Theta_1(s)}{R_m + s L_m} \qquad \text{(P2.20.5)}$$

The load torque is

$$T_L(t) = (J_m + J_1) \frac{d^2 \theta_1(t)}{dt^2} + B_C \frac{d(\theta_1(t) - \theta_2(t))}{dt} \qquad \text{(P2.20.6)}$$

If we suppose that the torque $T_d(t)$ equals the load torque, then

$$(J_m + J_1) \frac{d^2 \theta_1(t)}{dt^2} + B_C \frac{d(\theta_1(t) - \theta_2(t))}{dt} = k_m E_{in}(t) \qquad \text{(P2.20.7)}$$

$$-B_C \frac{d\left(\theta_2(t) - \theta_1(t)\right)}{dt} - k_1(\theta_2(t) - \theta_3(t)) = 0 \qquad \text{(P2.20.8)}$$

$$J_2 \frac{d^2\theta_3(t)}{dt^2} + k_1(\theta_3(t) - \theta_2(t)) = 0 \qquad \text{(P2.20.9)}$$

We apply again Laplace transform to the relationships (P2.20.7) through (P2.20.9):

$$[(J_m + J_1)s^2 + B_C s]\Theta_1(s) - B_C s\Theta_2(s) = k_m E_{in}(s) \qquad \text{(P2.20.10)}$$

$$-B_C s\Theta_1(s) + (B_C s + k_1)\Theta_2(s) - k_1\Theta_3(s) = 0 \qquad \text{(P2.20.11)}$$

$$0 - k_1\Theta_2(s) + (J_2 s^2 + k_1)\Theta_3(s) = 0 \qquad \text{(P2.20.12)}$$

We solve for $\Theta_3(s)$ and have

$$\Theta_3(s) = \frac{\begin{vmatrix} (J_m + J_1)s^2 + B_C s & -B_C s & k_m E_{in}(s) \\ -B_C s & B_C s + k_1 & 0 \\ 0 & -k_1 & 0 \end{vmatrix}}{\begin{vmatrix} (J_m + J_1)s^2 + B_C s & -B_C s & 0 \\ -B_C s & B_C s + k_1 & -k_1 \\ 0 & -k_1 & J_2 s^2 + k_1 \end{vmatrix}} \qquad \text{(P2.20.13)}$$

By solving (P2.20.13) we get

$$\Theta_3(s) = \frac{k_m E_{in}(s)}{(J_m + J_1)s^2} \left[\frac{1}{\dfrac{J_3 s^2}{k_1} + \dfrac{J_3 s}{B_C} + \dfrac{J_3}{J_m + J_1} + 1} \right] \qquad \text{(P2.20.14)}$$

but

$$E_o(s) = \frac{E_2\Theta_3(s)}{10 \cdot 2\pi} = \frac{E_2\Theta_3(s)}{62.8} \Rightarrow \Theta_3(s) = \frac{62.8}{E_2}E_o(s) \qquad \text{(P2.20.15)}$$

$$\text{(P2.20.15), (P2.20.16)} \Rightarrow \frac{E_o(s)}{E_{in}(s)} = \frac{E_2 k_m}{62.8(J_m + J_1)s^2} \left[\frac{1}{\dfrac{J_3 s^2}{k_1} + \dfrac{J_3 s}{B_C} + \dfrac{J_3}{J_m + J_1} + 1} \right] \qquad \text{(P2.20.16)}$$

2.21 Find the transfer function $G(s) = E_o(s)/Y(s)$ of the depicted system. Suppose that E_i is proportional to the difference between the displacements of the masses M_1 and M_2, more specifically $E_i = E(y - x)/l$. Finally assume that k_1 and D are approximately zero.

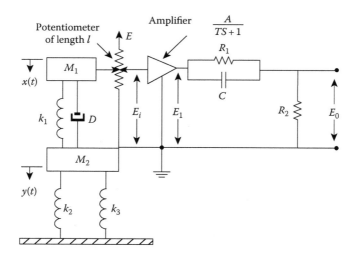

Solution

For the electrical circuit, we have

$$\frac{E_o(s)}{E_1(s)} = \frac{R_2}{R_2 + (R_1(1/sC_1))/(R_1 + (1/sC_1))} = \frac{R_2}{R_1 + R_2} \frac{sCR_1 + 1}{sC(R_1R_2/(R_1 + R_2)) + 1} \Rightarrow$$

$$\frac{E_o(s)}{E_1(s)} = a\frac{1 + T_1 s}{1 + aT_1 s}$$

(P2.21.1)

where

$$\left.\begin{aligned} a &= \frac{R_2}{R_1 + R_2} \\ T_1 &= CR_1 \end{aligned}\right\}$$

(P2.21.2)

Also

$$\frac{E_1(s)}{E_i(s)} = \frac{A}{Ts + 1}$$

(P2.21.3)

Consequently

$$\frac{E_o(s)}{E_i(s)} = \frac{E_o(s)}{E_1(s)} \cdot \frac{E_1(s)}{E_i(s)} = a\frac{1 + T_1 s}{1 + aT_1 s}\frac{A}{Ts + 1}$$

(P2.21.4)

or

$$E_o(s) = \frac{aA(1 + T_1 s)}{(1 + aT_1 s)(Ts + 1)}E_i(s)$$

(P2.21.5)

However,

$$E_i(s) = \frac{E}{l}(Y - X) \tag{P2.21.6}$$

$$(P2.21.5), (P2.21.6) \Rightarrow E_o(s) = aA\frac{E}{l}\frac{1 + T_1 s}{(1 + aT_1 s)(Ts + 1)}(Y(s) - X(s)) \tag{P2.21.7}$$

The difference $Y(s) - X(s)$ is computed from the mechanical system.
For the mass M_1, we have

$$M_1\frac{d^2 x}{dt^2} + k_1(x - y) + D\frac{d(x - y)}{dt} = 0 \tag{P2.21.8}$$

For the mass M_2, we have

$$M_2\frac{d^2 y}{dt^2} + k_1(y - x) + D\frac{d(y - x)}{dt} + 2k_2 y = 0 \tag{P2.21.9}$$

By supposing that $k_1 = D = 0$, we get

$$M_2\frac{d^2 y}{dt^2} + M_1\frac{d^2 x}{dt^2} + 2k_2 y = 0 \tag{P2.21.10}$$

Supposing zero initial conditions we apply Laplace transform and get

$$(M_2 s^2 + 2k_2)Y(s) + M_1 s^2 X(s) = 0 \tag{P2.21.11}$$

$$(P2.21.11) \Rightarrow X(s) = -\frac{M_2 s^2 + 2k_2}{M_1 s^2}Y(s) \Rightarrow$$

$$\tag{P2.21.12}$$

$$Y(s) - X(s) = \left[1 + \frac{M_2 s^2 + 2k_2}{M_1 s^2}\right]Y(s)$$

$$(P2.21.7), (P2.21.12) \Rightarrow G(s) = \frac{E_o(s)}{Y(s)} = aA\frac{E}{l}\frac{(1 + T_1 s)}{M_1 s^2}\frac{\left[(M_1 + M_2)s^2 + 2k_2\right]}{(1 + aT_1 s)(Ts + 1)} \tag{P2.21.13}$$

Equation P2.21.13 is the system transfer function.

2.22 In the hydraulic system of the figure below there is interaction between the two containers. Thus, the transfer function of the system is not equal to the product of the two first-degree transfer functions.

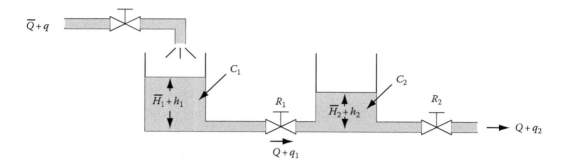

Assuming that the variables fluctuate around the steady-state values, the system equations are

$$C_1 \frac{dh_1(t)}{dt} = q(t) - q_1(t) \tag{P2.7}$$

$$q_1(t) = \frac{h_1(t) - h_2(t)}{R_1} \tag{P2.8}$$

$$C_2 \frac{dh_2(t)}{dt} = q_1(t) - q_2(t) \tag{P2.9}$$

$$q_2(t) = \frac{h_2(t)}{R_2} \tag{P2.10}$$

Assuming that all initial conditions are zero, compute

a. The transfer function of the system $G(s) = Q_2(s)/Q(s)$ and show that it can be expressed as $G(s) = 1/(s^2\tau_1\tau_2 + s(\tau_1 + \tau_2 + \tau_3) + 1)$ where τ_1, τ_2, τ_3 are constant system parameters.

b. For $\tau_1 = \tau_2 = \tau_3$, compute and plot the step response $q_2(t)$.

c. Plot the SFG of the system and repeat the first query by using Mason's gain formula.

Solution

a. We apply Laplace transform to the system equations and get

$$(\text{P2.7}) \Rightarrow C_1 \frac{dh_1(t)}{dt} = q(t) - q_1(t) \overset{\text{L.T.}}{\Rightarrow} C_1 s H_1(s) = Q(s) - Q_1(s) \tag{P2.22.1}$$

$$(\text{P2.8}) \Rightarrow q_1(t) = \frac{h_1(t) - h_2(t)}{R_1} \overset{\text{L.T.}}{\Rightarrow} Q_1(s) = \frac{H_1(s) - H_2(s)}{R_1} \tag{P2.22.2}$$

$$(\text{P2.9}) \Rightarrow C_2 \frac{dh_2(t)}{dt} = q_1(t) - q_2(t) \overset{\text{L.T.}}{\Rightarrow} C_2 s H_2(s) = Q_1(s) - Q_2(s) \tag{P2.22.3}$$

$$(\text{P2.10}) \Rightarrow q_2(t) = \frac{h_2(t)}{R_2} \overset{\text{L.T.}}{\Rightarrow} Q_2(s) = \frac{H_2(s)}{R_2} \tag{P2.22.4}$$

In order to find the transfer function $G(s) = Q_2(s)/Q(s)$, we proceed as follows:

$$G(s) = \frac{Q_2(s)}{Q(s)} \overset{(P2.22.1)}{=} \frac{Q_2(s)}{C_1 s H_1(s) + Q_1(s)} \overset{(P2.22.2)}{=} \frac{Q_2(s)}{C_1 s(Q_1(s)R_1 + H_2(s)) + Q_1(s)} =$$

$$\overset{(P2.22.3)}{=} \frac{Q_2(s)}{C_1 s((C_2 s H_2(s) + Q_2(s))R_1 + H_2(s)) + C_2 s H_2(s) + Q_2(s)} =$$

$$\overset{(P2.22.4)}{=} \frac{Q_2(s)}{C_1 s((C_2 s R_2 Q_2(s) + Q_2(s))R_1 + Q_2(s)R_2) + C_2 s Q_2(s)R_2 + Q_2(s)} =$$

$$= \frac{Q_2(s)}{Q_2(s)\left[C_1 s((C_2 s R_2 + 1)R_1 + R_2) + C_2 s R_2 + 1 \right]} \Rightarrow$$

$$G(s) = \frac{Q_2(s)}{Q(s)} = \frac{1}{s^2 R_1 R_2 C_1 C_2 + s R_1 C_1 + s R_2 C_1 + s R_2 C_2 + 1} \tag{P2.22.5}$$

By substituting $\tau_1 = R_1 C_1$, $\tau_2 = R_2 C_2$, and $\tau_3 = R_2 C_1$, we get

$$G(s) = \frac{Q_2(s)}{Q(s)} = \frac{1}{s^2 \tau_1 \tau_2 + s(\tau_1 + \tau_2 + \tau_3) + 1} \tag{P2.22.6}$$

b. For $\tau_1 = \tau_2 = \tau_3 = r$, we have

$$(P2.22.6) \Rightarrow G(s) = \frac{Q_2(s)}{Q(s)} = \frac{1}{s^2 \tau^2 + 3\tau s + 1} \Rightarrow Q_2(s) = \frac{1}{s^2 \tau^2 + 3\tau s + 1} Q(s) \tag{P2.22.7}$$

The step response is the system response to a unit step input, that is, $q(t) = u(t)$.

$$q(t) = u(t) \Rightarrow Q(s) = L\{q(t)\} = L\{u(t)\} = \frac{1}{s} \tag{P2.22.8}$$

$$(P2.22.7), (P2.22.8) \Rightarrow Q_2(s) = \frac{1}{s(s^2 \tau^2 + 3\tau s + 1)} = \frac{1}{s\tau^2(s^2 + (3/\tau)s + (1/\tau^2))}$$

$$\Rightarrow Q_2(s) = \frac{1/\tau^2}{s(s + (0.382/\tau))(s + (2.62/\tau))} = \frac{1}{\tau^2}\left(\frac{k_1}{s} + \frac{k_2}{s + (0.382/\tau)} + \frac{k_3}{s + (2.62/\tau)} \right) \tag{P2.22.9}$$

- Computation of k_1, k_2, k_3:

$$k_1 = \lim_{s \to 0} Q_2(s) \cdot s = \tau^2$$

$$k_2 = \lim_{s \to -0.382/\tau} Q_2(s) \cdot \left(s + \frac{0.382}{\tau} \right) = -\frac{\tau^2}{0.86}$$

$$k_3 = \lim_{s \to -2.62/\tau} Q_2(s) \cdot \left(s + \frac{2.62}{\tau} \right) = \frac{\tau^2}{5.86}$$

By substituting k_1, k_2, k_3 we get

$$(P2.22.9) \Rightarrow Q_2(s) = \frac{1}{\tau^2}\left(\frac{\tau^2}{s} - \frac{\frac{\tau^2}{0.86}}{s + \frac{0.382}{\tau}} + \frac{\frac{\tau^2}{5.86}}{s + \frac{2.62}{\tau}} \right) \overset{I.L.T}{\Rightarrow}$$

$$q_2(t) = \frac{1}{\tau^2}\left(\tau^2 u(t) - \frac{\tau^2}{0.86}e^{-(0.382/\tau)t} + \frac{\tau^2}{5.86}e^{-(2.62/\tau)t} \right) \Rightarrow$$

$$q_2(t) = \left(1 - 1.17e^{-(0.382/\tau)t} + 0.17e^{-(2.62/\tau)t}\right)u(t)$$

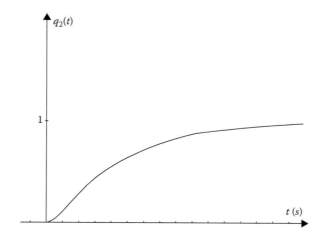

c. We plot the SFG of the system based on Equations P2.22.1′ through P2.22.4′.

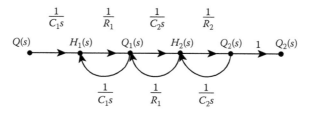

From Mason's gain formula we have

$$G(s) = \frac{Q_2(s)}{Q(s)} = \frac{T_1\Delta_1}{\Delta} \tag{P2.22.10}$$

where

$$T_1 = \frac{1}{C_1 s}\cdot\frac{1}{R_1}\cdot\frac{1}{C_2 s}\cdot\frac{1}{R_2} = \frac{1}{s^2 C_1 C_2 R_1 R_2}$$

$$\Delta = 1 - \sum L_1 + \sum L_2$$

and

$$\sum L_1 = \frac{1}{R_1}\cdot\left(-\frac{1}{C_1 s}\right) + \frac{1}{C_2 s}\cdot\left(-\frac{1}{R_1}\right) + \frac{1}{R_2}\cdot\left(-\frac{1}{C_2 s}\right) \Rightarrow$$

$$\sum L_1 = -\frac{1}{sR_1C_1} - \frac{1}{sR_1C_2} - \frac{1}{sR_2C_2}$$

$$\sum L_2 = \frac{1}{R_1}\cdot\left(-\frac{1}{C_1 s}\right)\cdot\frac{1}{R_2}\cdot\left(-\frac{1}{C_2 s}\right) \Rightarrow \sum L_2 = \frac{1}{s^2 C_1 C_2 R_1 R_2}$$

Hence,

$$\Delta = 1 + \frac{1}{sR_1C_1} + \frac{1}{sR_1C_2} + \frac{1}{sR_2C_2} + \frac{1}{s^2 C_1 C_2 R_1 R_2}$$

$$\Delta_1 = 1 - (0) = 1$$

$$(\text{P2.22.10}) \Rightarrow G(s) = \frac{1/s^2 C_1 C_2 R_1 R_2}{1 + (1/sR_1C_1) + (1/sR_1C_2) + (1/sR_2C_2) + (1/s^2 C_1 C_2 R_1 R_2)} \Rightarrow$$

$$G(s) = \frac{Q_2(s)}{Q(s)} = \frac{1}{s^2 C_1 C_2 R_1 R_2 + sR_1C_1 + sR_2C_1 + sR_2C_2 + 1} \Rightarrow$$

$$\Rightarrow G(s) = \frac{1}{s^2 \tau_1 \tau_2 + s(\tau_1 + \tau_2 + \tau_3) + 1}$$

2.23 The following figure represents a simple liquid level control system.

    a. Plot the SFG of the system.

    b. Find the transfer functions of the open-loop and of the closed-loop control system, that is, the transfer functions $G(s) = Y(s)/Q_i(s)$ and $F(s) = Y(s)/\Omega(s)$.

    c. Supposing that the system is activated by the inputs

       i. $\omega(t) = \delta(t)$

      ii. $\omega(t) = u(t)$

     iii. $\omega(t) = r(t) = t$

    plot each time the system response and comment upon them.

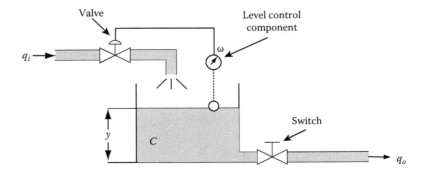

Solution

a. The mathematical model for the operation of the system is

$$C\frac{dy(t)}{dt} = q_i(t) - q_0(t) \tag{P2.23.1}$$

$$e(t) = \omega(t) - k_f y(t) \tag{P2.23.2}$$

$$p(t) = k_a e(t) \tag{P2.23.3}$$

$$q_i(t) = k_v p(t) \tag{P2.23.4}$$

$$y(t) = R q_o(t) \tag{P2.23.5}$$

Assuming zero initial conditions, we apply Laplace transform to these equations and get

$$(\text{P2.23.1}) \overset{\text{L.T.}}{\Rightarrow} CsY(s) = Q_i(s) - Q_0(s) \tag{P2.23.1'}$$

$$(\text{P2.23.2}) \overset{\text{L.T.}}{\Rightarrow} E(s) = \Omega(s) - k_f Y(s) \tag{P2.23.2'}$$

$$(\text{P2.23.3}) \overset{\text{L.T.}}{\Rightarrow} P(s) = k_a E(s) \tag{P2.23.3'}$$

$$(\text{P2.23.4}) \overset{\text{L.T.}}{\Rightarrow} Q_i(s) = k_v P(s) \tag{P2.23.4'}$$

$$(\text{P2.23.5}) \overset{\text{L.T.}}{\Rightarrow} Y(s) = RQ_o(s) \tag{P2.23.5'}$$

As usual, we plot the SFG of the system based on Equations P2.23.1′ through P2.23.5′.

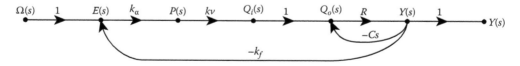

b. i. In order to compute the closed-loop transfer function, we apply Mason's gain formula:

$$F(s) = \frac{Y(s)}{\Omega(s)} = \frac{T_1 \Delta_1}{\Delta} \tag{P2.23.6}$$

where

$$T_1 = k_a k_v R \tag{P2.23.7}$$

$$\Delta = 1 - \sum L_1 \tag{P2.23.8}$$

and

$$L_1 = -k_a k_v k_f - RCs \tag{P2.23.9}$$

$$(P2.23.8),(P2.23.9) \Rightarrow \Delta = 1 + k_a k_v k_f R + RCs \tag{P2.23.10}$$

$$\Delta_1 = 1 - (0) = 1 \tag{P2.23.11}$$

Substituting (P2.23.7), (P2.23.10), and (P2.23.11) to (P2.23.6) we get

$$F(s) = \frac{Y(s)}{\Omega(s)} = \frac{k_a k_v R}{RCs + k_a k_v k_f R + 1} \tag{P2.23.12}$$

ii. In order to compute the open-loop transfer function, we apply Mason's gain formula:

$$G(s) = \frac{Y(s)}{Q_i(s)} = \frac{T_1 \Delta_1}{\Delta} \tag{P2.23.13}$$

where

$$T_1 = 1 \cdot R \cdot 1 = R \tag{P2.23.14}$$

$$\Delta = 1 - (-RCs) = 1 + RCs \tag{P2.23.15}$$

$$\Delta_1 = 1 - (0) = 1 \tag{P2.23.16}$$

Substitution (P2.23.14) through (P2.23.16) to (P2.23.13), we get

$$G(s) = \frac{Y(s)}{Q_i(s)} = \frac{R}{1 + RCs} \tag{P2.23.17}$$

c. i. For $\omega(t) = \delta(t) \Rightarrow \Omega(s) = 1$.
   Therefore,

$$(P2.23.12) \Rightarrow Y_\delta(s) = \frac{k_a k_v R}{1 + k_a k_v k_f R + RCs} \cdot \Omega(s) = \frac{k_a k_v R}{RCs + k_a k_v k_f R + 1} \tag{P2.23.18}$$

$$\left. \begin{aligned} \lim_{t \to 0} y_\delta(t) &= \lim_{s \to \infty} s Y_\delta(s) = \frac{k_a k_v}{C} \\ \lim_{t \to \infty} y_\delta(t) &= \lim_{s \to 0} s Y_\delta(s) = 0 \end{aligned} \right\} \tag{P2.23.19}$$

We plot the system impulse response $y_\delta(t)$.

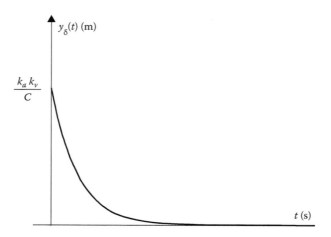

ii. For $\omega(t) = u(t) \Rightarrow \Omega(s) = 1/s$.
     Thus,

$$(\text{P2.23.12}) \Rightarrow Y_u(s) = \frac{k_a k_v R}{RCs + k_a k_v k_f R + 1} \cdot \Omega(s) = \frac{k_a k_v R}{s(1 + k_a k_v k_f R + RCs)} \qquad (\text{P2.23.20})$$

$$\left. \begin{array}{l} \lim_{t \to 0} y_u(t) = \lim_{s \to \infty} sY_u(s) = 0 \\[3mm] \lim_{t \to \infty} y_u(t) = \lim_{s \to 0} sY_u(s) = \dfrac{k_a k_v R}{1 + k_a k_v k_f R} \end{array} \right\} \qquad (\text{P2.23.21})$$

We plot the step response of the system $y_u(t)$.

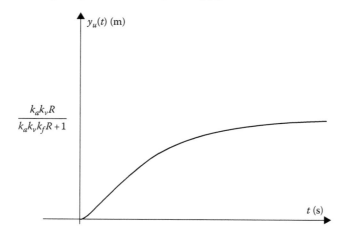

iii. For $\omega(t) = r(t) = t \Rightarrow \Omega(s) = 1/s^2$. Thus,

$$(\text{P2.23.12}) \Rightarrow Y_r(s) = \frac{k_a k_v R}{1 + k_a k_v k_f R + RCs} \cdot \Omega(s) = \frac{k_a k_v R}{s^2(1 + k_a k_v k_f R + RCs)} \qquad (\text{P2.23.22})$$

$$\left. \begin{aligned} \lim_{t \to 0} y_r(t) &= \lim_{s \to \infty} s Y_r(s) = 0 \\ \lim_{t \to \infty} y_r(t) &= \lim_{s \to 0} s Y_r(s) \to \infty \end{aligned} \right\}$$

(P2.23.23)

We plot the ramp response of the system $y_r(t)$.

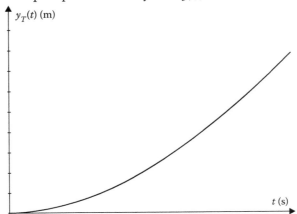

After examining the time response graphs for $y_\delta(t)$, $y_u(t)$, and $y_r(t)$, we conclude that the most suitable input is the step function, that is, $\omega(t) = u(t)$.

2.24 Consider the depicted displacement control system of a hydraulic amplifier.

a. Write down the equations that describe the system.

b. Plot the block diagram and the SFG of the system.

c. Compute the transfer function $Y(s)/E(s)$.

d. Find the steady-state impulse response and the steady-state step response of the system. Comment upon the results.

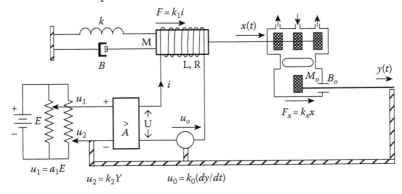

Solution

a. The equations that describe the displacement-control system of the hydraulic amplifier are

$$u_1(t) - u_2(t) = u_e(t) \qquad \text{(P2.24.1)}$$

$$u(t) = A u_e(t) \qquad \text{(P2.24.2)}$$

$$u(t) = Ri(t) + L\frac{di(t)}{dt} + u_o(t) \qquad (P2.24.3)$$

$$f(t) = k_1 i(t) \qquad (P2.24.4)$$

$$f(t) = M\frac{d^2x(t)}{dt^2} + B\frac{dx(t)}{dt} + kx(t) \qquad (P2.24.5)$$

$$f_x(t) = k_x x(t) \qquad (P2.24.6)$$

$$f_x(t) = M_o\frac{d^2y(t)}{dt^2} + B_o\frac{dy(t)}{dt} \qquad (P2.24.7)$$

$$u_o(t) = k_o\frac{dy(t)}{dt} \qquad (P2.24.8)$$

$$u_2(t) = k_2 y(t) \qquad (P2.24.9)$$

$$u_1(t) = a_1 E(t) \qquad (P2.24.10)$$

b. We apply Laplace transform to the system equations, assuming zero initial conditions, and get

$$V_1(s) - V_2(s) = V_e(s) \qquad (P2.24.1')$$

$$V(s) = AV_e(s) \qquad (P2.24.2')$$

$$V(s) = (R + Ls)I(s) + V_o(s) \qquad (P2.24.3')$$

$$F(s) = k_1 I(s) \qquad (P2.24.4')$$

$$F(s) = (Ms^2 + Bs + k)X(s) \qquad (P2.24.5')$$

$$F_x(s) = k_x X(s) \qquad (P2.24.6')$$

$$F_x(s) = (M_o s^2 + B_o s)Y(s) \qquad (P2.24.7')$$

$$V_o(s) = k_o s Y(s) \qquad (P2.24.8')$$

$$V_2(s) = k_2 Y(s) \qquad (P2.24.9')$$

$$V_1(s) = a_1 E(s) \qquad (P2.24.10')$$

Based on the previous set of equations, we plot the block diagram of the system:

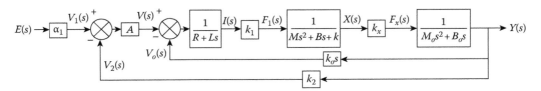

The SFG of the system is:

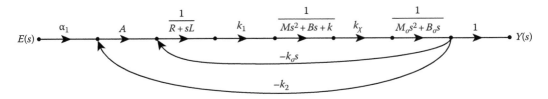

c. The transfer function $G(s) = Y(s)/E(s)$ is found with the use of Mason's gain formula

$$G(s) = \frac{Y(s)}{E(s)} = \frac{T_1 \Delta_1}{\Delta} \qquad \text{(P2.24.11)}$$

where

$$T_1 = \frac{a_1 A k_1 k_x}{(R+sL)(Ms^2 + Bs + k)(M_o s^2 + B_o s)} \qquad \text{(P2.24.12)}$$

$$\Delta = 1 - \sum L_1 \qquad \text{(P2.24.13)}$$

and

$$\sum L_1 = \frac{k_1 k_x k_o s}{(R+sL)(Ms^2 + Bs + k)(M_o s^2 + B_o s)} - \frac{A k_1 k_2 k_x}{(R+sL)(Ms^2 + Bs + k)(M_o s^2 + B_o s)} \qquad \text{(P2.24.14)}$$

$$\Delta_1 = 1 - (0) = 1 \qquad \text{(P2.24.15)}$$

After substituting (P2.24.12) through (P2.24.15) to (P2.24.11) we arrive at

$$G(s) = \frac{Y(s)}{E(s)}$$

$$= \frac{a_1 A k_1 k_x}{(R+sL)(Ms^2 + Bs + k)(M_o s^2 + B_o s) + k_1 k_x k_o s + A k_1 k_2 k_x} \qquad \text{(P2.24.16)}$$

d. The steady-state impulse response is

$$y_{\delta,ss}(t) = \lim_{t\to\infty} y_\delta(t) = \lim_{s\to 0} sY_\delta(s) \overset{(P2.24.16)}{=} 0$$

(P2.24.17)

$$(\text{Here } E(t) = \delta(t) \Rightarrow E(s) = L\{\delta(t)\} = 1)$$

The steady-state step response is

$$y_{u,ss}(t) = \lim_{t\to\infty} y_u(t) = \lim_{s\to 0} sY_u(s) \overset{(P2.24.16)}{=} \frac{a_1 A k_1 k_x}{A k_1 k_2 k_x} = \frac{a_1}{k_2}$$

(P2.24.18)

$$\left(\text{Here } E(t) = u(t) \Rightarrow E(s) = L\{u(t)\} = \frac{1}{s}\right)$$

After examining the time responses $y_{\delta,ss}(t) = 0$ and $y_{u,ss}(t) = a_1/k_2 \neq 0$ we conclude that the most suitable input is the step function.

2.25 Consider the depicted temperature control system.
   a. Describe the operation of the system.
   b. Plot the block diagram and the SFG.
   c. Compute the transfer function $G(s) = \Theta(s)/V_r(s)$.
   d. Find the steady-state system response $\theta_{ss}(t)$ if $v_r(t) = u(t)$.
   e. Compute the new steady-state response if a force $f_d(t) = u(t)$ acts as a disturbance and changes for a short time the temperature of the furnace.

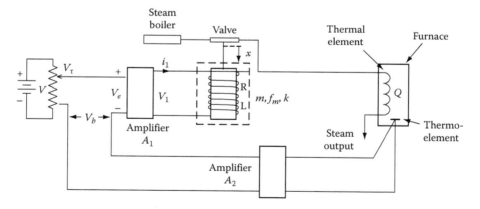

Solution
   a. A decrease in the temperature results in a decrease in the voltage at the ends of the thermo-element, which is in turn amplified by the gain amplifier A_2 and is then led to the input of the differential gain amplifier A_1. As $v_e(t) = v_r(t) - v_b(t)$ the error voltage is increased.

   In this way, the current $i_1(t)$ increases and it induces a force which displaces mass M, opens up the fuel supply control valve more widely, and leads to the desired increase of the furnace's temperature.

b. The equations that describe the system are

$$v_e(t) = v_r(t) - v_b(t) \qquad\qquad\text{(P2.25.1)}$$

$$v_1(t) = A_1 v_e(t) \qquad\qquad\text{(P2.25.2)}$$

$$v_1(t) = Ri_1(t) + L\frac{di_1(t)}{dt} \qquad\qquad\text{(P2.25.3)}$$

$$f_s(t) = k_s i_1(t) \qquad\qquad\text{(P2.25.4)}$$

$$f_s(t) = M\frac{d^2x(t)}{dt^2} + f_m\frac{dx(t)}{dt} + kx(t) \qquad\qquad\text{(P2.25.5)}$$

$$q(t) = k_1 x(t) \qquad\qquad\text{(P2.25.6)}$$

$$\theta(t) = k_2 q(t) \qquad\qquad\text{(P2.25.7)}$$

$$v_b(t) = A_2 v_\theta(t) \qquad\qquad\text{(P2.25.8)}$$

$$v_\theta(t) = k_\theta \theta(t) \qquad\qquad\text{(P2.25.9)}$$

Supposing zero initial conditions we apply Laplace transform and get

$$V_e(s) = V_r(s) - V_b(s) \qquad\qquad\text{(P2.25.1')}$$

$$V_1(s) = A_1 V_e(s) \qquad\qquad\text{(P2.25.2')}$$

$$V_1(s) = RI_1(s) + sLI_1(s) = (R + Ls)I_1(s) \qquad\qquad\text{(P2.25.3')}$$

$$F_s(s) = k_s I_1(s) \qquad\qquad\text{(P2.25.4')}$$

$$F_s(t) = (Ms^2 + f_m s + k)X(s) \qquad\qquad\text{(P2.25.5')}$$

$$Q(s) = k_1 X(s) \qquad\qquad\text{(P2.25.6')}$$

$$\Theta(s) = k_2 Q(s) \qquad\qquad\text{(P2.25.7')}$$

$$V_b(s) = A_2 V_\theta(s) \qquad\qquad\text{(P2.25.8')}$$

$$V_\theta(s) = k_\theta \Theta(s) \qquad\qquad\text{(P2.25.9')}$$

The block diagram of the system is:

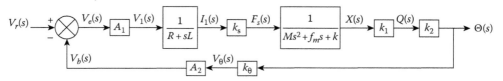

c. The transfer function is computed from the SFG (using Mason's gain formula):

$$G(s) = \frac{\Theta(s)}{V_r(s)} = \frac{T_1 \Delta_1}{\Delta} \qquad \text{(P2.25.10)}$$

where

$$T_1 = \frac{A_1 k_s k_1 k_2}{(R+sL)(Ms^2 + f_m s + k)} \qquad \text{(P2.25.11)}$$

$$\Delta = 1 - \sum L_1 \qquad \text{(P2.25.12)}$$

and

$$L_1 = -\frac{A_1 k_s k_1 k_2 A_2 k_\theta}{(R+sL)(Ms^2 + f_m s + k)} \qquad \text{(P2.25.13)}$$

$$\Delta_1 = 1 - (0) = 1 \qquad \text{(P2.25.14)}$$

By substituting (P2.25.11) through (P2.25.14) to (P2.25.10) it emerges that

$$G(s) = \frac{\Theta(s)}{V_r(s)} = \frac{A_1 k_s k_1 k_2}{(R+sL)(Ms^2 + f_m s + k) + A_1 k_s k_1 k_2 A_2 k_\theta} \qquad \text{(P2.25.15)}$$

d. The steady-state response for $v_r(t) = u(t)$ is

$$\theta_{ss}(t) = \lim_{t \to \infty} \theta(t) = \lim_{s \to 0} s\Theta(s) \overset{\text{(P2.25.15)}}{=}$$

$$= \lim_{s \to 0} G(s) V_r(s) s = \lim_{s \to 0} G(s) \frac{1}{s} s = \lim_{s \to 0} G(s) \Rightarrow \qquad \text{(P2.25.16)}$$

$$\theta_{ss}(t) \overset{\text{(P2.25.15)}}{=} \frac{A_1 k_s k_1 k_2}{Rk + A_1 k_s k_1 k_2 A_2 k_\theta}$$

e. According to the superposition principle, the output of the system is computed as

$$\Theta(s) = \Theta'(s)\big|_{F_d(s)=0} + \Theta''(s)\big|_{V_r(s)=0} \qquad \text{(P2.25.17)}$$

We have already computed $\Theta'(s)$. In order to compute $\Theta''(s)$ we plot the SFG, taking into account the effect of the disturbance.

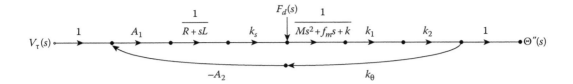

We have

$$\frac{\Theta''(s)}{F_d(s)} = \frac{T_1 \Delta_1}{\Delta} \tag{P2.25.18}$$

where

$$T_1 = \frac{k_1 k_2}{Ms^2 + f_m s + k} \tag{P2.25.19}$$

$$\Delta = 1 - \sum L_1 \tag{P2.25.20}$$

and

$$L_1 = -\frac{A_1 k_s k_1 k_2 A_2 k_\theta}{(R + sL)(Ms^2 + f_m s + k)} \tag{P2.25.21}$$

$$\Delta_1 = 1 - (0) = 1 \tag{P2.25.22}$$

Substituting (P2.25.19) through (P2.25.22) to (P2.25.18) we obtain

$$\frac{\Theta''(s)}{F_d(s)} = \frac{k_1 k_2 (R + sL)}{(R + sL)(Ms^2 + f_m s + k) + A_1 k_s k_1 k_2 A_2 k_\theta} \tag{P2.25.23}$$

From (P2.25.17) and considering also (P2.25.15) and (P2.25.23) we get

$$\Theta(s) = \frac{A_1 k_s k_1 k_2 V_r(s) + k_1 k_2 (R + sL) F_d(s)}{(R + sL)(Ms^2 + f_m s + k) + A_1 k_s k_1 k_2 A_2 k_\theta} \tag{P2.25.24}$$

For $v_r(t) = u(t)$ and $f_d(t) = u(t)$ the new steady-state response is

$$\theta_{ss}(t) = \lim_{t \to \infty} \theta(t) = \lim_{s \to 0} s\Theta(s) \overset{(P2.25.24)}{=} \frac{A_1 k_s k_1 k_2 + k_1 k_2 R}{Rk + A_1 k_s k_1 k_2 A_2 k_\theta} \tag{P2.25.25}$$

From relationship (P2.25.25) we conclude that the temperature increases by $k_1 k_2 R/(Rk + A_1 k_s k_1 k_2 A_2 k^\theta)$; an increase that is insignificant if we consider the practical values of the parameters of the system.

3

Control System Characteristics

3.1 System Sensitivity to Parameter Variations

Any observed environmental variation leads an open-loop system to a decrease in its output accuracy. On the contrary, a closed-loop system detects variations during its operation and corrects by itself the errors generated at the output. The sensitivity of a system to parameter variations is of special importance. Equally important is the ability to reduce the sensitivity of a closed-loop control system to parameter variations.

The sensitivity of a system is defined as the ratio of the change in the system transfer function to the change of a process transfer function (or parameter), for very small variations.

The **sensitivity** of a system with transfer function $F(s)$, in relation to a parameter m, is defined as

$$S_m^F = \frac{m}{F(s)} \left| \frac{\partial F(s)}{\partial m} \right| \tag{3.1}$$

A closed-loop control system (in contrast to an open-loop system) has the ability to reduce the influence of parameter variations by adding a feedback loop.

In open-loop control systems one must carefully choose the system transfer function $G(s)$, so that the performance requirements of the system are fully satisfied.

In closed-loop control systems, high accuracy is not required in defining $G(s)$, because the sensitivity of the system to possible variations or errors in $G(s)$ is reduced by the factor $1 + G(s)$.

3.2 Steady-State Error

The steady-state error $e_{ss}(t)$ is a factor that determines the operation of control systems. It is observed at the output of the system after the end of the transient response period. More specifically, the value of $e_{ss}(t)$ characterizes the final value of the error as a difference between the final value of the input $x(t)$ and the final value of the system response $y_{ss}(t)$.

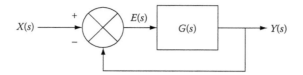

FIGURE 3.1
Closed-loop control system.

Generally speaking, the steady-state error of a stable closed-loop control system is much smaller than the associated error of an open-loop control system.

Consider the unity feedback system shown in Figure 3.1.

For the signal $E(s)$ it holds that

$$E(s) = X(s) - Y(s) \tag{3.2}$$

The quotient $\dfrac{E(s)}{X(s)}$ is called **error transfer function** and is defined by the following equation:

$$\frac{E(s)}{X(s)} = 1 - \frac{Y(s)}{X(s)} = \frac{1}{1+G(s)} \tag{3.3}$$

The error $e(t)$ is

$$e(t) = x(t) - y(t) \tag{3.4}$$

The steady-state error $e_{ss}(t)$ is computed by

$$e_{ss}(t) = \lim_{t \to \infty} e(t) = \lim_{s \to 0} sE(s) \overset{(3.3)}{=} \frac{sX(s)}{1+G(s)} \tag{3.5}$$

Relationship (3.5) is valid, provided that the roots of the equation $sE(s) = 0$ have negative real parts.

The steady-state error of a system depends on the input signal. We now introduce the basic cases.

1. The first case is that of a step input signal $x(t) = Au(t)$.
   In this case, the steady-state error is called *position error* and is given by

$$e_{ss}(t) = \frac{A}{1+\lim_{s \to 0} G(s)} = \frac{A}{1+k_p} \tag{3.6}$$

where the term k_p is called *position error constant* and is given by

$$k_p = \lim_{s \to 0} G(s) \tag{3.7}$$

2. The second case is that of a ramp input signal $x(t) = A \cdot t$
   In this case, the steady-state error is called *velocity error* and is given by

$$e_{ss}(t) = \frac{A}{\lim\limits_{s \to 0} sG(s)} = \frac{A}{k_v} \tag{3.8}$$

where the term k_v is called *velocity error constant* and is given by

$$k_v = \lim_{s \to 0} sG(s) \tag{3.9}$$

3. The third case is that of a parabolic input signal $x(t) = \frac{1}{2}At^2$
   In this case, the steady-state error is called *acceleration error* and is given by

$$e_{ss}(t) = \frac{A}{\lim\limits_{s \to 0} s^2 G(s)} = \frac{A}{k_a} \tag{3.10}$$

where the term k_a is called *acceleration error constant* and is given by

$$k_a = \lim_{s \to 0} s^2 G(s) \tag{3.11}$$

4. Finally, the steady-state error for any input signal $x(t)$ is computed as follows:
   From (3.3), we have

$$E(s) = \frac{1}{1+G(s)} \cdot X(s) \tag{3.12}$$

By denoting $F(s) = \dfrac{1}{1+G(s)}$, (3.12) becomes $E(s) = F(s)X(s)$. From the convolution property of Laplace transform, the error $e(t)$ is given by the convolution integral:

$$e(t) = \int_0^t f(\tau)x(t-\tau)d\tau \tag{3.13}$$

Applying Taylor series to (3.13) and taking into account that $e_{ss}(t) = \lim_{t \to \infty} e(t)$ we obtain a general expression for the steady-state error for any input signal

$$e_{ss}(t) = \sum_{k=0}^{\infty} \frac{c_k}{k!} x_{ss}^{(k)}(t) \tag{3.14}$$

where c_k are called generalized error constants and are given by the relationship

$$c_k = \lim_{s \to 0} F^{(k)}(s) \tag{3.15}$$

In the case of non-unity feedback systems like the one shown in the following figure, the output of the summing junction $E_a(s)$ is not the difference between the reference input and the actual output. The solution is to use block diagram manipulation in order to convert the system into an equivalent unity feedback system.

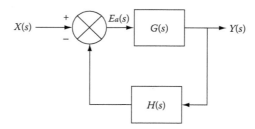

The equivalent system is depicted in the figure below. The transfer function of the inner loop is $H(s) - 1$ and the outer loop has unity feedback.

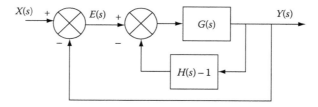

The transfer function $G_1(s)$ of the inner loop is given by

$$G_1(s) = \frac{G(s)}{1 + G(s)(H(s) - 1)} \tag{3.16}$$

and the equivalent unity feedback system is

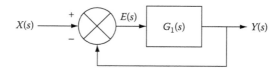

3.3 Types of Control Systems

Control systems are usually described in relation to the **system type**, and to the steady-state error constants k_p, k_v, and k_a. The open-loop transfer function $G(s)$ of a closed loop system can be written in the form:

$$G(s) = K \frac{\prod_{i=1}^{n}(T_i's+1)}{s^l \prod_{i=1}^{m}(T_is+1)} \tag{3.17}$$

The term s^l represents the number of multiple poles at the origin of the complex plane. Based on the number (l) of pure integrations in the denominator of the open-loop transfer function, control systems are defined as *system type l*. The usual values of l are 0, 1, and 2. Thus, a system is called a system type l if its open-loop transfer function has l poles at $s = 0$.

If $l > 0$, as s tends to 0 the denominator of (3.17) tends to 0 and the open-loop transfer tends to infinity.

Depending on the input signal (step/ramp/parabolic) the steady-state error can be zero, constant, or infinite. This is straightforward from Equations 3.7, 3.9, and 3.11.

The steady-state errors in relation to the system type and the input type are summarized in Table F3.1.

Formulas

TABLE F3.1

Steady-State Errors

		Steady-State Errors		
System Type	Error Constants	Step Input $x(t) = Au(t)$	Ramp Input $x(t) = A \cdot t$	Parabolic Input $x(t) = \frac{1}{2}A \cdot t^2$
0	Position: k_p const. Velocity: $k_v = 0$ Acceleration: $k_a = 0$	$\frac{A}{1+k_p}$	∞	∞
1	Position: $k_p \to \infty$ Velocity: k_v const. Acceleration: $k_a = 0$	0	$\frac{A}{k_v}$	∞
2	Position: $k_p \to \infty$ Velocity: $k_v \to \infty$ Acceleration: k_a const.	0	0	$\frac{A}{k_a}$

Problems

3.1 Study the sensitivity of the closed-loop control system shown in the figure below, in relation to the variations of $G(s)$ and $H(s)$.

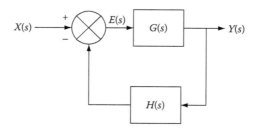

Solution

For the closed-loop control system it holds that

$$F(s) = \frac{Y(s)}{X(s)} = \frac{G(s)}{1+G(s)H(s)} \qquad \text{(P3.1.1)}$$

The sensitivity of the closed-loop control system to the variations of $G(s)$ is

$$S_G^F = \frac{G(s)}{F(s)}\left|\frac{\partial F(s)}{\partial G(s)}\right| = \frac{G(s)}{\dfrac{G(s)}{1+G(s)H(s)}} \cdot \frac{1}{\left(1+G(s)H(s)\right)^2} \Rightarrow$$

$$\qquad \text{(P3.1.2)}$$

$$S_G^F = \frac{1}{1+G(s)H(s)}$$

The factor $1+G(s)H(s)$ determines the characteristics of the closed-loop control system. From the relationship (P3.1.2), it follows that for $G(s)H(s) \gg 1$, the sensitivity of the system tends to zero $S_G^F \to 0$.

Thus, if $G(s)H(s)$ is sufficiently large, the influence of variations of $G(s)$ is not significant. The sensitivity of the system to variations of $H(s)$ is computed as

$$S_H^F = \frac{H(s)}{F(s)}\left|\frac{\partial F(s)}{\partial H(s)}\right| = \frac{H(s)}{\dfrac{G(s)}{1+G(s)H(s)}} \cdot \frac{G^2(s)}{\left(1+G(s)H(s)\right)^2} \Rightarrow$$

$$\qquad \text{(P3.1.3)}$$

$$S_H^F = \frac{G(s)H(s)}{1+G(s)H(s)} = \frac{1}{1+\dfrac{1}{G(s)H(s)}}$$

From the relationship (P3.1.3), it arises that for $G(s)H(s) \gg 1$ the system sensitivity tends to unity, that is, $S_G^F \to 1$. If $G(s)H(s)$ is sufficiently large, then the variations of $H(s)$ directly influence the output response.

Therefore, it is important to use feedback elements that are not affected from environmental variations.

3.2 Consider the block diagram of the closed-loop control system shown in the following figure, where J and F are real positive numbers.

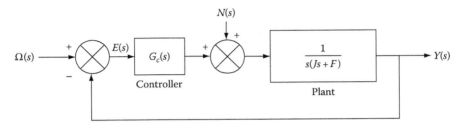

a. Compute the steady-state error if $G_c(s) = k_p > 0$, $\Omega(s) = 0$ and the disturbance is a step function with amplitude A.

b. Suppose that $G_c(s) = k_p\left(1 + \dfrac{1}{T_i s}\right)$, $T_i \neq 0$, $k_p > 0$, $\Omega(s) = 0$ and the disturbance is a step function with amplitude A. Compute the steady-state error if the closed-loop system is stable.

Discuss the results of the first and the second query in regards to the form of the controller and to the steady-state error.

c. Assuming that $G_c(s) = k_p\left[1 + \dfrac{1}{T_i s}\right]$, $T_i \neq 0$, find the region of values of k_p and T_i so that the effect of the disturbance $N(s) = 1$ is fully eliminated in the steady-state response, that is, $\lim_{t \to \infty} y_N(t) = 0$. Moreover, the output must follow the steady-state input signal $\Omega(s) = 1/s$, that is, $\lim_{t \to \infty}\{y(t) - \omega(t)\} = 0$.

Solution

a. From the depicted block diagram and for $\Omega(s) = 0$, we have

$$\frac{Y(s)}{N(s)} = \frac{\dfrac{1}{s(Js+F)}}{1 + G_c(s)\dfrac{1}{s(Js+F)}} = \frac{1}{Js^2 + Fs + k_p} \tag{P3.2.1}$$

and

$$\frac{E(s)}{N(s)} = -\frac{Y(s)}{N(s)} \overset{(P3.2.1)}{=} -\frac{1}{Js^2 + Fs + k_p} \tag{P3.2.2}$$

The second-order denominator has positive coefficients. Thus, the roots of the characteristic equation are in the left-half s-plane. Hence,

$$e_{ss}(t) = \lim_{t \to \infty} e(t) = \lim_{s \to 0} sE(s) \Rightarrow$$

$$e_{ss}(t) = \lim_{s \to 0}\left[s \cdot \left(-\frac{1}{Js^2 + Fs + k_p}\right) \cdot \frac{A}{s}\right] = -\frac{A}{k_p} \tag{P3.2.3}$$

b. It holds that

$$G_c(s) = k_p\left(1 + \frac{1}{T_i s}\right) \tag{P3.2.4}$$

We have

$$\frac{Y(s)}{N(s)} = \frac{\dfrac{1}{s(Js+F)}}{1 + k_p\left(1 + \dfrac{1}{T_i s}\right) \cdot \dfrac{1}{s(Js+F)}} = \frac{s}{Js^3 + Fs^2 + k_p s + \dfrac{k_p}{T_i}} \tag{P3.2.5}$$

and

$$\frac{E(s)}{N(s)} = -\frac{Y(s)}{N(s)} \overset{\text{(P3.2.5)}}{=} -\frac{s}{Js^3 + Fs^2 + k_p s + \dfrac{k_p}{T_i}} \tag{P3.2.6}$$

Supposing that the roots of the characteristic equation are in the left-half s-plane, we have

$$e_{ss}(t) = \lim_{t \to \infty} e(t) = \lim_{s \to 0} sE(s) \Rightarrow$$

$$e_{ss}(t) \overset{\text{(P3.2.6)}}{=} \lim_{s \to 0}\left[s \cdot \left(-\frac{s}{Js^3 + Fs^2 + k_p s + \dfrac{k_p}{T_i}}\right) \cdot \frac{A}{s}\right] = 0 \tag{P3.2.7}$$

We observe that if the controller is a gain k_p, the steady-state error is equal to $-A/k_p$. On the other hand, as the closed-loop system is stable, if the controller contains a gain and an integrator, the steady-state error tends to zero.

c. From the superposition property, the total output $Y(s)$ of the system is given by

$$Y(s) = Y_N(s) + Y_\Omega(s) \tag{P3.2.8}$$

From relationship (P3.2.5), we obtain

$$Y_N(s) = \frac{s}{Js^3 + Fs^2 + k_p s + \dfrac{k_p}{T_i}} N(s) \tag{P3.2.9}$$

For $N(s) = 0$ from the block diagram, we get

$$Y_\Omega(s) = \frac{k_p s + \dfrac{k_p}{T_i}}{Js^3 + Fs^2 + k_p s + \dfrac{k_p}{T_i}} \Omega(s) \tag{P3.2.10}$$

Substituting (P3.2.9) and (P3.2.10) to (P3.2.8), we get

$$Y(s) = \frac{s}{Js^3 + Fs^2 + k_p s + \dfrac{k_p}{T_i}} N(s) + \frac{k_p s + \dfrac{k_p}{T_i}}{Js^3 + Fs^2 + k_p s + \dfrac{k_p}{T_i}} \Omega(s) \qquad \text{(P3.2.11)}$$

The disturbance $N(t)$ is eliminated if

$$\lim_{t \to \infty} y_N(t) = 0 \qquad \text{(P3.2.12)}$$

The final value theorem can be applied only if the system is stable. We will use Routh's criterion to decide on the system stability; Routh's criterion is introduced analytically in Chapter 5.

Routh's tabulation is

$$
\begin{array}{c|ccc}
s^3 & J & k_p \\[2mm]
s^2 & F & k_p/T_i \\[4mm]
s^1 & \dfrac{Fk_p - \dfrac{Jk_p}{T_i}}{F} & 0 \\[4mm]
s^0 & k_p/T_i & 0
\end{array}
$$

From the last row, it follows that k_p και T_i must have the same sign.

As $k_p < 0$ leads to contradiction, $k_p > 0$ yields $T_i > \dfrac{J}{F}$. Thus,

$$\lim_{t \to \infty} y_N(t) = \lim_{s \to 0} s Y_N(s) \overset{\text{(P3.2.9)}}{=} \lim_{s \to 0} \left[s \left(\frac{s}{Js^3 + Fs^2 + k_p s + \dfrac{k_p}{T_i}} \right) \cdot 1 \right] = 0 \qquad \text{(P3.2.13)}$$

and

$$\lim_{t \to \infty} \left\{ y(t) - \omega(t) \right\} = \lim_{s \to 0} s \left(Y(s) - \Omega(s) \right) = \lim_{s \to 0} \left(s Y_\Omega(s) - 1 \right) \Rightarrow$$

$$\lim_{t \to \infty} \left\{ y(t) - \omega(t) \right\} = \lim_{s \to 0} \left[s \left(\frac{k_p s + \dfrac{k_p}{T_i}}{Js^3 + Fs^2 + k_p s + \dfrac{k_p}{T_i}} \right) \cdot \frac{1}{s} - 1 \right] = 0 \qquad \text{(P3.2.14)}$$

These relationships hold for any k_p and T_i, such that the roots of the characteristic equation are in the left-half s-plane.

3.3 The next figure depicts the block diagram of a position servomechanism with a PD controller.

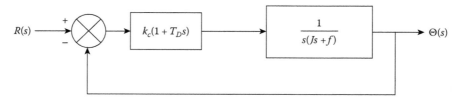

a. Compute the velocity error of the system and show how it can be minimized.

b. Compute the sensitivity of the system on the variations of k_c and J, and discuss the results.

Solution

a. The closed-loop transfer function of the system is

$$G(s)H(s) = \frac{k_c(1 + T_D s)}{s(Js + f)} \tag{P3.3.1}$$

The velocity error $e_{ss,v}$ is

$$e_{ss,v} = \frac{1}{k_v} = \frac{1}{\lim_{s \to 0} G(s)} = \frac{1}{k_c/f} = \frac{f}{k_c} \tag{P3.3.2}$$

Thus, in order to minimize $e_{ss,v}$, we need to choose a small f and a large k_c.

b. i.   The sensitivity of the system to variations of k_c is given by the relationship (P3.3.3):

$$S_{k_c}^F = \frac{k_c}{F(s)} \left| \frac{\partial F(s)}{\partial k_c} \right| \tag{P3.3.3}$$

However,

$$F(s) = \frac{\Theta(s)}{R(s)} = \frac{k_c(1 + T_D s)/s(Js + f)}{1 + \dfrac{k_c(1 + T_D s)}{s(Js + f)}} = \frac{k_c(1 + T_D s)}{Js^2 + s(f + k_c T_D) + k_c} \tag{P3.3.4}$$

and

$$\frac{\partial F(s)}{\partial k_c} = \frac{(1 + T_D s)\left[Js^2 + s(f + k_c T_D) + k_c \right] - k_c(1 + T_D s)(1 + T_D s)}{\left[Js^2 + s(f + k_c T_D) + k_c \right]^2} \tag{P3.3.5}$$

Thus, (P3.3.3) $\underset{(P3.3.5)}{\overset{(P3.3.4)}{\Rightarrow}}$ $S_{k_c}^F = \left| \dfrac{Js^2 + sf}{Js^2 + sf + sk_c T_D + k_c} \right| = \left| \dfrac{1}{1 + \dfrac{k_c(sT_D + 1)}{Js^2 + sf}} \right|$ (P3.3.6)

From relationship (P3.3.6), we conclude that if we choose a large k_c, then the sensitivity becomes $S_{k_c}^F \to 0$.

ii. The sensitivity of the system to the variations of the equivalent inertia constant J is

$$S_J^F = \left| \frac{Js^2 + sf}{Js^2 + sf + sk_cT_D + k_c} \right| = \frac{J}{F(s)} \left| \frac{\partial F(s)}{\partial J} \right| \qquad (P3.3.7)$$

and

$$\frac{\partial F(s)}{\partial J} = -\frac{k_c(1 + T_Ds)s^2}{\left[Js^2 + s(f + k_cT_D) + k_c \right]^2} \qquad (P3.3.8)$$

$$(P3.3.7) \overset{(P3.3.8)}{\underset{(P3.3.4)}{\Rightarrow}} S_J^F = \left| \frac{Js^2}{Js^2 + s(f + k_cT_D) + k_c} \right| = \left| \frac{1}{1 + \dfrac{s(f + k_cT_D) + k_c}{Js^2}} \right| \qquad (P3.3.9)$$

From relationship (P3.3.9), we see that if we choose a sufficiently large J, then the sensitivity becomes $S_J^F \to 1$.

We can also study the sensitivity S_J^F in relation to the variation of ω.

$$(P3.3.9) \overset{s=j\omega}{\Rightarrow} S_J^F = \frac{J\omega^2}{\left| -J\omega^2 + j\omega(f + k_cT_D) + k_c \right|} \Rightarrow$$

$$(P3.3.10)$$

$$S_J^F = \frac{J\omega^2}{\sqrt{(k_c - J\omega^2)^2 + \omega^2(f + k_cT_D)^2}} = \frac{J}{\sqrt{\left(\dfrac{k_c}{\omega^2} \right)^2 + \dfrac{1}{\omega^2}(f + k_cT_D)^2}}$$

From relationship (P3.3.10), we obtain that for $\omega \to 0 \, \text{rad/s} \Rightarrow S_J^F \to 0$ and for $\omega \to \infty \, \text{rad/s} \Rightarrow S_J^F \to 1$.

3.4 Calculate the steady-state errors (position, velocity, and acceleration errors) for the position control system of the following figure.

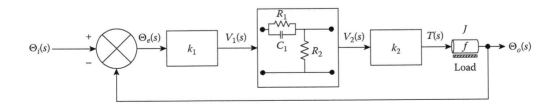

Solution

In order to find the steady-state errors, we must first compute the open-loop transfer function. For the electrical circuit of the system, we have

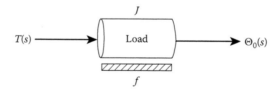

$$\frac{V_2(s)}{V_1(s)} = \frac{R_2 I(s)}{(Z(s) + R_2) I(s)} = \frac{R_2}{Z(s) + R_2} \tag{P3.4.1}$$

where

$$Z(s) = \frac{R_1 \dfrac{1}{sC_1}}{R_1 + \dfrac{1}{sC_1}} = \frac{R_1}{sR_1C_1 + 1} \tag{P3.4.2}$$

$$(P3.4.1) \overset{(P3.4.2)}{\Rightarrow} \frac{V_2(s)}{V_1(s)} = \frac{R_2}{\dfrac{R_1}{sR_1C_1 + 1} + R_2} = \frac{R_2(sR_1C_1 + 1)}{sR_1C_1 + R_1 + R_2} \tag{P3.4.3}$$

For the electrical load, we have:

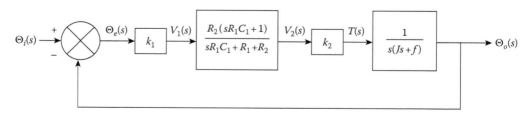

$$T(t) = J\frac{d^2\theta(t)}{dt^2} + f\frac{d\theta(t)}{dt} \tag{P3.4.4}$$

$$(P3.4.4) \underset{I.C.=0}{\overset{L.T.}{\Rightarrow}} T(s) = (Js + f)s\Theta(s) \Rightarrow \frac{\Theta(s)}{T(s)} = \frac{1}{(Js + f)s} \tag{P3.4.5}$$

Based on the relationships (P3.4.3) and (P3.4.5), the block diagram of the system becomes.

Hence, the open-loop transfer function is

$$G(s) = \frac{k_1 k_2 R_2 (s R_1 C_1 + 1)}{s(s R_1 C_1 + R_1 + R_2)(Js + f)} \qquad \text{(P3.4.6)}$$

The position error is given by

$$\theta e_{ss,p} = \frac{1}{1 + \lim_{s \to 0} G(s)} \to \infty \qquad \text{(P3.4.7)}$$

The velocity error is

$$\theta e_{ss,v} = \frac{1}{\lim_{s \to 0} s G(s)} \stackrel{\text{(P3.4.6)}}{=} \frac{k_1 k_2 R_2}{(R_1 + R_2) f} \qquad \text{(P3.4.8)}$$

The acceleration error is

$$\theta e_{ss,a} = \frac{1}{\lim_{s \to 0} s^2 G(s)} \stackrel{\text{(P3.4.6)}}{=} 0 \qquad \text{(P3.4.9)}$$

3.5 Calculate the steady-state errors of the control system shown in the following figure if the input is (i) a unit-step function, (ii) a unit-ramp function, and (iii) a unit-parabolic function.

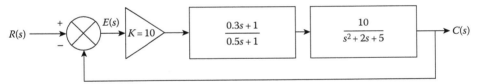

Solution

The open-loop transfer function of the system shown is

$$G(s) = \frac{100(0.3s + 1)}{(0.5s + 1)(s^2 + 2s + 5)} \qquad \text{(P3.5.1)}$$

i. For an input of the form $r(t) = u(t)$, the steady-state error is

$$e_{ss,p} = \frac{1}{1 + \lim_{s \to 0} G(s)} \stackrel{\text{(P3.5.1)}}{=} \frac{1}{21} \qquad \text{(P3.5.2)}$$

ii. For an input of the form $r(t) = t$, the steady-state error is

$$e_{ss,v} = \frac{1}{\lim_{s \to 0} s G(s)} \to \infty \qquad \text{(P3.5.3)}$$

iii. For an input of the form $r(t) = t^2$, the steady-state error is

$$e_{ss,a} = \frac{2}{\lim_{s \to 0} s^2 G(s)} \to \infty \qquad \text{(P3.5.4)}$$

3.6 Compute the error constants (position, velocity, and acceleration) for the unity feed-back systems with open-loop transfer functions:

a. $G(s) = \dfrac{10}{(0.1s+1)(0.5s+1)}$

b. $G(s) = \dfrac{20}{s(s+2)(s+5)}$

c. $G(s) = \dfrac{100(s+1)}{s^2(s^2+4s+5)}$

d. $G(s) = \dfrac{k(s+2)}{s^3(s^2+2s+5)}$

Solution

a. For $G(s) = \dfrac{10}{(0.1s+1)(0.5s+1)}$, the error constants are

$$k_p = \lim_{s\to0} G(s) = \lim_{s\to0}\frac{10}{(0.1s+1)(0.5s+1)} = 10 \qquad\qquad \text{(P3.6.1)}$$

$$k_v = \lim_{s\to0} sG(s) = \lim_{s\to0}\frac{10s}{(0.1s+1)(0.5s+1)} = 0 \qquad\qquad \text{(P3.6.2)}$$

$$k_a = \lim_{s\to0} s^2G(s) = \lim_{s\to0}\frac{10s^2}{(0.1s+1)(0.5s+1)} = 0 \qquad\qquad \text{(P3.6.3)}$$

b. For $G(s) = \dfrac{20}{s(s+2)(s+5)}$, the error constants are

$$k_p = \lim_{s\to0} G(s) \to \infty \qquad\qquad \text{(P3.6.4)}$$

$$k_v = \lim_{s\to0} sG(s) = 2 \qquad\qquad \text{(P3.6.5)}$$

$$k_a = \lim_{s\to0} s^2 G(s) \to 0 \qquad\qquad \text{(P3.6.6)}$$

c. For $G(s) = \dfrac{100(s+1)}{s^2(s^2+4s+5)}$, the error constants are

$$k_p = \lim_{s\to0} G(s) \to \infty \qquad\qquad \text{(P3.6.7)}$$

$$k_v = \lim_{s\to0} sG(s) \to \infty \qquad\qquad \text{(P3.6.8)}$$

$$k_a = \lim_{s\to0} s^2 G(s) = 20 \qquad\qquad \text{(P3.6.9)}$$

d. For $G(s) = \dfrac{k(s+2)}{s^3(s^2+2s+5)}$, the error constants are

$$k_p = \lim_{s \to 0} G(s) \to \infty \qquad \text{(P3.6.10)}$$

$$k_v = \lim_{s \to 0} sG(s) \to \infty \qquad \text{(P3.6.11)}$$

$$k_a = \lim_{s \to 0} s^2 G(s) \to \infty \qquad \text{(P3.6.12)}$$

3.7 Consider the unity feedback control systems (i.e., $H(s) = 1$) which have the following open-loop transfer functions:

i. $G(s)H(s) = G(s) = \dfrac{k}{s(s+1)}$

ii. $G(s)H(s) = G(s) = \dfrac{k}{s^2(s+5)(s+2)}$

Compute

  a. The error constants k_p, k_v, and k_a
  b. The position, velocity, and acceleration errors
  c. The general error constants and the general expression of the steady-state error for any excitation

Solution

  a. First, we consider the case where

$$G(s) = \frac{k}{s(s+1)} \qquad \text{(P3.7.1)}$$

Since the system is Type 1, we have

$$k_p = \infty$$

$$k_v = a \qquad \text{(P3.7.2)}$$

$$k_a = 0$$

For

$$G(s) = \frac{k}{s^2(s+5)(s+2)} \qquad \text{(P3.7.3)}$$

the system is Type 2; thus,

$$k_p = \infty$$

$$k_v = \infty \qquad \text{(P3.7.4)}$$

$$k_a = \lim_{s \to 0} s^2 G(s) \overset{(P3.7.3)}{=} \lim_{s \to 0} \frac{k}{(s+5)(s+2)} = \frac{k}{10}$$

b. For $G(s) = \dfrac{k}{s(s+1)}$, the steady-state errors are

$$e_{ss,p} = 0$$

$$e_{ss,v} = \frac{1}{k} \tag{P3.7.5}$$

$$e_{ss,a} = \infty$$

Similarly, for $G(s) = \dfrac{k}{s^2(s+5)(s+2)}$

$$e_{ss,p} = 0$$

$$e_{ss,v} = 0 \tag{P3.7.6}$$

$$e_{ss,a} = \frac{1}{k_a} = \frac{10}{k}$$

c. For $G(s) = \dfrac{k}{s(s+1)}$, the generalized error constants are computed by relationship (3.15) as

$$c_k = \lim_{s \to 0} F(s)^{(k)} \tag{P3.7.7}$$

where

$$F(s) = \frac{1}{1+G(s)} = \frac{s(s+1)}{s(s+1)+k} \tag{P3.7.8}$$

Consequently,

$$c_0 = \lim_{s \to 0} F(s) \overset{(P3.7.8)}{=} 0$$

$$c_1 = \lim_{s \to 0} \frac{dF(s)}{ds} \overset{(P3.7.8)}{=} \frac{1}{k} \tag{P3.7.9}$$

$$c_2 = \lim_{s \to 0} \frac{d^2F(s)}{ds^2} \overset{(P3.7.8)}{=} \frac{2(k-1)}{k^2}$$

The generalized expression of the steady-state error for any input signal is

$$e_{ss}(t) = \sum_{k=0}^{\infty} \frac{c_k}{k!} x_{ss}^{(k)}(t) \tag{P3.7.10}$$

From relationship (P3.7.10), it follows that

   i. For a unit-step input, that is, for $x(t) = u(t)$,

$$(P3.7.10) \Rightarrow e_{ss} = c_0 = 0 \qquad \text{(P3.7.11)}$$

   ii. For a unit-ramp input, that is, for $x(t) = t$,

$$(P3.7.10) \Rightarrow e_{ss} = \frac{1}{k} \qquad \text{(P3.7.12)}$$

   iii. For a unit-parabolic input, that is, for $x(t) = t^2$,

$$(P3.7.10) \Rightarrow e_{ss} \to \infty \qquad \text{(P3.7.13)}$$

As we expected, the steady errors computed in (P3.7.11), (P3.7.12), and (P3.7.13) are same to the ones computed in (P3.7.5). Finally the generalized expression of the steady-state error is

$$(P3.7.10) \Rightarrow e_{ss}(t) = c_0 x(t) + c_1 \frac{dx(t)}{dt} + c_2 \frac{d^2 x(t)}{dt^2} + \cdots \qquad \text{(P3.7.14)}$$

We repeat the same procedure for $G(s) = \dfrac{k}{s^2(s+5)(s+2)}$.

We have

$$F(s) = \frac{1}{1+G(s)} = \frac{s^2(s+5)(s+2)}{s^2(s+5)(s+2)+k} \qquad \text{(P3.7.15)}$$

The relationship (P3.7.7) provides the general error constants

$$c_0 = \lim_{s\to 0} F(s) \overset{(P3.7.15)}{=} 0$$

$$c_1 = \lim_{s\to 0} \frac{dF(s)}{ds} \overset{(P3.7.15)}{=} 0 \qquad \text{(P3.7.16)}$$

$$c_2 = \lim_{s\to 0} \frac{d^2 F(s)}{ds^2} \overset{(P3.7.15)}{=} \frac{10}{k}$$

From relationship (P3.7.16), it follows that

   i. For input $x(t) = u(t)$, we have

$$(P3.7.10) \Rightarrow e_{ss} = 0 \qquad \text{(P3.7.17)}$$

   ii. For input $x(t) = t$, we have

$$(P3.7.10) \Rightarrow e_{ss} = 0 \qquad \text{(P3.7.18)}$$

iii. For input $x(t) = t^2$, we have

$$(P3.7.10) \Rightarrow e_{ss} = \frac{10}{k} \tag{P3.7.19}$$

3.8 Calculate the position, velocity, and acceleration errors for the system depicted in the following figure.

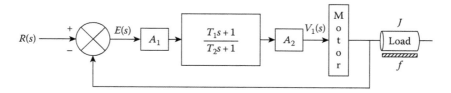

Solution

From the block diagram of the depicted system the following relationships hold:

$$V_1(s) = \frac{A_1 A_2 (T_1 s + 1)}{T_2 s + 1} E(s) \tag{P3.8.1}$$

$$E(s) = R(s) - \Theta(s) \tag{P3.8.2}$$

$$T_m(s) = J s^2 \Theta(s) + f s \Theta(s) \tag{P3.8.3}$$

$$T_m(s) = k V_1(s) + m s \Theta(s) \tag{P3.8.4}$$

By combining these relationships, we get

$$E(s) = \frac{\left[J s^2 + (f - m)s \right](T_2 s + 1) R(s)}{k A_1 A_2 (T_1 s + 1)\left[J s^2 + (f - m)s \right](T_2 s + 1)} \tag{P3.8.5}$$

The position error is

$$e_{ss,p} \stackrel{(P3.8.5)}{=} 0 \tag{P3.8.6}$$

The velocity error is

$$e_{ss,v} \stackrel{(P3.8.5)}{=} \frac{(f - m)R_1}{k A_1 A_2} \tag{P3.8.7}$$

The acceleration error is

$$e_{ss,a} \stackrel{(P3.8.5)}{=} \infty \tag{P3.8.8}$$

3.9 Consider the following block diagram of a closed-loop system,

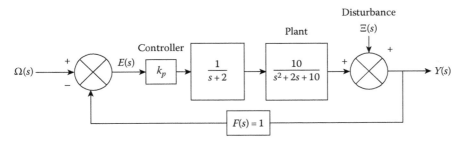

where k_p is a real number and the disturbance is $\Xi(s) = \dfrac{1}{s+2}$.

Compute the values of k_p (if they exist) for which the effect of the disturbance is eliminated and the steady-state output follows the unit-step input ($\Omega(s) = 1/s$), that is, $\lim_{t \to \infty} (y(t) - \omega(t)) = 0$.

Solution

We compute the total output of the system, based on the superposition principle. We have

$$Y(s) = Y_\omega(s) + Y_\Xi(s) \tag{P3.9.1}$$

In order to find $Y_\omega(s)$, we suppose that the disturbance $\Xi(s)$ is zero.

$$\frac{Y_\omega(s)}{\Omega(s)} = \frac{\dfrac{k_p}{(s+2)(s^2+2s+10)}}{1 + \dfrac{k_p}{(s+2)(s^2+2s+10)}} = \frac{k_p}{(s+2)(s^2+2s+10)+k_p} \tag{P3.9.2}$$

Similarly, we suppose that the system input $\Omega(s)$ is zero in order to find $Y_\Xi(s)$:

$$\frac{Y_\Xi(s)}{\Xi(s)} = \frac{1}{1 + \dfrac{k_p}{(s+2)(s^2+2s+10)}} = \frac{(s+2)(s^2+2s+10)}{(s+2)(s^2+2s+10)+k_p} \tag{P3.9.3}$$

Thus, the output of the system is

$$\overset{\text{(P3.9.2),(P3.9.3)}}{\text{(P3.9.1)}} \Rightarrow Y(s) = \frac{k_p}{(s+2)(s^2+2s+10)+k_p}\Omega(s) + \frac{s^2+2s+10}{(s+2)(s^2+2s+10)+k_p}\Xi(s) \tag{P3.9.4}$$

We have $\omega(t) = u(t) \Rightarrow \Omega(s) = 1/s$ and $\Xi(s) = 1/s + 2$.
By substituting these to (P3.9.4), we get

$$Y(s) = \frac{k_p}{(s+2)(s^2+2s+10)+k_p} \cdot \frac{1}{s} + \frac{s^2+2s+10}{(s+2)(s^2+2s+10)+k_p} \tag{P3.9.5}$$

In order to eliminate the steady-state disturbance, it must hold that

$$\lim_{t\to\infty} y_\Xi(t) = 0 \tag{P3.9.6}$$

We apply the final value theorem, which holds if $sY_\Xi(s)$ is stable. We will use Routh's criterion for the polynomial $P(s) = (s + 2)(s^2 + 2s + 10) + k_p = s^3 + 4s^2 + 14s + (k_p + 20)$. Routh's tabulation is

$$
\begin{array}{c|cc}
s^3 & 1 & 14 \\[2mm]
s^2 & 4 & k_p + 20 \\[2mm]
s^1 & \dfrac{36 - k_p}{4} & 0 \\[2mm]
s^0 & k_p + 20 & 0
\end{array}
$$

For the $P(s)$ to be stable it is sufficient that

$$k_p + 20 > 0 \tag{P3.9.7}$$

and

$$\frac{36 - k_p}{4} > 0 \tag{P3.9.8}$$

Thus,

$$(\text{P3.9.7}) \Rightarrow k_p > -20 \tag{P3.9.7'}$$

$$(\text{P3.9.8}) \Rightarrow k_p < 36 \tag{P3.9.8'}$$

From (P3.9.7) and (P3.9.8), we come up with the inequality

$$-20 < k_p < 36 \tag{P3.9.9}$$

We apply the final value theorem and get

$$\lim_{t\to\infty} y_\Xi(t) = \lim_{s\to0} sY_\Xi(s) \overset{(\text{P3.9.3})}{=} 0 \tag{P3.9.10}$$

Therefore, the influence of the disturbance goes to zero in the steady state for $-20 < k_p < 36$.

We examine the second requirement:

$$\lim_{t\to\infty}\left(y(t)-\omega(t)\right)=0 \qquad\text{(P3.9.11)}$$

Since the disturbance is eliminated for $-20 < k_p < 36$, it must hold that

$$\lim_{t\to\infty}\left(y(t)-\omega(t)\right)=\lim_{s\to0}\left(sY(s)-s\Omega(s)\right)\overset{\text{(P3.9.10)}}{=}\lim_{s\to0}\left(sY_\omega(s)-s\Omega(s)\right)\Rightarrow$$

$$\lim_{t\to\infty}\left(y(t)-\omega(t)\right)=\lim_{s\to0}\left(\frac{k_p}{(s+2)(s^2+2s+10)+k_p}-1\right)=0 \qquad\text{(P3.9.12)}$$

From Equation (P3.9.12), we get

$$\frac{k_p}{k_p+20}=1 \qquad\text{(P3.9.13)}$$

This requirement is not fulfilled for any k_p. Consequently, the output $y(t)$ cannot follow the step input in the steady state.

3.10 Consider the following closed-loop control system, where $J > 0$.

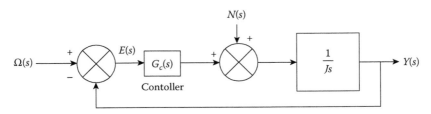

Suppose that the input $\Omega(s)$ is zero, and the disturbance $N(s)$ is a step function with amplitude A, that is, $N(s) = A/s$. Which one of the following controllers is the best for achieving (a) zero steady-state error, and (b) stability of the system.

i. $G_C(s) = k, \quad k > 0$

ii. $G_C(s) = \dfrac{k}{s}, \quad k > 0$

iii. $G_C(s) = k_p + \dfrac{k}{s}, \quad k > 0, k_p > 0$

Solution

i. Suppose that $G_c(s) = k$.
From the given block diagram for $\Omega(s) = 0$ we have

$$\frac{Y(s)}{N(s)}=\frac{1/Js}{1+k/Js}=\frac{1}{Js+k} \qquad\text{(P3.10.1)}$$

and

$$\frac{E(s)}{N(s)} = -\frac{Y(s)}{N(s)} = -\frac{1}{Js + k} \tag{P3.10.2}$$

The characteristic equation $Js + k = 0$ has one root in the left-half s-plane; thus, the system is stable.

Hence, for $N(s) = \dfrac{A}{s}$,

$$e_{ss} = \lim_{t \to \infty} e(t) = \lim_{s \to 0} sE(s) = \lim_{s \to 0} \left(s \cdot \left(-\frac{1}{Js+k} \right) \cdot \frac{A}{s} \right) = -\frac{A}{k} \neq 0 \tag{P3.10.3}$$

ii. Assume that $G_c(s) = \dfrac{k}{s}$

We have

$$\frac{Y(s)}{N(s)} = \frac{1/Js}{1 + k/Js^2} = \frac{s}{Js^2 + k} \tag{P3.10.4}$$

and

$$\frac{E(s)}{N(s)} = -\frac{Y(s)}{N(s)} = -\frac{s}{Js^2 + k} \tag{P3.10.5}$$

The characteristic equation $Js^2 + k = 0$ has the roots $s = \pm j\sqrt{\dfrac{k}{J}}$.

Therefore, the system is unstable. Also, the final value theorem cannot be applied. The error is

$$e(t) = L^{-1}[E(s)] = L^{-1}\left\{ \left(-\frac{s}{Js^2 + k} \right) \cdot \frac{A}{s} \right\} = -\frac{A}{\sqrt{kJ}} \sin\left(\sqrt{\frac{k}{J}}\, t \right) \tag{P3.10.6}$$

This means that we have a simple harmonic oscillation, with amplitude $A/\sqrt{mk}$. Accordingly, the steady-state error is not zero.

iii. Suppose that $G_c(s) = k_p + \dfrac{k}{s}$

We have

$$\frac{Y(s)}{N(s)} = \frac{\dfrac{1}{Js}}{1 + \dfrac{1}{Js} \cdot \dfrac{k_p s + k}{s}} = \frac{s}{Js^2 + k_p s + k} \tag{P3.10.7}$$

and

$$\frac{E(s)}{N(s)} = -\frac{Y(s)}{N(s)} = -\frac{s}{Js^2 + k_p s + k} \qquad \text{(P3.10.8)}$$

The final value theorem can be now applied, because the equation $Js^2 + k_p s + k = 0$ has roots in the left-half s-plane. Hence,

$$e_{ss} = \lim_{t \to \infty} e(t) = \lim_{s \to 0} sE(s) = \lim_{s \to 0} \left(s \cdot \left(-\frac{s}{Js^2 + k_p s + k} \right) \cdot \frac{A}{s} \right) = 0 \qquad \text{(P3.10.9)}$$

Consequently, the controller $G_c(s) = k_p + \dfrac{k}{s}$ is chosen, because the system is stable and the steady-state error becomes zero.

4

Time Response of First- and Second-Order Control Systems

4.1 Time Response

The **time response** $y(t)$ of a system describes the behavior of a system in relation to time, for a certain input. It consists of two parts:

    a. The **transient response** $y_t(t)$
    b. The **steady-state response** $y_{ss}(t)$

The following relationship holds:

$$y(t) = y_t(t) + y_{ss}(t) \tag{4.1}$$

Transient response is the response that follows right after the excitation of the system. It goes to zero after a certain period of time.

Steady-state response is the part of the time response, which remains after the transient part has faded away. Thus,

$$y_{ss}(t) = \lim_{t \to \infty} y(t) \tag{4.2}$$

The **design specifications** of a control system include, among other parameters, the time response to a specific input and the accuracy that must be preserved during the steady state. Specifications are usually given in terms of both the transient and the steady-state response. In order to evaluate them, we usually choose some **typical test (input) signals**.
   The most common typical test signals are

- The unit-step function
- The ramp function
- The Dirac function
- The parabolic function
- The sinusoidal function

The time response of a closed-loop control system is subject to the location of the poles of the transfer function in the complex field.

The relative positions of the poles of a system provide a graphical method for determining its behavior. The **poles** of a transfer function determine the form of the associated response, while the **zeros** determine the constant coefficients of the associated functions.

More specifically, the closer a zero is to a pole, the more reduced is the factor by which the function that corresponds to this pole participates in the response of the system.

4.2 First-Order Systems

Consider the block diagram of the first-order system shown below.

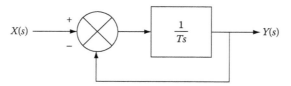

The transfer function of this system is

$$G(s) = \frac{Y(s)}{X(s)} = \frac{1}{Ts+1} \tag{4.3}$$

where T is called the time constant of the system.

- **Unit-step response**
  The system response to an input $x(t) = u(t)$, that is, the step response, is

$$y(t) = L^{-1}\{Y(s)\} = L^{-1}\left\{\frac{1}{Ts+1} \cdot \frac{1}{s}\right\} = 1 - e^{-\frac{t}{T}} \tag{4.4}$$

The next figure illustrates the curve of the time response $y(t)$, for the time constants $T_1 < T_2 < T_3$.

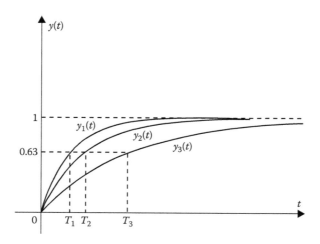

At time $t = T$, the system response $y(t)$ reaches approximately a 63% of its final value. This is straightforward from relationship (4.4) as

$$y(T) = 1 - e^{-1} \simeq 0.6321 \tag{4.5}$$

- **Unit-ramp response**
  The system response to an input $x(t) = r(t) = t$ is given by

$$y(t) = L^{-1}\{Y(s)\} = L^{-1}\left\{\frac{1}{Ts+1} \cdot \frac{1}{s^2}\right\} = t - T + Te^{-\frac{t}{T}}, \quad t \geq 0 \tag{4.6}$$

The corresponding curve is shown in the following figure.

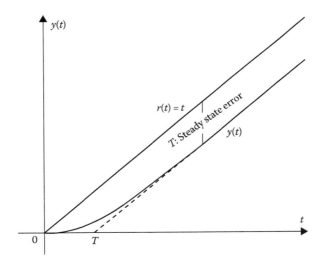

The signal error $e(t)$ is

$$e(t) = r(t) - y(t) = T\left(1 - e^{-\frac{t}{T}}\right), \quad t \geq 0 \tag{4.7}$$

The steady-state error is

$$e_{ss} = \lim_{t \to \infty} e(t) \overset{(4.7)}{=} T \tag{4.8}$$

- **Unit impulse response**
  The system response to an input $x(t) = \delta(t)$ is given by

$$y(t) = L^{-1}\{Y(s)\} = L^{-1}\left\{\frac{1}{Ts+1}\cdot 1\right\} = \frac{1}{T}e^{-\frac{t}{T}} \qquad (4.9)$$

The impulse response curve is shown in the following figure.

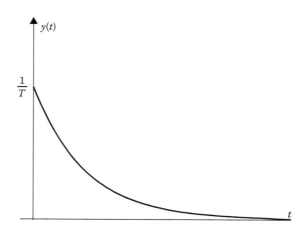

4.3 Second-Order Systems

Consider the block diagram of the second-order system depicted in the following figure.

The transfer function of the system is

$$G(s) = \frac{Y(s)}{X(s)} = \frac{\omega_n^2}{s^2 + 2J\omega_n s + \omega_n^2} \qquad (4.10)$$

The characteristic equation of the system is

$$s^2 + 2J\omega_n s + \omega_n^2 = 0 \qquad (4.11)$$

The poles of $G(s)$ are the roots of the characteristic equation, that is,

$$s_{1,2} = -J\omega_n \pm \omega_n\sqrt{J^2 - 1} \qquad (4.12)$$

where J is called the **damping ratio** of the system.

- **Unit-step response**
  a. **The system is underdamped**, that is, $0 < J < 1$.
     If $0 < J < 1$, then the characteristic equation has two complex conjugate roots:

$$s_{1,2} = -J\omega_n \pm j\omega_n\sqrt{1-J^2} \tag{4.13}$$

The time response is

$$y(t) = L^{-1}\{Y(s)\} = 1 - \frac{e^{-J\omega_n t}}{\sqrt{1-J^2}}\sin(\omega_d t + \varphi) \tag{4.14}$$

where

$$\varphi = \tan^{-1}\frac{\sqrt{1-J^2}}{J} \tag{4.15}$$

and

$$\omega_d = \omega_n\sqrt{1-J^2} \tag{4.16}$$

The frequency ω_d is called **damped natural frequency** of the system, while the frequency ω_n is called the **undamped natural frequency**.

  b. **The system is critically damped**, that is, $J = 1$.
     In case $J = 1$, then $G(s)$ has a real double pole, that is,

$$s_1 = s_2 = -\omega_n \tag{4.17}$$

The time response is

$$y(t) = 1 - e^{-\omega_n t}(1 + \omega_n t) \tag{4.18}$$

  c. **The system is overdamped**, that is, $J > 1$.
     If $J > 1$, then the characteristic equation has two real roots:

$$s_{1,2} = -J\omega_n \pm \omega_n\sqrt{J^2-1} \tag{4.19}$$

The time response is written as

$$y(t) = 1 - e^{J\omega_n t}\left[\cosh\left(\omega_n\sqrt{J^2-1}\cdot t\right) + \frac{1}{\sqrt{J^2-1}}\sinh\left(\omega_n\sqrt{J^2-1}\cdot t\right)\right] \tag{4.20}$$

  d. **The system is undamped**, that is, $J = 0$.
     If $J = 0$, then the poles of $G(s)$ are imaginary and conjugates, that is,

$$s_1 = s_2 = \pm j\omega_n \tag{4.21}$$

The time response can be written as

$$y(t) = 1 - \cos\omega_n t \tag{4.22}$$

For practical applications, only cases for which the damping ratio is $0 < J < 1$ are meaningful, since they correspond to stable control systems.

The next figure depicts the time response of a second-order system to the unit-step function, for various values of the damping ratio J.

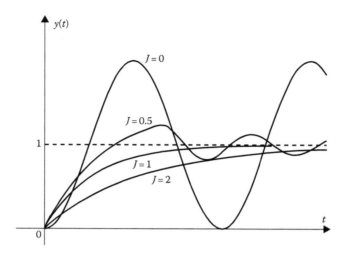

4.4 Transient Response

The time response of a system can be described in relation to two factors:

- The response speed, as it is expressed by the rise time and the peak time
- The degree of convergence between the real and the desired response of the system, as it is expressed by the percent overshoot and the settling time

These factors clash with each other; therefore, the proper trade-offs must be taken into consideration.

In most practical applications the transient response exhibits damped oscillations before it reaches the steady state.

The next figure presents the values of t_d, t_r, t_p, t_s, y_m and M_p. The definitions and the relationships by which these values can be computed are the following:

a. **Delay time** t_d is the time required for the response to reach a 50% of the amplitude of the step input, that is, of its final value, for the first time.

b. **Rise time** t_r is the time required for the response to rise from a 10% to a 90% of its final value. It is given by the relationship

$$t_r = \frac{1}{\omega_d} \tan^{-1}\left(-\frac{\sqrt{1-J^2}}{J}\right) \qquad (4.23)$$

c. **Peak time** t_p is the time required for the response to reach the first peak of the curve. The relationship for the peak time is

$$t_p = \frac{\pi}{\omega_n\sqrt{1-J^2}} = \frac{\pi}{\omega_d} \qquad (4.24)$$

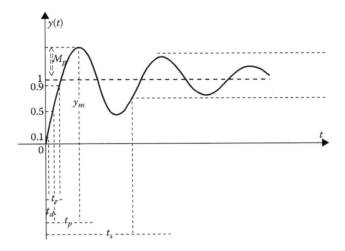

d. **Settling time** t_s is the time required for the output of a system to reach and stay within (usually) ±2% or ±5% of the steady value.

For the 2% requirement $(e^{-J\omega_n t_s} < 0.02)$, settling time is computed as follows:

$$t_s = \frac{4}{J\omega_n} \tag{4.25}$$

For the 5% requirement, settling time is given by

$$t_s = \frac{3}{J\omega_n} \tag{4.26}$$

e. **Maximum response value** y_m is the value of the response that corresponds to the first peak of the curve. It is given by

$$y_m = 1 + e^{\frac{-J\pi}{\sqrt{1-J^2}}} \tag{4.27}$$

f. **Maximum overshoot** M_p is the difference between the maximum response value y_m and the final value y_{ss} of $y(t)$.

g. The **maximum percent overshoot** is defined as

$$M_p\% = \frac{y_m - y_{ss}}{y_{ss}} \cdot 100\% \tag{4.28}$$

where

$$y_{ss} = \lim_{t \to \infty} y(t) \tag{4.29}$$

Supposing that $y_{ss} = 1$ from relationship (4.27), the percent overshoot is

$$M_p\% = e^{\frac{-J\pi}{\sqrt{1-J^2}}} \cdot 100\% \tag{4.30}$$

Formulas

TABLE F4.1

Typical Test Signals

S/N	Name	Graph	Equation	Laplace Transform
1.	Unit step function $u(t)$		$u(t) = \begin{cases} 1, & t > 0 \\ 0, & t < 0 \end{cases}$	$U(s) = \dfrac{1}{s}$
2.	Unit impulse function $\delta(t)$		$\delta(t) = \begin{cases} \infty, & t = 0 \\ 0, & t \neq 0 \end{cases}$	$D(s) = 1$
3.	Unit ramp function $r(t) = t$		$r(t) = \begin{cases} t, & t \geq 0 \\ 0, & t < 0 \end{cases}$	$R(s) = \dfrac{1}{s^2}$

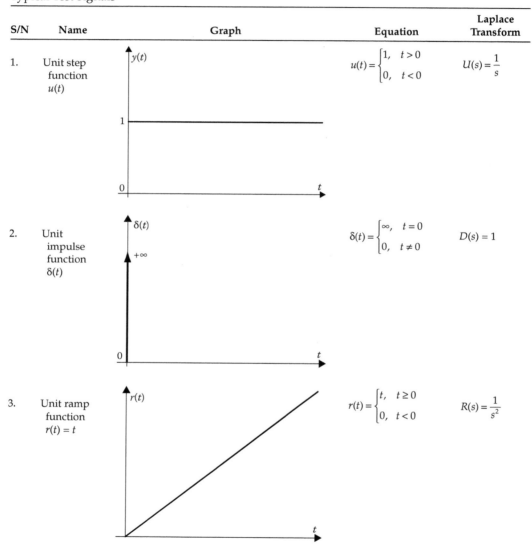

TABLE F4.1 (continued)

Typical Test Signals

S/N	Name	Graph	Equation	Laplace Transform
4.	Rectangular function $g(t)$		$g(t) = u(t) - u(t-T) \Rightarrow$ $g(t) = \begin{cases} 1, & 0 < t < T \\ 0, & \text{elsewhere} \end{cases}$	$G(s) = \dfrac{1}{s} - \dfrac{1}{s}e^{-Ts}$
5.	Exponential function $f(t)$		$f(t) = e^{at}$	$F(s) = \dfrac{1}{s-a}$
6.	Sinusoidal function $s(t)$		$s(t) = \sin \omega t$	$S(s) = \dfrac{\omega}{s^2 + \omega^2}$

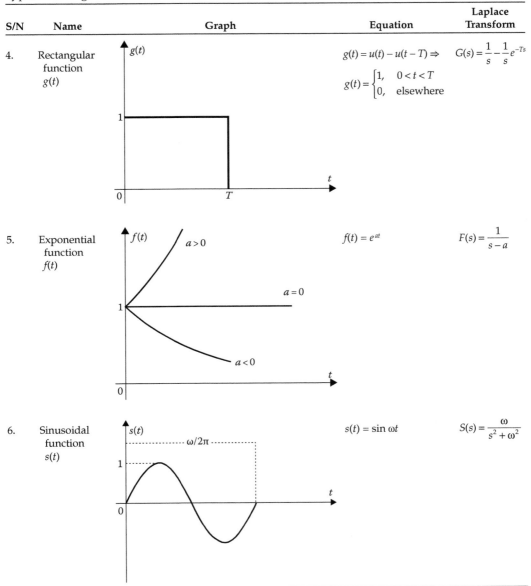

TABLE F4.2

Time Response of First-Order Systems

S/N	Input	Time Response	
1.	Unit step $x(t) = u(t) = 1$	$y(t) = 1 - e^{-\frac{t}{T}}$	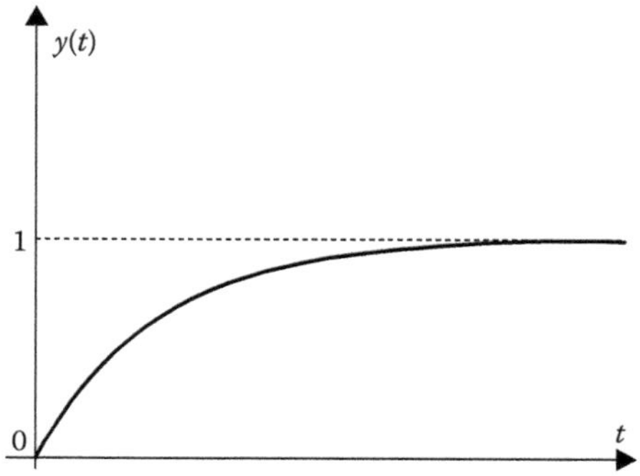
2.	Unit ramp $x(t) = r(t) = t$	$y(t) = t - T + Te^{-\frac{t}{T}}$	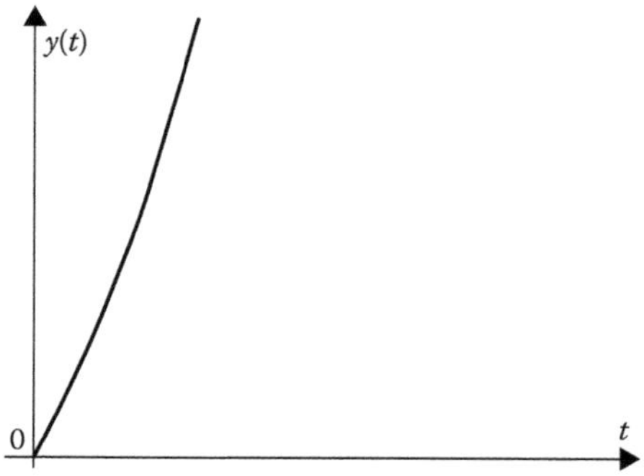
3.	Unit impulse $x(t) = \delta(t)$	$y(t) = \frac{1}{T}e^{-\frac{t}{T}}$	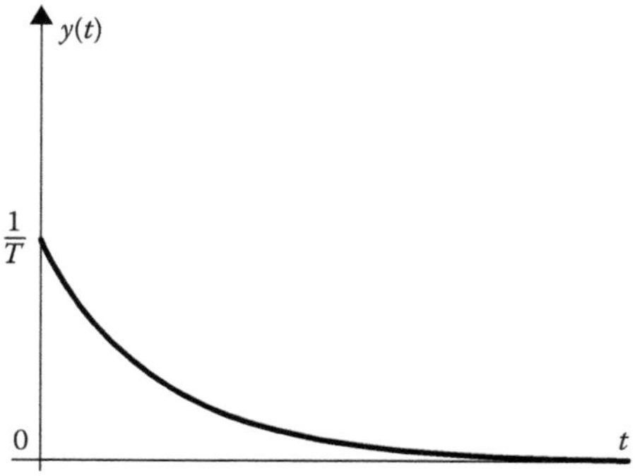

TABLE F4.3

Time Response of Second-Order Systems

Input $x(t)$	Damping Ratio J	Time Response $y(t)$
Unit-step function $x(t) = u(t)$	$0 < J < 1$ under-damping	$y(t) = 1 - \dfrac{e^{-J\omega_n t}}{\sqrt{1-J^2}} \sin\left(\omega_n \sqrt{1-J^2}\, t + \tan^{-1} \dfrac{\sqrt{1-J^2}}{J}\right)$
	$J = 1$ critical damping	$y(t) = 1 - e^{-\omega_n t}(1 + \omega_n t)$
	$J > 1$ over-damping	$y(t) = 1 - e^{-J\omega_n t}\cosh\left[\left(\omega_n\sqrt{J^2-1}\; t\right) + \right.$ $\left. + \dfrac{J}{\sqrt{J^2-1}}\sinh\left(\omega_n\sqrt{J^2-1}\; t\right)\right]$
	$J = 0$ zero damping	$y(t) = 1 - \cos\omega_n t$
Unit-ramp function $x(t) = r(t) = t$	$0 < J < 1$ under-damping	$y(t) = t - \dfrac{2J}{\omega_n} + \dfrac{e^{-J\omega_n t}}{\omega_n\sqrt{1-J^2}}\sin\left(\omega_n\sqrt{1-J^2}\; t + \right.$ $\left. + 2\tan^{-1}\dfrac{\sqrt{1-J^2}}{J}\right)$
	$J = 1$ critical damping	$y(t) = t - \dfrac{2J}{\omega_n} + \dfrac{2}{\omega_n}e^{-\omega_n t}\left(1 + \dfrac{\omega_n t}{2}\right)$
	$J > 1$ over-damping	$y(t) = t - \dfrac{2J}{\omega_n} + \dfrac{2J^2 - 1 - 2J\sqrt{J^2-1}}{2\omega_n\sqrt{J^2-1}}e^{-\left(J+\sqrt{J^2-1}\right)\omega_n t} +$ $+ \dfrac{2J^2 - 1 + 2J\sqrt{J^2-1}}{2\omega_n\sqrt{J^2-1}}e^{-\left(J-\sqrt{J^2-1}\right)\omega_n t}$
Unit-impulse function $x(t) = \delta(t)$	$0 < J < 1$ under-damping	$y(t) = \dfrac{\omega_n}{\sqrt{1-J^2}}e^{-J\omega_n t}\cdot\sin\omega_n\sqrt{1-J^2}\; t$
	$J = 1$ critical damping	$y(t) = \omega_n^2 t e^{-\omega_n t}$
	$J > 1$ over-damping	$y(t) = \dfrac{\omega_n}{2\sqrt{J^2-1}}e^{-\left(J-\sqrt{J^2-1}\right)\omega_n t} -$ $- \dfrac{\omega_n}{2\sqrt{J^2-1}}e^{-\left(J+\sqrt{J^2-1}\right)\omega_n t}$

TABLE F4.4

Time Domain Specifications

S/N	Mathematical Relationship	Symbol-Type-Name
1.	$t_r = \dfrac{1}{\omega_n\sqrt{1-J^2}}\tan^{-1}\left(-\dfrac{\sqrt{1-J^2}}{J}\right)$	Rise time
2.	$t_p = \dfrac{\pi}{\omega_n\sqrt{1-J^2}}$	Peak time
3.	$t_s = \dfrac{\pi}{J\omega_n}$ for (±2%) or $t_s = \dfrac{3}{J\omega_n}$ for (±5%)	Settling time
4.	$M_p = \dfrac{y_m - y_f}{y_f}\cdot 100\%$ or $M_p\% = 100e^{-\frac{J\pi}{\sqrt{1-J^2}}}$	Percent overshoot
5.	$y_m = 1 - e^{-\frac{J\pi}{\sqrt{1-J^2}}}$	Maximum response value
6.	$J = \dfrac{\left|\ln(M_p - 1)\right|}{\sqrt{\pi^2 + \ln^2(M_p - 1)}}$	Relationship between J and M_p
7.	$\omega_n = \dfrac{\pi}{t_p\sqrt{1-J^2}}$	Relationship between ω_n and t_p
8.	$\omega_n = \dfrac{\pi}{t_p\sqrt{\dfrac{\pi^2 - 2\ln^2(M_p - 1)}{\pi^2 - \ln^2(M_p - 1)}}}$	Relationship between ω_n and M_p
9.	$s_{1,2} = -J\omega_n \pm j\omega_n\sqrt{1-J^2} =$ $= \dfrac{\pi}{t_p}\left(-\dfrac{\ln(M_p - 1)}{\sqrt{\pi^2 - 2\ln^2(M_p - 1)}} \pm j\right)$	Roots of the characteristic equation

Problems

4.1 The input of the system that is depicted in the following figure is a unit-step function. Given that $J = 0.6$ and $\omega_n = 5\,\text{rad/s}$, compute ω_d, t_r, t_p, M_p, and t_s.

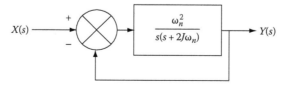

Solution

The transfer function of the system is

$$G(s) = \frac{Y(s)}{X(s)} = \frac{\omega_n^2}{s^2 + 2J\omega_n s + \omega_n^2} \tag{P4.1.1}$$

It is a second-order system with a unit-step input; hence,

$$\omega_d = \omega_n \sqrt{1-J^2} = 4 \text{ rad/s} \tag{P4.1.2}$$

$$t_r = \frac{1}{\omega_d} \tan^{-1}\left(-\sqrt{1-J^2}/J\right) = \frac{\pi - \tan^{-1}\left(\sqrt{1-J^2}/J\right)}{\omega_d} = 0.553 \text{ s} \tag{P4.1.3}$$

$$t_p = \frac{\pi}{\omega_d} = 0.785 \text{ s} \tag{P4.1.4}$$

$$M_p = e^{-\frac{J\pi}{\sqrt{1-J^2}}} = 0.095 \tag{P4.1.5}$$

The percent overshoot and the settling time are computed as follows:

$$M_p\% = 9.5\% \tag{P4.1.6}$$

$$t_s = \frac{4}{J\omega_n} = 1.33 \text{ s} \quad \text{(for 2\%)}$$

or

$$t_s = \frac{3}{J\omega_n} = 1 \text{ s} \quad \text{(for 5\%)}$$

4.2 The transfer function of a control system is

$$G(s) = \frac{k}{s^2 + 10s + k} = \frac{Y(s)}{X(s)}$$

where k is the gain of the system. Suppose that the input signal is the unit-step function and compute for $k = 10$, 100, and 1000

a. The undamped natural frequency ω_n.
b. The damping ratio J.
c. The damped natural frequency ω_d.
d. The roots of the characteristic equation $s_{1,2}$.
e. The maximum value of the gain k in order to have real negative roots of the characteristic equation.
f. The maximum percent overshoot $M_p\%$.
g. The time response of the system $y(t)$.
h. Discuss the influence of the amplifier gain k upon the specifications of the system.

Solution

The transfer function of the system is

$$G(s) = \frac{Y(s)}{X(s)} = \frac{k}{s^2 + 10s + k} \qquad\text{(P4.2.1)}$$

It is a second-order system with a unit-step input; therefore,

$$
\text{a. } \omega_n = \sqrt{k} \Rightarrow
\begin{array}{l}
\text{For } k = 10, \quad \omega_n = 3.16 \text{ rad/s} \\
\text{For } k = 100, \quad \omega_n = 10 \text{ rad/s} \\
\text{For } k = 1000, \quad \omega_n = 31.6 \text{ rad/s}
\end{array}
\qquad\text{(P4.2.2)}
$$

$$
\text{b. } J = \frac{10}{2\omega_n} \Rightarrow
\left.\begin{array}{l}
\text{For } k = 10, \quad J = 1.58 \\
\text{For } k = 100, \quad J = 0.5 \\
\text{For } k = 1000, \quad J = 0.158
\end{array}\right\}
\qquad\text{(P4.2.3)}
$$

$$
\text{c. } \omega_d = \omega_n\sqrt{1 - J^2} \Rightarrow
\left.\begin{array}{l}
\text{For } k = 10, \quad \omega_d \text{ is not defined as } J > 1 \\
\text{For } k = 100, \quad \omega_d \simeq 8.66 \text{ rad/s} \\
\text{For } k = 1000, \quad \omega_d \simeq 31.2 \text{ rad/s}
\end{array}\right\}
\qquad\text{(P4.2.4)}
$$

$$
\text{d. } s_{1,2} = -J\omega_n \pm \omega_n\sqrt{J^2 - 1} \Rightarrow
\left.\begin{array}{l}
\text{For } k = 10, \quad
\begin{cases}
s_1 = -1.125 \\
s_2 = -8.875
\end{cases} \\
\text{For } k = 100, \quad s_{1,2} = -5 \pm j8.66 \\
\text{For } k = 1000, \quad s_{1,2} = -5 \pm j31.2
\end{array}\right\}
\qquad\text{(P4.2.5)}
$$

e. From the characteristic equation, we get

$$s^2 + 10s + k = 0 \Rightarrow s_{1,2} = \frac{-10 \pm \sqrt{100 - 4k}}{2} \qquad\text{(P4.2.6)}$$

For real and negative roots, it must hold that

$$100 - 4k \geq 0 \Rightarrow k \leq 25 \Rightarrow k_{max} = 25 \qquad\text{(P4.2.7)}$$

$$
\text{f. } M_p\% = 100 \cdot e^{-\frac{J\pi}{\sqrt{1-J^2}}} \Rightarrow
\left.\begin{array}{l}
\text{For } k = 10, \quad M_p\% \text{ is not defined} \\
\text{For } k = 100, \quad M_p\% \simeq 16.3\% \\
\text{For } k = 1000, \quad M_p\% \simeq 60\%
\end{array}\right\}
\qquad\text{(P4.2.8)}
$$

g. For $k = 10$, and since $J > 1$, the time response is

$$y(t) = 1 - e^{-5t}(\cosh 3.88t + 1.29\sinh 3.88t) \qquad \text{(P4.2.9)}$$

For $k = 100$, and since $0 < J < 1$, the time response is

$$y(t) = 1 - 1.15e^{-5t}\sin(8.66t + 60°) \qquad \text{(P4.2.10)}$$

For $k = 1000$, the time response is

$$y(t) = 1 - 1.05e^{-5t}\sin(31.2t + 80.24°) \qquad \text{(P4.2.11)}$$

The graph of $y(t)$ is illustrated in the following figure for the three cases.

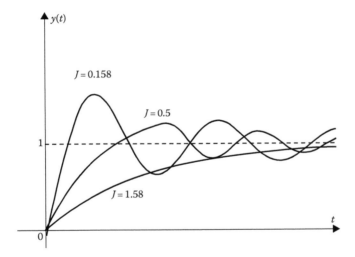

h. i. If k increases, then the damping ratio decreases and the percent overshoot increases.

   ii. The settling time for the 2% requirement is

$$t_s = \frac{4}{J\omega_n} = \begin{cases} 0.80115 \text{ s} & \text{for } k = 10 \\ 0.8 \text{ s} & \text{for } k = 100 \\ 0.80115 \text{ s} & \text{for } k = 1000 \end{cases}$$

   The fastest settling time is for $k = 100$. In general, the settling time is minimized for $J = 0.707$.

4.3 Figure (a) illustrates a mechanical system in longitudinal motion. If a force $f(t) = 8.9\,\text{N}$ is applied to the system, the mass oscillates as shown in Figure (b). Calculate the mass M, the spring constant k, and the viscous-friction coefficient B.

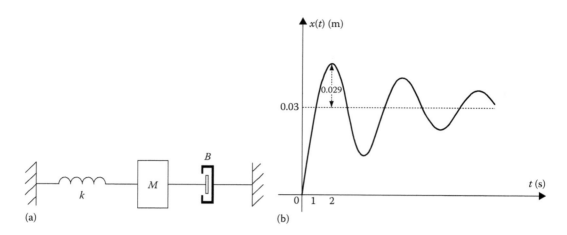

(a)   (b)

Solution

The transfer function $G(s)$ of the mechanical system is

$$G(s) = \frac{X(s)}{F(s)} = \frac{1}{Ms^2 + Bs + k} \tag{P4.3.1}$$

However,

$$f(t) = 8.9u(t) \Rightarrow F(s) = L\{8.9u(t)\} = \frac{8.9}{s} \tag{P4.3.2}$$

Solving the relationship (P4.3.1) for $X(s)$, we get

$$X(s) = \frac{8.9}{s(Ms^2 + Bs + k)} \tag{P4.3.3}$$

From Figure (b), we observe that

$$\lim_{t \to \infty} x(t) = 0.03 \Rightarrow \lim_{s \to 0} sX(s) = 0.03 \overset{(3)}{\Rightarrow} \frac{8.9}{k} = 0.03$$

Thus, k is

$$k = \frac{8.9}{0.03} \simeq 297 \text{ N/m} \tag{P4.3.4}$$

The percent overshoot is

$$M_p\% = \frac{0.0329 - 0.03}{0.03} \cdot 100\% = 9.67\% \tag{P4.3.5}$$

We know that $e^{-\frac{J\pi}{\sqrt{1-J^2}}} = 0.0967$. Hence,

$$J = 0.6 \tag{P4.3.6}$$

From the transfer function of relationship (P4.3.1), we get

$$\omega_n^2 = \frac{k}{M} \Rightarrow M = \frac{k}{\omega_n^2} \qquad\qquad (P4.3.7)$$

$$2J\omega_n = \frac{B}{M} \Rightarrow B = 2J\omega_n M \qquad\qquad (P4.3.8)$$

The peak time t_p is 2 s; consequently,

$$t_p = \frac{\pi}{\omega_n\sqrt{1-J^2}} = 2 \Rightarrow \omega_n = \frac{\pi}{t_p\sqrt{1-J^2}} = 1.96 \text{ rad/s} \qquad\qquad (P4.3.9)$$

By substituting ω_n to (P4.3.7) and (P4.3.8), we get

$$(P4.3.7) \Rightarrow M = \frac{297}{1.96^2} \simeq 77 \text{ kg} \qquad\qquad (P4.3.10)$$

$$(P4.3.8) \Rightarrow B = 2 \cdot 0.6 \cdot 1.96 \cdot 77 \Rightarrow B = 181.2 \text{ N s/m} \qquad\qquad (P4.3.11)$$

4.4 For the block diagram shown in the following figure:
   a. Calculate ω_n, J, and ω_d.
   b. Compute and plot the time response of the system.

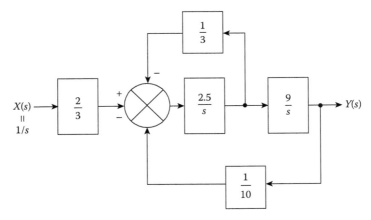

Solution
   a. The signal flow diagram of the system is

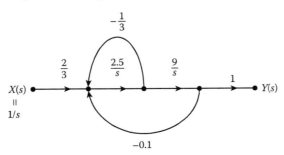

Applying Mason's formula, the transfer function form is

$$G(s) = \frac{Y(s)}{X(s)} = \frac{T_1 \Delta_1}{\Delta} \tag{P4.4.1}$$

where

$$T_1 = \frac{2}{3} \cdot \frac{2.5}{s} \cdot \frac{9}{s} \cdot 1 = \frac{15}{s^2} \tag{P4.4.2}$$

$$\Delta = 1 - \sum L_1 \tag{P4.4.3}$$

$$\sum L_1 = -\frac{2.5}{3s} - \frac{2.5 \cdot 9 \cdot 0.1}{s^2} \tag{P4.4.4}$$

$$(P4.4.3), (P4.4.4) \Rightarrow \Delta = 1 + \frac{2.5}{3s} - \frac{22.5}{s^2} = \frac{s^2 + 0.83s + 22.5}{s^2} \tag{P4.4.5}$$

$$\Delta = 1 - (0) = 1 \tag{P4.4.6}$$

$$(P4.4.1) \underset{(P4.4.5),(P4.4.6)}{\overset{(P4.4.2)}{\Rightarrow}} G(s) = \frac{Y(s)}{X(s)} = \frac{15}{s^2 + 0.835s + 22.5} \tag{P4.4.7}$$

From relationship (P4.4.7), we conclude that the system under consideration is a second-order system, that is,

$$G(s) = \frac{k\omega_n^2}{s^2 + 2J\omega_n s + \omega_n^2} \tag{P4.4.8}$$

Hence,

$$\omega_n = \sqrt{22.5} = 1.5 \text{ rad/s} \tag{P4.4.9}$$

and

$$J = \frac{0.83}{2\omega_n} = 0.28 < 1 \tag{P4.4.10}$$

$$\omega_d = \omega_n \sqrt{1 - J^2} = 1.44 \text{ rad/s} \tag{P4.4.11}$$

b. As $0 < J < 1$, the time response is

$$y(t) = 1 - \frac{e^{-J\omega_n t}}{\sqrt{1 - J^2}} \sin(\omega_d t + \varphi) \tag{P4.4.12}$$

where

$$\varphi = \tan^{-1} \frac{\sqrt{1 - J^2}}{J} = 1.28 \text{ rad} = 73.7° \tag{P4.4.13}$$

$$(P4.4.12), (P4.4.13) \Rightarrow y(t) = 1 - 1.085 e^{-0.42t} \sin(1.44t + 1.28) \tag{P4.4.14}$$

4.5 The block diagram of a position control system is depicted in the following figure. Suppose that the percent overshoot is 20% and the peak time is 0.2 s. Compute k_ω and k_θ.

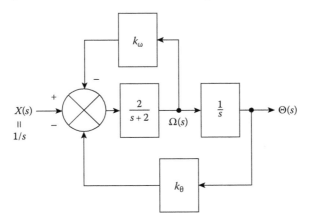

Solution

The signal flow diagram of the system is

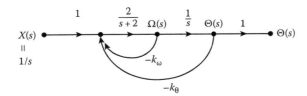

The transfer function is computed from Mason's gain formula:

$$G(s) = \frac{\Theta(s)}{X(s)} = \frac{T_1 \Delta_1}{\Delta} \qquad \text{(P4.5.1)}$$

where

$$T_1 = \frac{2}{s(s+2)} \qquad \text{(P4.5.2)}$$

$$\Delta = 1 - \sum L_1 = 1 - \left(-\frac{2k_\omega}{s+2} - \frac{2k_\theta}{s(s+2)} \right) \qquad \text{(P4.5.3)}$$

$$\Delta = 1 - (0) = 1 \qquad \text{(P4.5.4)}$$

$$(\text{P4.5.1}) \underset{(\text{P4.5.4})}{\overset{(\text{P4.5.2}),(\text{P4.5.3})}{\Rightarrow}} G(s) = \frac{\Theta(s)}{X(s)} = \frac{2}{s^2 + s(2 + 2k_\omega) + 2k_\theta} \qquad \text{(P4.5.5)}$$

The system under consideration is a second-order system. Moreover, the input signal is a unit-step function. Therefore,

$$\omega_n^2 = 2k_\theta \Rightarrow k_\theta = \frac{\omega_n^2}{2} \tag{P4.5.6}$$

$$2J\omega_n = 2 + 2k_\omega \Rightarrow k_\omega = J\omega_n - 1 \tag{P4.5.7}$$

The percent overshoot is 20%; thus,

$$M_p\% = 100 \cdot e^{-\frac{J\pi}{\sqrt{1-J^2}}} = 20 \Rightarrow e^{-\frac{J\pi}{\sqrt{1-J^2}}} = 0.2 \Rightarrow -\frac{J\pi}{\sqrt{1-J^2}} = \ln 0.2 \Rightarrow$$

$$J\pi = 1.61\sqrt{1-J^2} \Rightarrow J^2\pi^2 = 2.59(1-J^2) \Rightarrow J^2(\pi^2 + 2.59) = 2.59 \Rightarrow$$

$$J = \sqrt{\frac{2.59}{\pi^2 + 2.59}} = 0.456 < 1 \tag{P4.5.8}$$

The peak time is 0.2 s; hence,

$$t_p = 0.2 \Rightarrow \frac{\pi}{\omega_n\sqrt{1-J^2}} = 0.2 \Rightarrow \omega_n = \frac{\pi}{0.2\sqrt{1-J^2}} = 17.6 \text{ rad/s} \tag{P4.5.9}$$

$$(\text{P4.5.6}) \overset{(\text{P4.5.9})}{\Rightarrow} k_\theta \simeq 156$$

$$(\text{P4.5.7}) \overset{(\text{P4.5.8}),(\text{P4.5.9})}{\Rightarrow} k_\omega \simeq 7$$

4.6 A position servomechanism is depicted in the following figure.
   a. Explain how the system operates.
   b. Given that the input is $\theta_\omega(t) = u(t)$ and $L_a \approx 0 \text{H}$, compute the undamped natural frequency ω_n of the system and the damping ratio J.

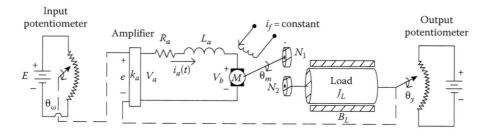

Solution
   a. The system shown is designed so that every change in the angular position $\theta_\omega(t)$ of the input is followed by a corresponding change in the angular position $\theta_y(t)$ of the output. If $\theta_\omega(t) = \theta_y(t)$, then the error voltage $u_e(t)$ is equal to $k_p(\theta_\omega(t) - \theta_y(t)) = 0$.

Consequently, the motor is not activated and the system is immobilized. Every change in $\theta_\omega(t)$ generates an electrical signal through the input potentiometer. If this signal is amplified, it excites the motor. The motor rotates the load axis at an angle of $\theta_y(t) = N\theta_m(t)$, where $N = N_1/N_2$. If $u_e(t) \neq 0$, then the motor rotates the axis right or left until $u_e(t) = 0$.

b. The equations that describe the system are

1. $u_e(t) = k_p\left(\theta_\omega(t) - \theta_y(t)\right) \overset{\text{L.T.}}{\Rightarrow} V_e(s) = k_p\left(\Theta_\omega(s) - \Theta_y(s)\right)$ (P4.6.1')

2. $u_a(t) = k_a u_e(t) \overset{\text{L.T.}}{\Rightarrow} V_a(s) = k_a V_e(s)$ (P4.6.2')

3. $u_a(t) = R_a i_a + L_a \dfrac{di_a(t)}{dt} + u_b(t) \overset{\text{L.T.}}{\Rightarrow}$

   $\Rightarrow V_a(s) = (R_a + sL_a)I_a(s) + V_b(s)$ (P4.6.3')

4. $u_b(t) = k_b\omega_m(t) \overset{\text{L.T.}}{\Rightarrow} V_b(s) = k_b\Omega_m(s)$ (P4.6.4')

5. $J_m^*\dfrac{d\omega_m(t)}{dt} + B_m^*\omega_m(t) = T_m(t) \overset{\text{L.T.}}{\Rightarrow} \left(J_m^*s + B_m^*\right)\Omega_m(s) = T_m(s)$ (P4.6.5')

6. $T_m(t) = k_1 i_a(t) \overset{\text{L.T.}}{\Rightarrow} T_m(s) = k_1 I_a(s)$ (P4.6.6')

7. $\omega_m(t) = \dfrac{d\theta_m(t)}{dt} \overset{\text{L.T.}}{\Rightarrow} \Omega_m(s) = s\Theta_m(s)$ (P4.6.7')

8. $\theta_y(t) = N\theta_m(t) \overset{\text{L.T.}}{\Rightarrow} \Theta_y(s) = N\Theta_m(s)$ (P4.6.8')

Supposing that the gear arrangement presents no losses, the constants J_m^* and B_m^* are computed by the following relationships:

$$J_m^* = J_m + N^2 J_L$$ (P4.6.9)

$$B_m^* = B_m + N^2 B_L$$ (P4.6.10)

The constant J_m^* is the equivalent inertial torque constant and the constant B_m^* is the viscous-friction coefficient of the subsystem (which consists of the motor, the gears and the load).

The signal flow diagram of the system, for $L_a \simeq 0$, is

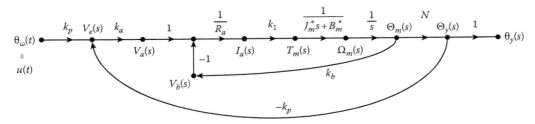

The transfer function $G(s) = \dfrac{\Theta_y(s)}{\Theta_\omega(s)}$ of the system is

$$G(s) = \frac{\Theta_y(s)}{\Theta_\omega(s)} = \frac{T_1 \Delta_1}{\Delta} \qquad \text{(P4.6.11)}$$

where

$$T_1 = \frac{k_p k_a k_1 N}{s R_a \left(J_m^* s + B_m^* \right)} \qquad \text{(P4.6.12)}$$

$$\Delta = 1 - \sum L_1 \qquad \text{(P4.6.13)}$$

$$\sum L_1 = -\frac{k_1 k_b}{R_a \left(J_m^* s + B_m^* \right)} - \frac{k_p k_a k_1 N}{s R_a \left(J_m^* s + B_m^* \right)} \qquad \text{(P4.6.14)}$$

$$\text{(P4.6.13), (P4.6.14)} \Rightarrow \Delta = 1 + \frac{k_1 k_b}{R_a \left(J_m^* s + B_m^* \right)} + \frac{k_p k_a k_1 N}{s R_a \left(J_m^* s + B_m^* \right)} \Rightarrow$$

$$\Delta = \frac{s R_a \left(J_m^* s + B_m^* \right) + s k_1 k_b + k_p k_a k_1 N}{s R_a \left(J_m^* s + B_m^* \right)} \qquad \text{(P4.6.15)}$$

$$\Delta_1 = 1 - (0) = 1 \qquad \text{(P4.6.16)}$$

$$\text{(P4.6.11)} \overset{\text{(P4.6.12),(P4.6.15)}}{\underset{\text{(P4.6.16)}}{\Rightarrow}} G(s) = \frac{\Theta_y(s)}{\Theta_\omega(s)} = \frac{k_p k_a k_1 N}{s^2 R_a J_m^* + s \left(R_a B_m^* + k_1 k_b \right) + k_p k_a k_1 N} \qquad \text{(P4.6.17)}$$

or

$$G(s) = \frac{\Theta_y(s)}{\Theta_\omega(s)} = \frac{\dfrac{k_p k_a k_1 N}{R_a J_m^*}}{s^2 + \dfrac{s \left(R_a B_m^* + k_1 k_b \right)}{R_a J_m^*} + \dfrac{k_p k_a k_1 N}{R_a J_m^*}} \qquad \text{(P4.6.18)}$$

We conclude that the system is a second-order system. Thus,

$$G(s) = k \cdot \frac{\omega_n^2}{s^2 + 2 J \omega_n s + \omega_n^2} \qquad \text{(P4.6.19)}$$

The following relationships hold:

$$\omega_n = \sqrt{\frac{k_p k_a k_1 N}{R_a J_m^*}} \ \text{rad/s} \qquad \text{(P4.6.20)}$$

and

$$J = \frac{\dfrac{R_a B_m^* + k_1 k_b}{R_a J_m^*}}{2\omega_n} \tag{P4.6.21}$$

4.7 The next figure depicts the block diagram of a position servomechanism with a PD controller. Verify that the damped natural frequency of the system is given by the relationship $\omega_d = \dfrac{1}{2J}\sqrt{4k_c J - (f + k_c T_D)^2}$.

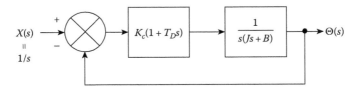

Solution

The transfer function of the position servomechanism is

$$G(s) = \frac{k_c(1 + T_D s)\dfrac{1}{s(Js + B)}}{1 + \dfrac{k_c(1 + T_D s)}{s(Js + B)}} = \frac{\dfrac{k_c}{J}(1 + T_D s)}{s^2 + \dfrac{f + k_c T_D}{J} + \dfrac{k_c}{J}} \tag{P4.7.1}$$

It is a second-order system; thus, it holds that

$$\omega_n = \sqrt{\frac{k_c}{J}} \tag{P4.7.2}$$

and

$$J = \frac{f + k_c T_D}{2\omega_n} = \frac{f + k_c T_D}{2J\sqrt{k_c/J}} \Rightarrow J = \frac{f + k_c T_D}{2\sqrt{k_c J}} \tag{P4.7.3}$$

The damped natural frequency is

$$\omega_d = \omega_n\sqrt{1 - J^2} = \sqrt{\frac{k_c}{J}} \cdot \sqrt{1 - \left(\frac{f + k_c T_D}{2\sqrt{k_c J}}\right)^2} = \sqrt{\frac{k_c}{J}\left[1 - \left(\frac{f + k_c T_D}{4k_c J}\right)^2\right]} \Rightarrow \tag{P4.7.4}$$

$$\omega_d = \sqrt{\frac{1}{4J^2}\left[4k_c J - (f + k_c T_D)^2\right]} = \frac{1}{2J}\sqrt{4k_c J - (f + k_c T_D)^2}$$

4.8 Consider the system shown in the following figure.

a. Compute the value of k so that the system has a 2% maximum overshoot percentage if the switch is in the ON position.

b. Study the time–domain behavior of the system (for the value of k computed in the first query) if the switch is first in the ON position and then in the OFF position.

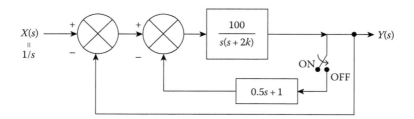

Solution

a. The switch is in the ON position. In this case the block diagram of the system is

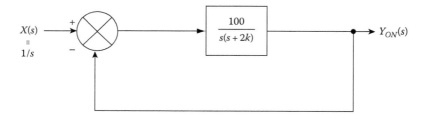

The transfer function $G_{ON}(s)$ of the system is

$$G_{ON}(s) = \frac{Y_{ON}(s)}{X(s)} = \frac{100}{s^2 + 2ks + 100} \equiv \frac{\omega_n^2}{s^2 + 2J\omega_n s + \omega_n^2} \qquad \text{(P4.8.1)}$$

It follows that

$$\omega_n = \sqrt{100} = 10 \text{ rad/s} \qquad \text{(P4.8.2)}$$

and

$$J = \frac{2k}{2\omega_n} = \frac{k}{\omega_n} = \frac{k}{10} \qquad \text{(P4.8.3)}$$

However,

$$M_p\% = 2\% \Rightarrow e^{\frac{-J\pi}{\sqrt{1-J^2}}} = 0.02 \Rightarrow \frac{-J\pi}{\sqrt{1-J^2}} = \ln 0.02 \Rightarrow$$

$$\frac{-J\pi}{\sqrt{1-J^2}} = -3.91202 \Rightarrow J^2\pi^2 = (1-J^2)15.3039 \Rightarrow J^2 = \frac{15.3039}{\pi^2 + 15.3039} \Rightarrow \quad \text{(P4.8.4)}$$

$$J = 0.78 < 1$$

Hence,

$$(\text{P4.8.3}) \overset{(\text{P4.8.4})}{\Rightarrow} k = 10J = 7.8 \qquad \text{(P4.8.5)}$$

b. 1. The switch is in the ON position.
    Since $0 < J < 1$, the system performs damped oscillations. The time response is

$$y_{ON}(t) = 1 - 1.6e^{-7.8t} \sin(6.26t + 38.7°) \qquad \text{(P4.8.6)}$$

Also

$$y_p = y_{max} = 1 + 0.02 = 1.02 \qquad \text{(P4.8.7)}$$

$$t_p = \frac{\pi}{\omega_n\sqrt{1-J^2}} \simeq 0.5 \text{ s} \qquad \text{(P4.8.8)}$$

$$T_s = \frac{4}{J\omega_n} = 0.512 \text{ s} \qquad \text{(P4.8.9)}$$

2. The switch is in the OFF position.
   We plot the associated block diagram.

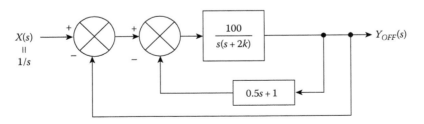

$$G_{OFF}(s) = \frac{Y_{OFF}(s)}{X(s)} = \frac{G(s)}{1+G(s)} \qquad \text{(P4.8.10)}$$

where

$$G(s) = \cfrac{\cfrac{100}{s(s+15.6)}}{1+\cfrac{100}{s(s+15.6)} \cdot (0.5s+1)} = \frac{100}{s^2 + 65.6s + 100} \qquad (P4.8.11)$$

$$(P4.8.10) \overset{(P4.8.11)}{\Rightarrow} G_{OFF}(s) = \frac{100}{s^2 + 65.6s + 200} \equiv \frac{\omega_n^2}{s^2 + 2J\omega_n s + \omega_n^2} \qquad (P4.8.12)$$

It follows that

$$\omega_n = \sqrt{200} = 14.14 \text{ rad/s} \qquad (P4.8.13)$$

and

$$J = \frac{65.6}{2\omega_n} = \frac{65.6}{2 \cdot 14.14} = 2.32 > 1 \qquad (P4.8.14)$$

Since $J > 1$, the system is overdamped and its time response is given by

$$y_{OFF}(t) = 1 - e^{-32.8t}[\cosh(29.6t) + 1.11\sinh(29.6t)] \qquad (P4.8.15)$$

The graphs of the time response for the two positions of the switch (ON and OFF) are depicted in the following figure.

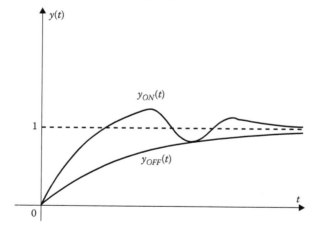

4.9 The loop transfer function of a unity-feedback control system is $G(s) = \dfrac{k}{s(sT+1)}$.

a. By which factor must the gain k be multiplied so that the damping ratio is increased from 0.2 to 0.8?

b. By which factor must the time constant T be multiplied so that the damping ratio is reduced from 0.6 to 0.3?

c. Verify that $\dfrac{4Tk_1 - 1}{4Tk_2 - 1} \simeq 10$, where k_1 and k_2 are the values of k for a 60% and a 20% percent overshoot, respectively.

Solution

The loop transfer function of the control system is

$$G(s) = \frac{k}{s(sT+1)} \tag{P4.9.1}$$

Since it is a unity-feedback system, its transfer function can be written as

$$F(s) = \frac{G(s)}{1+G(s)} = \frac{k}{Ts^2+s+k} = \frac{k/T}{s^2 + \frac{1}{T}s + \frac{k}{T}} \tag{P4.9.2}$$

This is a second-order system; thus,

$$\omega_n = \sqrt{\frac{k}{T}} \tag{P4.9.3}$$

and

$$J = \frac{1}{2T\omega_n} = \frac{1}{2\sqrt{kT}} \tag{P4.9.4}$$

a. Given that the damping ratios are $J_1 = 0.2$ and $J_2 = 0.8$ and the time constant is T, from relationship (P4.9.4), we get

$$\left. \begin{array}{l} J_1 = \dfrac{1}{2\sqrt{k_1 T}} \\[3mm] J_2 = \dfrac{1}{2\sqrt{k_2 T}} \end{array} \right\} \Rightarrow \frac{J_1}{J_2} = \sqrt{\frac{k_2}{k_1}} = \frac{0.2}{0.8} = \frac{1}{4} \Rightarrow k_2 = \frac{k_1}{16} \tag{P4.9.5}$$

b. For damping ratios $J_1 = 0.6$ and $J_2 = 0.3$, and an amplifier gain k, we have

$$\left. \begin{array}{l} J_1 = \dfrac{1}{2\sqrt{k T_1}} \\[3mm] J_2 = \dfrac{1}{2\sqrt{k T_2}} \end{array} \right\} \Rightarrow \frac{J_1}{J_2} = \sqrt{\frac{T_2}{T_1}} = \frac{0.6}{0.3} = 2 \Rightarrow T_2 = 4T_1 \tag{P4.9.6}$$

c. From the 60% and a 20% specifications and since the input is a unit step function, we have

$$60 = 100e^{-\frac{J_1 \pi}{\sqrt{1-J_1^2}}} \tag{P4.9.7}$$

$$20 = 100e^{-\frac{J_2 \pi}{\sqrt{1-J_2^2}}} \tag{P4.9.8}$$

$$(\text{P4.9.7}) \Rightarrow \frac{J_1\pi}{\sqrt{1-J_1^2}} = 0.51 \left. \right\}$$
$$(\text{P4.9.8}) \Rightarrow \frac{J_2\pi}{\sqrt{1-J_2^2}} = 1.61 \left. \right\} \Rightarrow \frac{\dfrac{J_2}{\sqrt{1-J_2^2}}}{\dfrac{J_1}{\sqrt{1-J_1^2}}} = \frac{1.61}{0.51} = 3.16 \Rightarrow$$

$$\frac{J_2}{J_1}\frac{\sqrt{1-J_1^2}}{\sqrt{1-J_2^2}} = 3.16 \Rightarrow \sqrt{\frac{k_1}{k_2}} \cdot \frac{\sqrt{1-\dfrac{1}{4k_1T}}}{\sqrt{1-\dfrac{1}{4k_2T}}} = 3.16 \Rightarrow$$

$$\sqrt{\frac{k_1}{k_2}} \cdot \sqrt{\frac{(4k_1T-1)k_2}{(4k_2T-1)k_1}} = 3.16 \Rightarrow \sqrt{\frac{4k_1T-1}{4k_2T-1}} = 3.16 \qquad (\text{P4.9.9})$$

or

$$\frac{4k_1T-1}{4k_2T-1} \simeq 10 \qquad (\text{P4.9.10})$$

4.10 The signal flow diagram of a position-control servomechanism is depicted in the figure below. Moreover,

$$J = 0.4\,\text{N}\cdot\text{m s}^2/\text{rad}, \quad f = 2\,\text{N}\cdot\text{m s}/\text{rad}, \quad k_p = 0.6\,\text{V}/\text{rad}, \quad k_T = 2\,\text{N}\cdot\text{m}$$

a. Assuming that the switch is in position 1, compute the value of the gain k so that the undamped natural frequency of the system is $10\,\text{rad/s}$.

b. Supposing that the switch is in position 2, compute the value of k_t for $k_A = 5$ so that the system becomes critically damped.

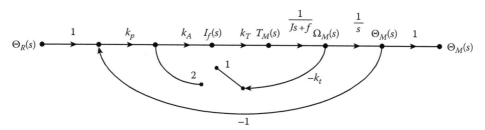

Solution

a. **The switch is in position 1.**

The tachometer generator is not in the loop; therefore the open-loop transfer function is

$$G(s)H(s) = k_p k_A k_T \cdot \frac{1}{Js+f} \cdot \frac{1}{s} = \frac{k_p k_A k_T}{s(Js+f)} \qquad (\text{P4.10.1})$$

The characteristic equation of the system is

$$1 + G(s)H(s) = 0 \Rightarrow s(Js + f) + k_p k_A k_T = 0 \Rightarrow$$

$$Js^2 + fs + k_p k_A k_T = 0 \tag{P4.10.2}$$

Hence,

$$\omega_n^2 = \frac{k_p k_A k_T}{J} \tag{P4.10.3}$$

and

$$k_A = \frac{J\omega_n^2}{k_p k_T} = \frac{0.4 \cdot 100}{0.6 \cdot 2} = 33.3 \text{ A/V} \tag{P4.10.4}$$

b. **The switch is in position 2.**
The open-loop transfer function is

$$G(s)H(s) = k_p \frac{k_A k_T \cdot \dfrac{1}{Js + f}}{1 + k_A k_T \cdot \dfrac{1}{Js + f} k_t} \cdot \frac{1}{s} \Rightarrow$$

$$G(s)H(s) = \frac{k_p k_A k_T}{s(Js + f + k_A k_T k_t)} \tag{P4.10.5}$$

where k_t is the tachometer constant.
The characteristic equation of the system is

$$1 + G(s)H(s) = 0 \Rightarrow s(Js + f + k_A k_T k_t) + k_p k_A k_T = 0 \Rightarrow$$

$$Js^2 + (f + k_A k_T k_t)s + k_p k_A k_T = 0 \tag{P4.10.6}$$

The following relationships hold:

$$\omega_n^2 = k_p k_A k_T \tag{P4.10.7}$$

and

$$2J\omega_n = \frac{f + k_A k_T k_t}{J} \tag{P4.10.8}$$

$$(P4.10.7), (P4.10.8) \Rightarrow J = \frac{f + k_A k_T k_t}{J} \cdot \frac{1}{2} \cdot \sqrt{\frac{J}{k_p k_A k_T}} \tag{P4.10.9}$$

For critical damping, $J = 1$. Consequently,

$$2J\sqrt{k_p k_A k_T} = \sqrt{J}(f + k_A k_T k_t) \Rightarrow$$

$$k_t = \frac{2\sqrt{k_p k_A k_T} - f}{k_A k_T} = 0.11 \text{ V/rad/s} \tag{P4.10.10}$$

5

Stability of Control Systems

5.1 Introduction

A system is **bounded-input bounded-output (BIBO)** stable, if every bounded input results in a bounded output. The output of a stable system is kept within the admissible boundaries (Figure 5.1), whereas the output of an unstable system can, theoretically, increase to infinity (Figure 5.2).

A linear time-invariant system is **stable**, if the poles of the closed-loop system are in the **left-half s-plane**, that is, the poles have real negative parts (Figure 5.3). On the other hand, if at least one pole is in the right-half s-plane, the system is unstable (Figure 5.4).

The response of an automatic control system is related to the **roots** of the characteristic equation (poles) of the transfer function:

- If the poles are in the left-half s-plane, then the response of the system to various disturbance signals is decreasing.
- If there are poles on the imaginary $j\omega$ axis or in the right-half s-plane, then the response of the system to a disturbance input is constant or increasing.

If the characteristic equation of the system has simple roots on the imaginary $j\omega$ axis, and all of its other roots are in the left-half s-plane, then its steady-state output is a bounded function. In this case, the system is called **marginally stable**. In the next section, we introduce some methods useful for defining if a system is stable.

5.2 Algebraic Stability Criteria

5.2.1 Routh's Stability Criterion

Routh's stability criterion determines the number of the poles of the transfer function, which are in the right-half s-plane. Recall that a pole in the right-half s-plane results in system instability.

Suppose that the characteristic equation of the system is

$$a_0 s^5 + a_1 s^4 + a_2 s^3 + a_3 s^2 + a_4 s + a_5 = 0 \tag{5.1}$$

where $a_i \in R$.

To ensure that the roots of Equation 5.1 have no positive real parts, all coefficients a_i must be of the same sign.

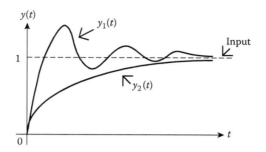

FIGURE 5.1
Output of a stable system.

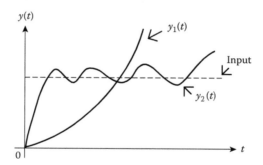

FIGURE 5.2
Output of an unstable system.

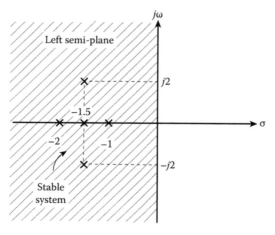

FIGURE 5.3
Poles of a stable system.

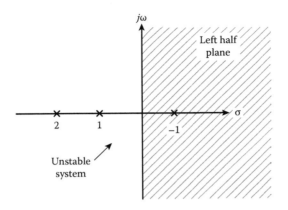

FIGURE 5.4
Poles of an unstable system.

Next, we introduce the procedure.

The Routh array is constructed as follows:

5.2.1.1 Routh's Tabulation

$$
\begin{array}{c|ccc}
s^5 & a_0 & a_2 & a_4 \\
s^4 & a_1 & a_3 & a_5 \\
s^3 & b_1 & b_2 & \\
s^2 & c_1 & c_2 & \\
s^1 & d_1 & & \\
s^0 & e_1 & &
\end{array}
\tag{5.2}
$$

where the terms $b_1, b_2, c_1, c_2, d_1, e_1$ are computed as follows:

$$
b_1 = -\frac{\begin{vmatrix} a_0 & a_2 \\ a_1 & a_3 \end{vmatrix}}{a_1}, \quad b_2 = -\frac{\begin{vmatrix} a_0 & a_4 \\ a_1 & a_5 \end{vmatrix}}{a_1}
$$

$$
c_1 = -\frac{\begin{vmatrix} a_1 & a_3 \\ b_1 & b_2 \end{vmatrix}}{b_1}, \quad c_2 = -\frac{\begin{vmatrix} a_1 & a_5 \\ b_1 & 0 \end{vmatrix}}{b_1} = a_5
\tag{5.3}
$$

$$
d_1 = -\frac{\begin{vmatrix} b_1 & b_2 \\ c_1 & c_2 \end{vmatrix}}{c_1}, \quad e_1 = -\frac{\begin{vmatrix} c_1 & c_2 \\ d_1 & 0 \end{vmatrix}}{d_1} = c_2
$$

According to Routh's stability criterion, a system is stable if the terms of the first column of Routh's tabulation (i.e., $a_0, a_1, b_1, c_1, d_1, e_1$) are of the **same sign**.

The number of roots of the characteristic equation, which are in the right-half *s*-plane, is equal to the number of sign changes of the coefficients in the first column of Routh's tabulation.

If a system satisfies Routh's criterion, which means it is absolutely stable, then it is desirable to define its relative stability. It follows that the relative damping of every root of the characteristic equation must be examined. The greater the distance of a pole from the *jω* axis, the greater is its relative stability.

5.2.1.2 Special Cases

a. If **a term of the first column is zero**, while all the other terms of the row are nonzero or do not exist, then the zero term is replaced by a very small number, which is of the same sign as the previous terms of the first column. The procedure continues as normal. Alternatively, the characteristic polynomial can be multiplied by $s + m$, where $m > 0$ and $-m$ is not a root of the characteristic equation.

b. If **all terms of a row in Routh's array are zero**, then the array is completed by replacing the zero terms by the terms of the differentiated auxiliary equation of the previous row.

c. If **two (or more) rows have zero terms**, then the system is unstable and the characteristic polynomial has two real poles with multiplicity 2.

d. In order to find the **marginal value of K that yields stability**, it is sufficient to suppose that the term of the s^1 row is zero and to solve the equation for $K = K_c$.

e. In order to find the **critical frequency of oscillation of the system**, it is sufficient to solve the auxiliary equation of the s^2 row for $\omega = \omega_c$.

Routh's stability criterion allows determining the region of values of a system parameter that leads to **closed-loop stability**.

If a system fulfills Routh's criterion, that is, it is absolutely stable, then it is desirable to define **relative stability**. The relative stability of a system is computed by the corresponding real parts of each pole or of each couple of poles as to the *jω* axis.

It can be defined with the help of the damping ratios *J*, which correspond to each couple of complex conjugate poles.

Hence, it is defined in relation to the percent overshoot and the response speed of the system. The separate examination of every pole is very important, as the position of the poles of a closed-loop system in the *s*-plane determines the behavior of the system.

5.2.2 Hurwitz Stability Criterion

The **Hurwitz stability criterion** determines whether there are any poles of the characteristic equation in the right-half *s*-plane or on the $s = j\omega$ axis. However, it does not determine the number of these poles.

Consider the following characteristic equation:

$$a_n s^n + a_{n-1} s^{n-1} + \cdots + a_1 s + a_0 = 0 \tag{5.4}$$

The Hurwitz criterion can be applied if the Hurwitz determinants are carried out:

$$D_0 = a_n$$

$$D_1 = a_{n-1}$$

$$D_2 = \begin{vmatrix} a_{n-1} & a_{n-3} \\ a_n & a_{n-2} \end{vmatrix}$$

$$D_3 = \begin{vmatrix} a_{n-1} & a_{n-3} & a_{n-5} \\ a_n & a_{n-2} & a_{n-4} \\ 0 & a_{n-1} & a_{n-3} \end{vmatrix}$$

$$\vdots$$

$$D_n = \begin{vmatrix} a_{n-1} & a_{n-3} & \cdots & \begin{bmatrix} a_0 & \text{for} & n = 2k+1 \\ a_1 & \text{for} & n = 2k \end{bmatrix} & 0 & \cdots & 0 \\ a_n & a_{n-2} & \cdots & \begin{bmatrix} a_1 & \text{for} & n = 2k+1 \\ a_0 & \text{for} & n = 2k \end{bmatrix} & 0 & \cdots & 0 \\ 0 & a_{n-1} & \cdots & a_{n-3} & 0 & \cdots & 0 \\ 0 & a_n & \cdots & a_{n-2} & 0 & \cdots & 0 \\ \vdots & \vdots & \vdots & \vdots & \vdots & \vdots & \vdots \\ 0 & 0 & 0 & 0 & 0 & \cdots & a \end{vmatrix}$$

(5.5)

According to the Hurwitz stability criterion, a system is stable if $D_i > 0$, $i = 1, 2, \ldots .. n$.

5.2.3 Continued Fraction Stability Criterion

The stability criterion of continued fractions determines whether the poles of the characteristic equation are in the right-half s-plane or on the $s = j\omega$ axis.

The characteristic polynomial is divided into two polynomials as follows:

$$\left. \begin{array}{l} \Pi_1(s) = a_n s^n + a_{n-2} s^{n-2} + a_{n-4} s^{n-4} + \cdots \\ \Pi_2(s) = a_{n-1} s^{n-1} + a_{n-3} s^{n-3} + a_{n-5} s^{n-5} + \cdots \end{array} \right\}$$

(5.6)

Subsequently, the quotient $\Pi_1(s)/\Pi_2(s)$ is formed and developed into a continued fraction:

$$\frac{\Pi_1(s)}{\Pi_2(s)} = P_1 s + \cfrac{1}{P_2 s + \cfrac{1}{P_3 s + \cfrac{1}{\ddots + \cfrac{1}{P_n s}}}}$$

(5.7)

According to the criterion of the continued fractions, a system is stable if $P_j > 0$, $j = 1, 2, \ldots .. n$

5.2.4 Nyquist Stability Criterion

The **Nyquist stability criterion** is based on the plot of the open-loop transfer function $G(s)H(s)$ for a special closed path in the complex frequency plane. It provides information not only for the stability of closed-loop systems, but also for their **relative stability**. The special closed path is called **Nyquist path**, and it includes the whole of the right-half s-plane.

The Nyquist plot includes the whole $j\omega$ axis from $\omega = -\infty$ to $\omega = +\infty$, and a semicircle path with an infinite radius in the right-half s-plane. It is traversed in clockwise direction as shown in the figure below.

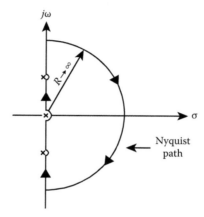

If Z is the number of the roots of the characteristic equation $1 + G(s)H(s) = 0$ in the right-half s-plane, N is the number of encirclements of the point $-1 + j0$ in the clockwise direction of $G(j\omega)H(j\omega)$, and P is the number of the poles of $G(s)H(s)$ in the right-half s-plane, then the following relationship holds:

$$Z = N + P \tag{5.8}$$

According to the Nyquist criterion, a closed-loop system is stable, if $Z = 0$, that is if,

$$N = -P \tag{5.9}$$

This means that the outline of $G(j\omega)H(j\omega)$ encircles the point $-1 + j0$ P times in the counterclockwise direction.

Remarks

1. If $P = 0$, that is, the open-loop transfer function $G(s)H(s)$ has no poles in the right-half s-plane, it is enough to plot the diagram in the GH plane, which is the region of the positive frequencies ($\omega \to 0$ to $\omega \to \infty$). The stability of this system can be tested by examining whether the point $-1 + j0$ is encircled by the Nyquist plot of $G(j\omega)H(j\omega)$. The following figures illustrate a stable and an unstable system, respectively. The point $(-1, 0) = -1 + j0$ is called **critical point**.

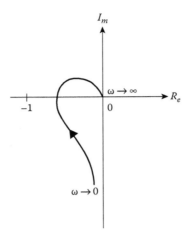

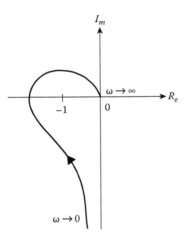

2. The Nyquist criterion provides information about the absolute stability and it can be used for determining the relative stability of a system. The proximity of the curve $GH(j\omega)$ to the critical point of stability is an indicator of the relative stability of a system.

Gain margin k_g is the inverse value of gain $|GH(j\omega)|$ at a frequency for which the phase angle tends to $-180°$. It is given by

$$k_g(\text{dB}) = -20\log_{10}|G(j\omega_c)H(j\omega_c)| \qquad (5.10)$$

where ω_c is the frequency, where the Nyquist diagram of $G(s)H(s)$ intersects the $\text{Re}\{GH\}$ axis.

A closed-loop control system is stable if

$$k_g > 0 \qquad (5.11)$$

The gain margin is a factor by which the gain of a system can be increased, so that the relevant Nyquist plot intersects the critical point $-1 + j0$.

Phase margin φ_α is the angle by which the diagram of $GH(j\omega)$ must be rotated, so that the unit magnitude point $|GH(j\omega)| = 1$ intersects the point $-1 + j0$. It is given by

$$\varphi_\alpha = 180° + \varphi \qquad (5.12)$$

where
   φ is the angle of $G(j\omega_1)H(j\omega_1)$
   ω_1 is the frequency for which $|G(j\omega_1)H(j\omega_1)|$ is equal to unity

A closed-loop control system is stable if

$$\varphi_\alpha > 0 \qquad (5.13)$$

The larger φ_α and k_g are, the larger becomes the relative stability of the closed-loop control system.

Remark

The Nyquist stability criterion can be also used for determining the influence of a time delay when examining the relative stability of a system.
   Time delay is the time that passes between the commencement of an event at an operation point of a system and the emerging effect at another operation point of the same system.
   The mathematical expression of a transfer function that yields time delay T is

$$G_d(s) = e^{-sT} \qquad (5.14)$$

The term e^{-sT} does not introduce any additional poles or zeros to the system; thus, the Nyquist criterion is still valid. The term e^{-sT} results in a **phase shift** $\varphi(\omega) = -\omega T$ of the relevant frequency response. The magnitude of the frequency response does not change. This phase shift is added to the phase angle (in radians), which corresponds to $G(j\omega)$.
   In general, the presence of a term e^{-sT} in an automatic control system introduces an additional phase lag, which results to a less stable system. Therefore, it is often required to decrease the system gain in order to retain system stability. On the other hand, decreasing the gain constant yields an increase of the steady-state error.

Formulas

TABLE F5.1
Relation between Poles and Transient Response

S/N	Pole Position	Transient Response
1.	Negative real pole	Ae^{-P_1t}: stable

| 2. | Complex conjugate poles with negative real part | $Ae^{-\sigma_1 t}\sin\omega t$: stable |

| 3. | Positive real pole | Ae^{P_1t}: unstable |

(*continued*)

TABLE F5.1 (continued)

Relation between Poles and Transient Response

S/N	Pole Position	Transient Response
4.	Complex conjugate poles with positive real part 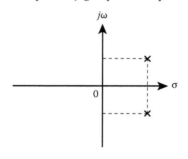	$Ae^{\sigma_1 t}\sin\omega t$: unstable 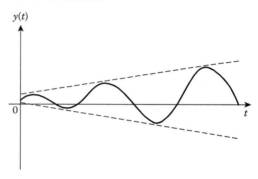
5.	Simple pole at zero 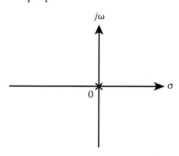	A: critically stable 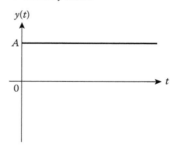
6.	Complex imaginary poles 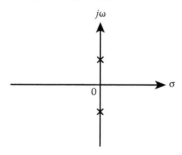	$A\sin\omega t$: critically stable 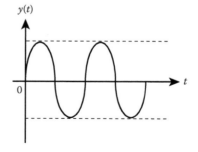

TABLE F5.1 (continued)

Relation between Poles and Transient Response

S/N	Pole Position	Transient Response
7.	Double pole at zero	At: unstable
8.	Double conjugate imaginary poles	$(A_1 + A_2t)\sin \omega t$: unstable

TABLE F5.2

Algebraic Stability Criteria

S/N	C.E.: $a_n s^n + a_{n-1}s^{n-1} + a_{n-2}s^{n-2} + \cdots + a_1 s + a_0 = 0$

1.　　**Routh's Criterion**

$$
\begin{array}{c|cccc}
s^n & a_n & a_{n-2} & a_{n-4} & \cdots \\
s^{n-1} & a_{n-1} & a_{n-3} & a_{n-5} & \cdots \\
s^{n-2} & b_1 & b_2 & b_3 & \cdots \\
s^{n-3} & c_1 & c_2 & c_3 & \cdots \\
\vdots & \vdots & \vdots & \vdots & \vdots
\end{array}
$$

$$
b_1 = -\frac{\begin{vmatrix} a_n & a_{n-2} \\ a_{n-1} & a_{n-3} \end{vmatrix}}{a_{n-1}}, \quad b_2 = -\frac{\begin{vmatrix} a_n & a_{n-4} \\ a_{n-1} & a_{n-5} \end{vmatrix}}{a_{n-1}}, \cdots
$$

$$
c_1 = -\frac{\begin{vmatrix} a_{n-1} & a_{n-3} \\ b_1 & b_2 \end{vmatrix}}{b_1}, \quad c_2 = -\frac{\begin{vmatrix} a_{n-1} & a_{n-5} \\ b_1 & b_3 \end{vmatrix}}{b_1}, \cdots
$$

The system is stable if $a_n, a_{n-1}, b_1, c_1, \ldots > 0$

2.　　**Hurwitz Criterion**

$$D_0 = a_n$$

$$D_1 = a_{n-1}$$

$$D_2 = \begin{vmatrix} a_{n-1} & a_{n-3} \\ a_n & a_{n-2} \end{vmatrix}$$

$$D_3 = \begin{vmatrix} a_{n-1} & a_{n-3} & a_{n-5} \\ a_n & a_{n-2} & a_{n-4} \\ 0 & a_{n-1} & a_{n-3} \end{vmatrix}$$

$$\vdots$$

$$
D_n = \begin{vmatrix}
a_{n-1} & a_{n-3} & \cdots & \begin{bmatrix} a_0 & \text{for } n = 2k+1 \\ a_1 & \text{for } n = 2k \end{bmatrix} & 0 & \cdots & 0 \\
a_n & a_{n-2} & \cdots & \begin{bmatrix} a_1 & \text{for } n = 2k+1 \\ a_0 & \text{for } n = 2k \end{bmatrix} & 0 & \cdots & 0 \\
0 & a_{n-1} & \cdots & a_{n-3} & 0 & \cdots & 0 \\
0 & a_n & \cdots & a_{n-2} & 0 & \cdots & 0 \\
\vdots & \vdots & \vdots & \vdots & \vdots & \vdots & \vdots \\
0 & 0 & 0 & 0 & 0 & \cdots & a
\end{vmatrix}
$$

The system is stable if $D_i > 0, \ i = 1, 2, \ldots n$

3.　　**Continued Fraction Criterion**

$$\Pi_1(s) = a_n s^n + a_{n-2}s^{n-2} + a_{n-4}s^{n-4} + \cdots$$

$$\Pi_2(s) = a_{n-1}s^{n-1} + a_{n-3}s^{n-3} + a_{n-5}s^{n-5} + \cdots$$

$$\frac{\Pi_1(s)}{\Pi_2(s)} = P_1 s + \cfrac{1}{P_2 s + \cfrac{1}{P_3 s + \cfrac{1}{\ddots + \cfrac{1}{P_n s}}}}$$

The system is stable if $P_j > 0, \ j = 1, 2, \ldots n$

TABLE F5.3

Nyquist Criterion

Open-Loop Transfer Function $G(s)H(s)$	Closed-Loop Transfer Function $\dfrac{G(s)}{1+G(s)H(s)}$

- The system is stable if $N = -P$

  where

      N is the number of encirclements of the point $-1 + j0$ in the clockwise direction of $G(j\omega)H(j\omega)$

      P is the number of the poles of $G(s)H(s)$ in the right-half s-plane

- **Gain margin**

  $k_g(db) = -20\log_{10}|G(j\omega_c)H(j\omega_c)|$

  where, ω_c is the frequency for which the Nyquist plot of $G(s)H(s)$ intersects the axis Re{GH}

- **Phase margin**

  $\varphi_a = 180° + \varphi$

  where

      $\varphi = \sphericalangle(G(j\omega_1)H(j\omega_1))$

      ω_1 is the frequency where $|G(j\omega_1)H(j\omega_1)| = 1$

  The system is stable if $k_g > 0$ $\varphi_\alpha > 0$

Problems

5.1 Determine the stability of the systems with the following characteristic equations:

  a. $s^5 + 3s^4 + 7s^3 + 20s^2 + 6s + 15 = 0$

  b. $2s^4 + s^3 + 3s^2 + 5s + 10 = 0$

  c. $s^5 + 2s^4 + 2s^3 + 4s^2 + 11s + 10 = 0$

  d. $s^5 + 2s^4 + 24s^3 + 48s^2 - 25s - 50 = 0$

Solution

  a. The characteristic equation is

$$s^5 + 3s^4 + 7s^3 + 20s^2 + 6s + 15 = 0 \qquad (P5.1.1)$$

Routh's tabulation is

s^5	1	7	6
s^4	3	20	15
s^3	1/3	1	
s^2	11	15	
s^1	6/11		
s^0	15		

There is no change of sign in the first column of Routh's tabulation. Hence, the system is stable.

b. The characteristic equation is

$$2s^4 + s^3 + 3s^2 + 5s + 10 = 0 \qquad (P5.1.2)$$

Routh's tabulation is written as

$$
\begin{array}{c|ccc}
s^4 & 2 & 3 & 10 \\
s^3 & 1 & 5 \\
s^2 & -7 & 10 \\
s^1 & 6.43 \\
s^0 & 10
\end{array}
$$

There are two sign changes in the first column of Routh's tabulation ($1 \rightarrow -7 \rightarrow 6.43$); hence, the characteristic equation has two roots in the right-half s-plane, and the system is unstable.

a. The characteristic equation is

$$s^5 + 2s^4 + 2s^3 + 4s^2 + 11s + 10 = 0 \qquad (P5.1.3)$$

Routh's tabulation is written as

$$
\begin{array}{c|ccc}
s^5 & 1 & 2 & 11 \\
s^4 & 2 & 4 & 10 \\
s^3 & \varepsilon & 6 \\
s^2 & \dfrac{4\varepsilon - 12}{\varepsilon} & 10 \\
s^1 & 6 + \dfrac{10}{12}\varepsilon^2 \\
s^0 & 10
\end{array}
$$

The first term in row s^3 was zero and it is replaced by a very small number ε (where, $\varepsilon > 0$ and $\lim \varepsilon \rightarrow 0$).

We have $4\varepsilon - 12/\varepsilon < 0$ and $6 + (10/12)\varepsilon^2 > 0$.

Hence, the system is unstable, and the characteristic equation has two roots in the right-half s-plane.

b. The characteristic equation is

$$s^5 + 2s^4 + 24s^3 + 48s^2 - 25s - 50 = 0 \qquad (P5.1.4)$$

The system is unstable, because the polynomial $s^5 + 2s^4 + 24s^3 + 48s^2 - 25s - 50$ has two coefficients of different sign. By applying Routh's criterion we obtain the same conclusion.

Routh's tabulation is

$$
\begin{array}{c|ccc}
s^5 & 1 & 24 & -25 \\
s^4 & 2 & 48 & -50 \\
s^3 & 8 & 96 & \\
s^2 & 24 & -50 & \\
s^1 & 112.7 & & \\
s^0 & -50 & &
\end{array}
$$

All coefficients of row s^3 were zero, thus, they have been replaced by the terms of the differentiated auxiliary equation of row s^4.
We have

$$
2s^4 + 48s^2 - 50 = 0 \Rightarrow \frac{d}{dt}(2s^4 + 48s^2 - 50) = 0 \Rightarrow 8s^3 + 96s = 0 \qquad \text{(P5.1.5)}
$$

At the first column of Routh's tabulation, there is a change of sign; therefore, the characteristic equation has one root in the right-half s-plane. It can be computed by solving the auxiliary equation $2s^4 + 48s^2 - 50 = 0$.
We have $s_{1,2}^2 = 1$ and $s_{3,4}^2 = -25$. Thus,

$$
\Rightarrow s_{1,2} = \pm 1 \quad \text{and} \quad s_{3,4} = \pm j5 \qquad \text{(P5.1.6)}
$$

Hence, the initial equation of relationship (P5.1.4) is written as

$$
(s+1)(s-1)(s+j5)(s-j5)(s+2) = 0
$$

Notice that the root $s = 1$ is at the right-half s-plane.

5.2 Determine the region of values for the parameter k so that the systems with the following characteristic equations are stable.
   For each case, compute the critical frequency of oscillation ω_c:
   a. $s^4 + 7s^3 + 15s^2 + (25+k)s + 2k = 0$
   b. $s^3 + 1{,}040s^2 + 48{,}500s + 400{,}000k = 0$
   c. $s^3 + 3ks^2 + (k+2)s + 4 = 0$

Solution
   a. The characteristic equation is

$$
s^4 + 7s^3 + 15s^2 + (25+k)s + 2k = 0 \qquad \text{(P5.2.1)}
$$

Routh's tabulation is

$$
\begin{array}{c|ccc}
s^4 & 1 & 15 & 2k \\
s^3 & 7 & 25+k & \\
s^2 & \dfrac{80-k}{7} & 2k & \\
s^1 & \dfrac{(80-k)(25+k)-98k}{80-k} & & \\
s^0 & 2k & &
\end{array}
$$

For the system to be stable, it must hold that

1. $\dfrac{80-k}{7} > 0 \Rightarrow k < 80$ (P5.2.2)

2. $\dfrac{(80-k)(25+k)-98k}{80-k} > 0 \Rightarrow -71.1 < k < 28.1$ (P5.2.3)

3. $2k > 0 \Rightarrow k > 0$ (P5.2.4)

By combining inequalities (P5.2.2), (P5.2.3), and (P5.2.4), we get

$$
0 < k < 28.1 \tag{P5.2.5}
$$

For $k = 28.1 = k_c$, the characteristic equation has a couple of imaginary roots, while for $k = 0$ there is no response ($y(t) = 0$). The angular frequency of oscillation if $k = 28.1 = k_c$ is

$$
\frac{80-k_c}{7}s^2 + 2k_c = 0 \Rightarrow s^2 = -7.58 \Rightarrow s = \pm j2.75 \Rightarrow \omega_c = 2.75 \text{ rad/s} \tag{P5.2.6}
$$

b. The characteristic equation is

$$
s^3 + 1{,}040s^2 + 48{,}500s+ = 4 \cdot 10^5 k = 0 \tag{P5.2.7}
$$

Routh's tabulation is

$$
\begin{array}{c|cc}
s^3 & 1 & 48{,}500 \\
s^2 & 1{,}040 & 4 \cdot 10^5 k \\
s^1 & \dfrac{5{,}044 \cdot 10^4 - 4 \cdot 10^5 k}{1{,}040} & \\
s^0 & 4 \cdot 10^5 k &
\end{array}
$$

The conditions for the stability of the system are

1. $4 \cdot 10^5 k > 0 \Rightarrow k > 0$ $\qquad$ (P5.2.8)

2. $\dfrac{5044 \cdot 10^4 - 4 \cdot 10^5 k}{1040} > 0 \Rightarrow k < 126.1$ $\qquad$ (P5.2.9)

$$(P5.2.8),(P5.2.9) \Rightarrow 0 < k < 126.1 \qquad (P5.2.10)$$

In order to compute ω_c, we substitute $k = 126.1 = k_c$ in the auxiliary equation of row s^2 and get:

$$1040 s^2 + 4 \times 10^5 \times 126.1 = 0 \Rightarrow s = \pm j220.23 \Rightarrow \omega_c = 220.23 \text{ rad/s} \qquad (P5.2.11)$$

c. The characteristic equation is

$$s^3 + 3k s^2 + (k+2)s + 4 = 0 \qquad (P5.2.12)$$

Routh's tabulation is

$$
\begin{array}{c|cc}
s^3 & 1 & k+2 \\
s^2 & 3k & 4 \\
s^1 & \dfrac{3k^2 + 6k - 4}{3k} & \\
s^0 & 4 &
\end{array}
$$

The conditions for the stability of the system are

1. $3k > 0 \Rightarrow k > 0$ $\qquad$ (P5.2.13)

2. $\dfrac{3k^2 + 6k - 4}{3k} > 0 \Rightarrow 3k^2 + 6k - 4 > 0 \Rightarrow \begin{cases} k < -2.528 \\ \text{or} \\ k > 0.528 \end{cases}$ $\qquad$ (P5.2.14)

$$(P5.2.13),(P5.2.14) \Rightarrow k > 0.528 \qquad (P5.2.15)$$

From the auxiliary equation of row s^2, we get

$$3k_c s^2 + 4 = 0 \Rightarrow s^2 = -2.525 \Rightarrow s = \pm j1.59 \Rightarrow \omega_c = 1.59 \text{ rad/s} \qquad (P5.2.16)$$

5.3 Compute the maximum value of k so that the system shown in following figure is stable.

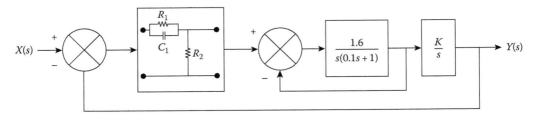

Solution

First we compute the transfer function of the phase lead circuit shown in the figure below:

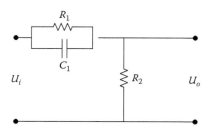

We have

$$G(s) = \frac{V_o(s)}{V_i(s)} = \frac{R_2}{R_1 \cdot (1/sC_1)/R_1 + (1/sC_1) + R_2} = \frac{R_2}{R_1/(sR_1C_1 + 1) + R_2} \Rightarrow$$

(P5.3.1)

$$G(s) = \frac{R_2(sR_1C_1 + 1)}{sR_1R_2C_1 + R_1 + R_2} = \frac{R_2}{R_1 + R_2} \cdot \frac{sR_1C_1 + 1}{s(R_1R_2C_1/R_1 + R_2) + 1}$$

Substituting the values of R_1, R_2, and C_1 in (P5.3.1), we get

$$G(s) = \frac{5 \cdot 10^4}{5 \cdot 10^5 + 5 \cdot 10^4} \cdot \frac{s \cdot 5 \cdot 10^5 \cdot 10^{-6} + 1}{s \cdot (5 \cdot 10^5 \cdot 5 \cdot 10^4 \cdot 10^{-6}/5 \cdot 10^5 + 5 \cdot 10^4) + 1} \simeq 0.1 \frac{s \cdot 0.5 + 1}{s \cdot 0.05 + 1}$$

(P5.3.2)

We plot the signal flow diagram of the system:

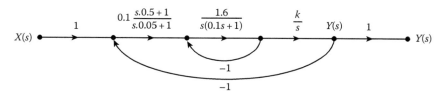

The total transfer function $F(s)$, from Mason's gain formula, is

$$F(s) = \frac{Y(s)}{X(s)} = \frac{T_1\Delta_1}{\Delta}$$

(P5.3.3)

where

$$T_1 = 1 \cdot 0.1 \frac{s \cdot 0.5 + 1}{s \cdot 0.05 + 1} \cdot \frac{1.6}{s(0.1s + 1)} \cdot \frac{k}{s} \cdot 1$$

(P5.3.4)

$$\Delta = 1 - \left(-\frac{1.6}{s(0.1s + 1)} - \frac{0.1 \cdot 1.6k(0.5s + 1)}{s^2(0.05s + 1)(0.1s + 1)} \right)$$

(P5.3.5)

$$\Delta_1 = 1 - (0) = 1$$

(P5.3.6)

By substituting (P5.3.4), (P5.3.5), and (P5.3.6) to (P5.3.3), we get the characteristic equation:

$$0.005s^4 + 0.15s^3 + 1.08s^2 + (1.6 + 0.08k)s + 0.16k = 0$$

Routh's tabulation is

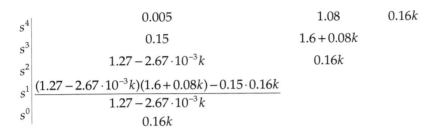

For the system to be stable, it must hold that

1. $1.27 - 2.67 \cdot 10^{-3}k > 0 \Rightarrow k < 475.6$ (P5.3.7)
2. $-2.136 \cdot 10^{-4} \cdot k^2 + 0.054k + 1.6432 > 0 \Rightarrow -27.45 < k < 280.3$ (P5.3.8)
3. $0.16k > 0 \Rightarrow k > 0$ (P5.3.9)

Hence, it suffices that

$$0 < k < 280.3 \qquad\qquad\text{(P5.3.10)}$$

The maximum value of k is 280.3.

5.4 Calculate the critical gain k_c and the critical frequency of oscillation ω_c for the position-velocity control system depicted in the figure, if the switch is in positions 1 and 2. What conclusions can be drawn about the stability of the system?

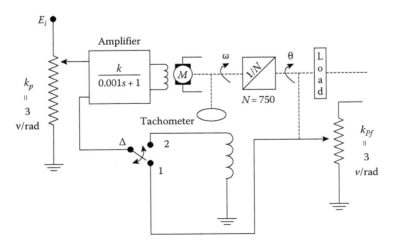

Suppose that $GM(s) = 10/(s(0.5s + 1)(0.001s + 1))$ is the transfer function of the motor, and $G_T(s) = 0.002s$ is the transfer function of the tachometer.

Solution

a. We plot the block diagram of the position control system for the switch in position 1.

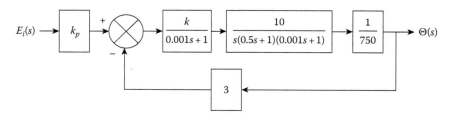

The characteristic equation of the system is

$$1 + \frac{k}{0.001s+1} \cdot \frac{10}{s(0.5s+1)(0.001s+1)} \cdot \frac{1}{750} \cdot 3 = 0 \Rightarrow$$

$$750 \cdot s(0.001s+1)(0.5s+1)(0.001s+1) + 30k = 0 \Rightarrow \qquad \text{(P5.4.1)}$$

$$3.75 \cdot 10^{-4}s^4 + 0.75075s^3 + 376.5s^2 + 750s + 30k = 0$$

Routh's tabulation is

$$
\begin{array}{c|ccc}
s^4 & 3.75 \cdot 10^{-4} & 376.5 & 30k \\
s^3 & 0.75,075 & 750 & \\
s^2 & 376.12 & 30k & \\
s^1 & \dfrac{282,090 - 22.5225k}{376.12} & & \\
s^0 & 30k & &
\end{array}
$$

For the system to be stable, it must hold that

1. $\dfrac{282,090 - 22.5225k}{376.12} > 0 \Rightarrow k < 12,524.9$ \hfill (P5.4.2)

2. $30k > 0 \Rightarrow k > 0$ \hfill (P5.4.3)

Hence, it is sufficient that

$$0 < k < 12,524.9 \qquad \text{(P5.4.4)}$$

The critical gain is

$$k_c = 12,524.9 \qquad \text{(P5.4.5)}$$

We compute ω_c:

$$376.12s^2 + 30k_c = 0 \Rightarrow s^2 = -999 \Rightarrow \omega_c = 31.6\,\text{rad/s} \qquad \text{(P5.4.6)}$$

b. We plot the block diagram of the velocity control system for the switch in position 2.

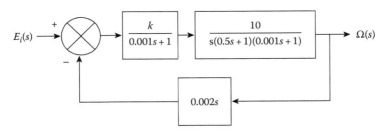

The characteristic equation of the system is

$$1+\frac{k}{0.001s+1}\cdot\frac{10}{s(0.5s+1)(0.001s+1)}\cdot 0.002s = 0 \Rightarrow$$

$$(0.001s+1)(0.5s+1)(0.001s+1)+0.02k = 0 \Rightarrow \qquad (P5.4.7)$$

$$5\cdot10^{-7}s^3 +1.001\cdot10^{-3}s^2 +0.502s+1+0.02k = 0$$

Routh's tabulation is

$$
\begin{array}{c|cc}
s^3 & 5\cdot10^{-7} & 0.502 \\
s^2 & 1.001\cdot10^{-3} & 1+0.02k \\
s^1 & \dfrac{5.02002\cdot10^{-4}-10^{-8}k}{1.001\cdot10^{-3}} & \\
s^0 & 1+0.02k &
\end{array}
$$

The condition for the stability of the system is

1. $\dfrac{5.02002\cdot10^{-4}-10^{-8}k}{1.001\cdot10^{-3}} > 0 \Rightarrow k < 50,200.2$ \qquad (P5.4.8)

2. $1+0.02k > 0 \Rightarrow k > -50$ \qquad (P5.4.9)

Hence, it is sufficient that

$$0 < k < 50,200.2 \qquad (P5.4.10)$$

The critical value of gain is

$$k_c = 50,200.2 \qquad (P5.4.11)$$

We compute ω_c:

$$1.001\cdot10^{-3}s^2 +0.02k_c = 0 \Rightarrow s^2 = -1,004,000 \Rightarrow \omega_c = 1,001.998\,\text{rad/s} \qquad (P5.4.12)$$

5.5 The open-loop transfer function of a unity-feedback system is

$$G(s) = \frac{50}{s(1+0.05s)(1+0.2s)}$$

Moreover, $R_1 = 1M\Omega$, $C_1 = 1\mu F$, $C_2 = 10\mu F$.

a. Demonstrate that the system is stable.

b. Examine the stability of the system if the electrical circuit depicted in the following figure is connected in series with the system.

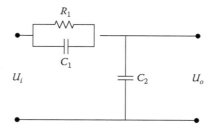

Solution

a. The loop transfer function is

$$G(s) = \frac{50}{s(1+0.05s)(1+0.2s)} \tag{P5.5.1}$$

The characteristic equation is

$$1+G(s) = 0 \Rightarrow s(1+0.05s)(1+0.2s)+50 = 0 \Rightarrow$$

$$0.01s^3 + 0.25s^2 + s + 50 = 0 \tag{P5.5.2}$$

Routh's tabulation is

$$
\begin{array}{c|cc}
s^3 & 0.01 & 1 \\
s^2 & 0.25 & 50 \\
s^1 & -1 & \\
s^0 & 50 &
\end{array}
$$

The system is **unstable** because there is a change of sign in the first column of Routh's tabulation.

b. We connect in series with the system the electrical circuit shown in the above figure. The new system is shown below.

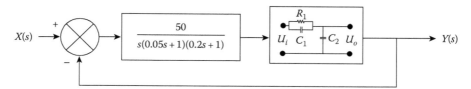

The transfer function of the electrical circuit is

$$G_c(s) = \frac{V_o(s)}{V_i(s)} = \frac{1/sC_2}{(R_1 \cdot (1/sC_1)/R_1 + (1/sC_1)) + 1/sC_2} = \frac{1/sC_2}{(R_1/(sR_1C_1 + 1)) + 1/sC_2} \Rightarrow$$

(P5.5.3)

$$G_c(s) = \frac{sR_1C_1 + 1}{sR_1(C_1 + C_2) + 1}$$

The new characteristic equation is

$$1 + G(s)G_c(s) = 0 \Rightarrow s(1 + 0.05s)(1 + 0.2s)[sR_1(C_1 + C_2) + 1] + 50(sR_1C_1 + 1) = 0 \Rightarrow$$
$$s(1 + 0.05s)(1 + 0.2s)(10.5s + 1) + 25s + 50 = 0$$

$$0.105s^4 + 2.635s^3 + 10.75s^2 + 26s + 50 = 0 \qquad \text{(P5.5.4)}$$

Routh's tabulation is

s^4	0.105	10.75	50
s^3	2.635	26	
s^2	9.714	50	
s^1	12.437		
s^0	50		

We observe that after connecting the electrical circuit, the system becomes **stable**.

5.6 Determine the values of the amplifier gain k so that the system shown in the figure is stable:

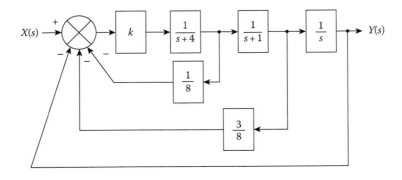

Solution

We gradually reduce the initial block diagram. We have

$$G_1(s) = \frac{k/(s+4)}{1 + (k/s+4) \cdot \frac{1}{8}} = \frac{8k}{8(s+4)+k} = \frac{8k}{8s+32+k} \qquad \text{(P5.6.1)}$$

The block diagram becomes

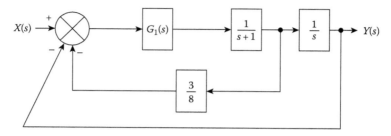

Also,

$$G_2(s) = \frac{G_1(s) \cdot (1/(s+1))}{1 + G_1(s) \cdot (1/(s+1)) \cdot \dfrac{3}{8}} = \frac{8k}{8s^2 + 40s + ks + 32 + 4k} \tag{P5.6.2}$$

The block diagram after the previous reduction becomes

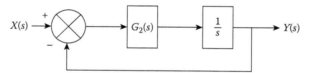

The characteristic equation of the system is

$$1 + G_2(s) \cdot \frac{1}{s} = 0 \Rightarrow s(8s^2 + 40s + ks + 32 + 4k) = 0 \Rightarrow$$

$$8s^3 + (40 + k)s^2 + (32 + 4k)s + 8k = 0 \tag{P5.6.3}$$

Routh's tabulation is

$$
\begin{array}{c|cc}
s^3 & 8 & 32 + 4k \\
s^2 & 40 + k & 8k \\
s^1 & \dfrac{4k^2 + 128k + 1280}{40 + k} & \\
s^0 & 8k &
\end{array}
$$

For the system to be stable, it must hold that

1. $40 + k > 0 \Rightarrow k > -40$ \hfill (P5.6.4)

2. $\dfrac{4k^2 + 128k + 1280}{40 + k} > 0 \Rightarrow 4k^2 + 128k + 1280 > 0$ \hfill (P5.6.5)

3. $8k > 0 \Rightarrow k > 0$ \hfill (P5.6.6)

The equation $4k^2 + 128k + 1280 = 0$ has complex roots; therefore, the system is **stable** for every $k > 0$.

5.7 Consider the following signal flow diagram of the system:

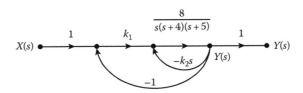

a. Determine the values of the parameters k_1 and k_2 for which the system has a critical frequency of oscillation $\omega_c = 9\,\text{rad/s}$.

b. Find the values of k_1 and k_2 for which the system has poles at the points $-2 \pm j2$. Plot the poles in the s-plane and discuss the stability of the system.

Solution

a. The transfer function of the system, from Mason's gain formula, is

$$G(s) = \frac{Y(s)}{X(s)} = \frac{T_1 \Delta_1}{\Delta} \tag{P5.7.1}$$

$$T_1 = 1 \cdot k_1 \cdot \frac{8}{s(s+4)(s+5)} \cdot 1 = \frac{8k_1}{s(s+4)(s+5)} \tag{P5.7.2}$$

$$\Delta = 1 - \left(-\frac{8}{s(s+4)(s+5)} \cdot k_2 s - \frac{8k_1}{s(s+4)(s+5)} \right) \Rightarrow$$

$$\Delta = \frac{s(s+4)(s+5) + 8k_2 s + 8k_1}{s(s+4)(s+5)} \tag{P5.7.3}$$

$$\Delta_1 = 1 - (0) = 1 \tag{P5.7.4}$$

By substituting to (P5.7.1), we get

$$G(s) = \frac{Y(s)}{X(s)} = \frac{8k_1}{s(s+4)(s+5) + 8k_2 s + 8k_1} \tag{P5.7.5}$$

The characteristic equation of the system is

$$s(s+4)(s+5) + 8k_2 s + 8k_1 = 0 \Rightarrow s^3 + 9s^2 + s(20 + 8k_2) + 8k_1 = 0 \tag{P5.7.6}$$

Routh's tabulation is

$$
\begin{array}{c|cc}
s^3 & 1 & 20 + 8k_2 \\
s^2 & 9 & 8k_1 \\
s^1 & \dfrac{180 + 72k_2 - 8k_1}{9} & \\
s^0 & 8k_1 & \\
\end{array}
$$

From the auxiliary equation of row s^2, we have

$$9s^2 + 8k_{1c} = 0 \Rightarrow s^2 = -\frac{8k_{1c}}{9} \Rightarrow s = \pm j\sqrt{\frac{8k_{1c}}{9}} \Rightarrow$$

$$\omega_c = \sqrt{\frac{8k_{1c}}{9}} = 9 \Rightarrow k_{1c} = 91.125 \tag{P5.7.7}$$

From row s^1, we have

$$\frac{180 + 72k_{2c} - 8k_{1c}}{9} = 0 \Rightarrow k_{2c} = 7.625 \tag{P5.7.8}$$

b. The characteristic equation of the system is a third order; hence, the system has three poles p_1, p_2, p_3. We know that $p_1 = 2 + j2$, $p_2 = 2 - j2$ (conjugate pole); therefore,

$$(s - p_1)(s - p_2) = (s + 2 + j2)(s + 2 - j2) = s^2 + 4s + 8 \tag{P5.7.9}$$

In order to find the third pole p_3, we divide the polynomial $s^3 + 9s^2 + (20 + 8k_2)s + 8k_1$ by the polynomial $s^2 + 4s + 8$:

$$
\begin{array}{r|l}
s^3 + 9s^2 + (20 + 8k_2)s + 8k_1 & s^2 + 4s + 8 \\
-s^3 - 4s^2 - 8s & \\
\hline
5s^2 + s(12 + 8k_2) + 8k_1 & s + 5 \\
-5s^2 - 20s - 40 & \\
\hline
-8s + 8k_2s + 8k_1 - 40 &
\end{array}
$$

The third pole is $p_3 = -5$. The remainder of the division must be zero; thus, it must hold that

$$-8 + 8k_2 = 0 \Rightarrow k_2 = 1 \tag{P5.7.10}$$

$$-8k_1 - 40 = 0 \Rightarrow k_1 = 5 \tag{P5.7.11}$$

We plot the poles in the s-plane

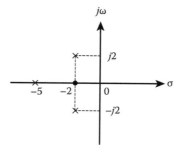

Apparently, the system is **stable**, because all of its poles are located in the left-half s-plane.

5.8 Consider the loop transfer functions:

a. $G(s)H(s) = \dfrac{1}{s^2 + a^2}$, $\quad a > 0$

b. $G(s)H(s) = \dfrac{1}{s^4(s+a)}$, $\quad a > 0$

Sketch the Nyquist paths and the Nyquist plots. What conclusions can be drawn about the stability of the systems from the Nyquist plots?

Solution

a. We have the loop transfer function:

$$G(s)H(s) = \frac{1}{s^2 + a^2}, \quad a > 0 \tag{P5.8.1}$$

The Nyquist path is

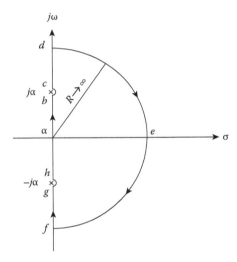

In order to sketch the Nyquist plot, we have

Part *ab*: $s = j\omega$, $0 \le \omega < \alpha$. Thus,

$$GH(j\omega) = \left(\frac{1}{a^2 - \omega^2}\right)^{\angle(0°)} \tag{P5.8.2}$$

For $\omega = 0$,

$$GH(j0) = \left(\frac{1}{a^2}\right)^{\angle(0°)} \tag{P5.8.3}$$

And for $\omega \to \alpha^-$,

$$\lim_{\omega \to \alpha^-} (GH(j\omega)) = \infty^{\angle(0^\circ)} \tag{P5.8.4}$$

Part *be*: $s = j\alpha + \rho e^{j\theta}$, $-90^\circ \le \theta \le 90^\circ$. Hence,

$$\lim_{\rho \to 0}\left(GH\left(j\alpha + \rho e^{j\theta}\right)\right) = \lim_{\rho \to 0}\left\{\frac{1}{\rho e^{j\theta}\left(\rho e^{j\theta} + 2ja\right)}\right\} = -j\infty^{\angle(-j\theta)} = \infty^{\angle(-\theta^\circ - 90^\circ)} \tag{P5.8.5}$$

For $\theta = -90^\circ$ the limit is $\infty^{\angle(0^\circ)}$, for $\theta = 0^\circ$ it is $\infty^{\angle(-90^\circ)}$, and for $\theta = 90^\circ$ it is $\infty^{\angle(-180^\circ)}$.

Part *cd*: $s = j\omega$, $\omega > a$. Thus,

$$\lim_{\omega \to \alpha^+} (GH(j\omega)) = \infty^{\angle(180^\circ)} \tag{P5.8.6}$$

and

$$\lim_{\omega \to \infty}(GH(j\omega)) = 0^{\angle(180^\circ)} \tag{P5.8.7}$$

The **part *fgha*** is symmetrical to ***abcd*** with respect to the Re{GH} axis. The **part *def*** is depicted on the axes origin.
    The Nyquist plot is

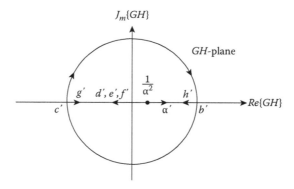

The closed-loop system is **unstable** as

$$N = 1, \quad P = 0 \quad \text{so} \quad N \ne -P \tag{P5.8.8}$$

b. The loop transfer function is

$$G(s)H(s) = \frac{1}{s^4(s+a)} \tag{P5.8.9}$$

The Nyquist path is

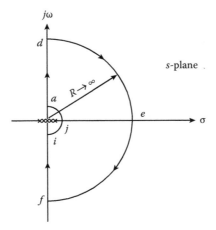

Part ad: $s = j\omega$, $0 < \omega < \infty$. Thus,

$$GH(j\omega) = \frac{1}{(j\omega)^4(j\omega + a)} = \frac{1}{\omega^4(\omega^2 + a^2)} {}^{\angle(-\tan^{-1}(\omega/a))} \tag{P5.8.10}$$

$$\lim_{\omega \to 0^+}(GH(j\omega)) = \infty^{\angle(0^\circ)} \tag{P5.8.11}$$

$$\lim_{\omega \to \infty}(GH(j\omega)) = 0^{\angle(-90^\circ)} \tag{P5.8.12}$$

As ω increases from 0 to ∞, the angle remains negative and decreases to -90°. The magnitude decreases steadily to zero.

Part def: $s = \lim_{R \to \infty} Re^{j\theta}$, where the angle θ increases from $+90^\circ$ to -90°. The magnitude of $GH(s)$ tends to zero and we can depict the part def on the axes origin.

The **part fi** is symmetrical to ad with respect to the Re{GH} axis.

Part ija: $s = \lim_{\rho \to 0} \rho e^{j\theta}$, $-90^\circ \leq \theta \leq 90^\circ$. Thus,

$$\lim_{\rho \to 0} GH(\rho e^{j\theta}) = \lim_{\rho \to 0} \left\{ \frac{1}{(\rho e^{j\theta})^4(\rho e^{j\theta} + \beta)} \right\} = \infty^{\angle(-4\theta)} \tag{P5.8.13}$$

For $\theta = -90^\circ$, the limit is $\infty^{\angle(360^\circ)}$.

For $\theta = +90^\circ$, the limit is $\infty^{\angle(-360^\circ)}$.

Consequently, the total angle is $-4 \cdot 90^\circ - 4 \cdot 90^\circ = -4 \cdot 180^\circ$, which corresponds to four "infinite" semicircles. The number of these semicircles is equal to the number that indicates the type of the system (here, system type 4). As the Nyquist path in the s-plane "rotates clockwise" at 90° in i, the Nyquist plot does the same at the i in the GH-plane.

The Nyquist plot is

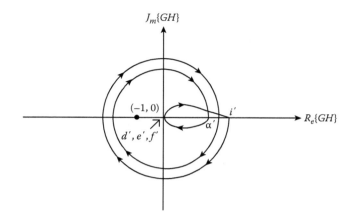

We have

$$N = 2, \quad P = 0 \quad \text{so} \quad N \neq -P \tag{P5.8.14}$$

Hence, the system is **unstable**.

5.9 A control system has the following loop transfer function:

$$G(s)H(s) = \frac{10}{s^2(1+0.2s)(1+0.1s)}$$

a. Prove that the system is unstable, by the use of the Nyquist method.
b. Examine how the stability of the system is influenced by the addition of a zero in the point $s = -1$ of the function.

Solution

a.
$$G(s)H(s) = \frac{10}{s^2(1+0.2s)(1+0.1s)} \tag{P5.9.1}$$

$$GH(j\omega) \overset{\text{(P5.9.1)}}{=} \frac{10}{(j\omega)^2(1+0.2j\omega)(1+0.1j\omega)} \tag{P5.9.2}$$

$$\lim_{\omega \to 0} GH(j\omega) = \infty^{\angle(-180°)} \tag{P5.9.3}$$

and

$$\lim_{\omega \to \infty} GH(j\omega) = 0^{\angle(-360°)} \tag{P5.9.4}$$

The two boundary points (for $\omega \to 0$ and for $\omega \to \infty$) are not sufficient for drawing the Nyquist plot. Therefore, we have to compute the amplitude and the phase of $GH(j\omega)$ for various values of ω:

$$
\left.\begin{aligned}
GH(j1) &= 9.7^{\angle(-197°)} \\[6pt]
GH(j2) &= 2.27^{\angle(-213°)} \\[6pt]
GH(j5) &= 1.25^{\angle(-257°)} \\[6pt]
GH(j8) &= 0.64^{\angle(-217°)} \\[6pt]
GH(j10) &= 0.32^{\angle(-288°)}
\end{aligned}\right\} \tag{P5.9.5}
$$

The Nyquist plot is depicted in the figure below:

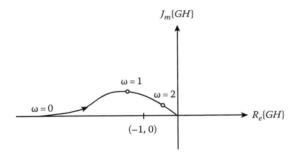

Apparently the system is **unstable**, since traversing the Nyquist plot as ω increases, the point $(-1, 0)$ is at the right-hand side of the diagram.

b. The loop transfer function with the addition of a zero at $s = -1$ is

$$
G(s)H(s) = \frac{10(s+1)}{s^2(1+0.2s)(1+0.1s)} \tag{P5.9.6}
$$

$$
GH(j\omega) \overset{(P5.9.6)}{=} \frac{10(j\omega+1)}{(j\omega)^2(1+0.2j\omega)(1+0.1j\omega)} \tag{P5.9.7}
$$

$$
\lim_{\omega \to 0} GH(j\omega) = \infty^{\angle(-180°)} \tag{P5.9.8}
$$

and

$$
\lim_{\omega \to \infty} GH(j\omega) = 0^{\angle(-270°)} \tag{P5.9.9}
$$

The intermediate points of the diagram for various values of ω are

$$GH(j1) = 14^{\angle(-152°)}$$

$$GH(j2) = 5.1^{\angle(-150°)}$$

$$GH(j5) = 1.3^{\angle(-172°)}$$ (P5.9.10)

$$GH(j8) = 0.6^{\angle(-194°)}$$

$$GH(j10) = 0.3^{\angle(-204°)}$$

The Nyquist plot is sketched approximately in

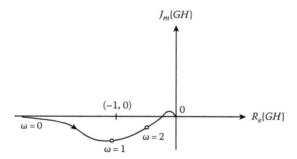

We observe that the system is **stable**, since the point $(-1, 0)$ is at the left-hand side of the diagram, as ω increases.

5.10 Assume two closed loop systems with the following loop transfer functions:

1. $G(s)H(s) = \dfrac{100(s+2)}{s(s+1)(s+10)}$

2. $G(s)H(s) = \dfrac{100(s+2)^2}{s(s-1)(s+10)}$

a. Draw the Nyquist plots and draw conclusions about the stability of the closed-loop systems.

b. Compute the gain margin and the phase margin.

Solution

a. 1. $$G(s)H(s) = \dfrac{100(s+2)}{s(s+1)(s+10)}$$ (P5.10.1)

We set $s = j\omega$ and

$$GH(j\omega) = \dfrac{100(j\omega+2)}{j\omega(j\omega+1)(j\omega+10)}$$ (P5.10.2)

Amplitude

$$(P5.10.2) \Rightarrow |GH(j\omega)| = \frac{100\sqrt{4+\omega^2}}{\omega\sqrt{1+\omega^2}\sqrt{100+\omega^2}} \tag{P5.10.3}$$

Phase

$$(P5.10.2) \Rightarrow \sphericalangle(GH(j\omega)) = \tan^{-1}\frac{\omega}{2} - \tan^{-1}\frac{\omega}{0} - \tan^{-1}\frac{\omega}{1} - \tan^{-1}\frac{\omega}{10} \Rightarrow$$
$$\sphericalangle(GH(j\omega)) = \tan^{-1}\frac{\omega}{2} - 90° - \tan^{-1}\omega - \tan^{-1}\frac{\omega}{10} \tag{P5.10.4}$$

$$\lim_{\omega \to 0^+} GH(j\omega) = \infty^{\sphericalangle(-\pi/2)} \tag{P5.10.5}$$

$$\lim_{\omega \to +\infty} GH(j\omega) = 0^{\sphericalangle(-\pi)} \tag{P5.10.6}$$

For intermediate values of ω, we get

$$\left.\begin{array}{l} GH(j1) = 15.7^{\sphericalangle(-114.1°)} \\[2mm] GH(j2) = 6.2^{\sphericalangle(-119.7°)} \\[2mm] GH(j5) = 1.89^{\sphericalangle(-127.1°)} \\[2mm] GH(j10) = 0.72^{\sphericalangle(-140.6°)} \end{array}\right\} \tag{P5.10.7}$$

The Nyquist plot is sketched below:

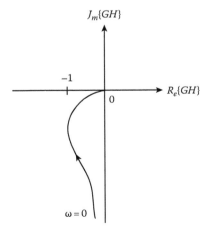

The system is evidently stable since the point $-1 + j0 = (-1, 0)$ is at the left-hand side of the Nyquist plot, at the direction that ω increases.

2.
$$G(s)H(s) = \frac{100(s+2)^2}{s(s-1)(s+10)} \Rightarrow \qquad \text{(P5.10.8)}$$

$$GH(j\omega) \overset{(P5.10.8)}{=} \frac{100(j\omega+2)^2}{j\omega(j\omega-1)(j\omega+10)} \qquad \text{(P5.10.9)}$$

Magnitude Computation:

$$(P5.10.9) \Rightarrow |GH(j\omega)| = \frac{100(4+\omega^2)}{\omega\sqrt{1+\omega^2}\sqrt{100+\omega^2}} \qquad \text{(P5.10.10)}$$

Phase Computation:

$$(P5.10.9) \Rightarrow \sphericalangle(GH(j\omega)) = 2\tan^{-1}\frac{\omega}{2} - \tan^{-1}\frac{\omega}{0} - \left[180° - \tan^{-1}\frac{\omega}{1}\right] - \tan^{-1}\frac{\omega}{10} \Rightarrow$$

$$\sphericalangle(GH(j\omega)) = 2\tan^{-1}\frac{\omega}{2} - 270° + \tan^{-1}\omega - \tan^{-1}\frac{\omega}{10} \qquad \text{(P5.10.11)}$$

$$\lim_{\omega \to 0^+} GH(j\omega) = \infty^{\sphericalangle(-270°)} \qquad \text{(P5.10.12)}$$

$$\lim_{\omega \to +\infty} GH(j\omega) = 0^{\sphericalangle(-90°)} \qquad \text{(P5.10.13)}$$

For intermediate values of ω, we have

$$\left.\begin{array}{l} GH(j0.5) = 75.9^{\sphericalangle(-218.2°)} \\[2mm] GH(j1) = 35.2^{\sphericalangle(-177.6°)} \\[2mm] GH(j5) = 10.17^{\sphericalangle(-81.5°)} \\[2mm] GH(j10) = 7.3^{\sphericalangle(-73.3°)} \\[2mm] GH(j100) = 0.995^{\sphericalangle(-87.2°)} \\[2mm] GH(j150) = 0.67^{\sphericalangle(-88.2°)} \end{array}\right\} \qquad \text{(P5.10.14)}$$

The Nyquist plot is sketched below:

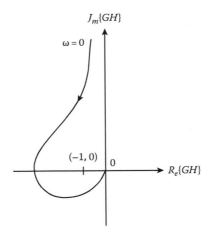

From the relationship (P5.10.11) it suffices that $\omega_c = 0.965\,\text{rad/s}$, so that $\sphericalangle GH(j\omega) = 180°$. The point $(-1, 0)$ is at the left-hand side of the Nyquist plot, as ω increases; hence, the system is **stable**.

We come to the same conclusion since

$$N = -1 = -P \qquad\qquad (P5.10.15)$$

b. 1. The **gain margin** of the system with the loop transfer function of the relationship (P5.10.1) is

$$k_g = -20\log_{10}\left|GH(j\omega_c)\right| \qquad\qquad (P5.10.16)$$

From the relationship (P5.10.4) that provides the phase, we get $\omega_c \to \infty$.

$$(P5.10.16) \Rightarrow k_g = -20\log_{10}\left|GH(j\omega_c)\right| \to \infty \qquad\qquad (P5.10.17)$$

For the **phase margin**, we have

$$\varphi_a = 180° + \sphericalangle(G(j\omega_1)H(j\omega_1)) \qquad\qquad (P5.10.18)$$

where ω_1 is the frequency for which $\left|GH(j\omega_1)\right| = 1 \overset{(P5.10.3)}{\Rightarrow}$

$$\frac{100\sqrt{4+\omega_1^2}}{\omega_1\sqrt{1+\omega_1^2}\sqrt{100+\omega_1^2}} = 1 \Rightarrow 10,000\left(4+\omega_1^2\right) = \omega_1^2\left(1+\omega_1^2\right)\left(100+\omega_1^2\right) \Rightarrow$$

$$10,000\omega_1^2 + 40,000 = \omega_1^2(\omega_1^4 + 101\omega_1^2 + 100) \Rightarrow \qquad (P5.10.19)$$

$$10,000\omega_1^2 + 40,000 = \omega_1^6 + 101\omega_1^4 + 100\omega_1^2 \Rightarrow$$

$$\omega_1^6 + 101\omega_1^4 - 9,900\omega_1^2 + 40,000 = 0$$

Let us suppose that

$$\omega_1{}^2 = x \tag{P5.10.20}$$

$$(\text{P5.10.19}) \overset{(\text{P5.10.20})}{\Rightarrow} x^3 + 101x^2 - 9{,}900x + 40{,}000 = 0 \Rightarrow x = 4.2 \tag{P5.10.21}$$

$$(\text{P5.10.20}) \overset{(\text{P5.10.21})}{\Rightarrow} \omega_1 = \sqrt{4.2} \Rightarrow \omega_1 = 2.05\,\text{rad/s} \tag{P5.10.22}$$

We have

$$\sphericalangle(GH(j\omega_1)) = \sphericalangle(GH(j2{,}05)) \overset{(\text{P5.10.4})}{=} 45.7° - 90° - 64° - 11.6° \simeq -120° \tag{P5.10.23}$$

$$(\text{P5.10.18}) \overset{(\text{P5.10.22})}{\Rightarrow} \varphi_a = 180° - 120° = 60° > 0 \tag{P5.10.24}$$

The gain and phase margins are positive; therefore, the system is stable.

2. The **gain margin** of the system with the loop transfer function of the relationship (P5.10.8) is

$$k_g = -20\log_{10}|GH(j\omega_c)| \tag{P5.10.25}$$

From the relationship (P5.10.11) that computes the phase, we get $\omega_c = 0.965\,\text{rad/s}$.

$$|GH(j\omega_c)| \overset{(\text{P5.10.10})}{=} 36.6 \tag{P5.10.26}$$

$$(\text{P5.10.25}) \overset{(\text{P5.10.26})}{\Rightarrow} k_g = -20\log_{10} 36.6 = -31.27\,\text{db}$$

or

$$k_g = \frac{1}{36.6} = 0.027 \tag{P5.10.27}$$

For the **phase margin**, we have

$$\varphi_a = 180° + \sphericalangle(G(j\omega_1)H(j\omega_1)) \tag{P5.10.28}$$

where ω_1 is the frequency for which $|GH(j\omega_1)| = 1$.
From the relationship (P5.10.14), we get

$$\omega_1 \simeq 100\,\text{rad/s} \tag{P5.10.29}$$

$$\sphericalangle(GH(j\omega_1)) = \sphericalangle(GH(j100)) = -87.2° \tag{P5.10.30}$$

$$(P5.10.28) \overset{(P5.10.30)}{\Rightarrow} \quad \phi_a = 180° - 87.2° = 92.8° > 0 \qquad (P5.10.31)$$

The gain and phase margin are positive; hence, the system is stable.

5.11 The following figure illustrates a control system. By using the Nyquist plot, examine
a. The stability of the inner loop
b. The stability of the closed-loop system, if $|G_1| = 100$

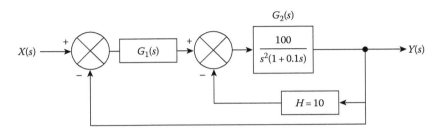

Solution

a. For the inner loop of the depicted system, we have

$$G_2(s)H(s) = \frac{1000}{s^2(1+0.1s)} \qquad (P5.11.1)$$

$$G_2H(j\omega) \overset{(P5.11.1)}{=} \frac{1000}{-\omega^2(1+0.1j\omega)} \qquad (P5.11.2)$$

We have

$$\left. \begin{array}{l} \lim_{\omega \to +\infty} G_2H(j\omega) = 0^{\angle(-270°)} \\[8pt] \lim_{\omega \to 0^+} G_2H(j\omega) = \infty^{\angle(-180°)} \\[8pt] \lim_{\omega \to -\infty} G_2H(j\omega) = 0^{\angle(270°)} \\[8pt] \lim_{\omega \to 0^-} G_2H(j\omega) = \infty^{\angle(180°)} \end{array} \right\} \qquad (P5.11.3)$$

In order to find the magnitude and the phase of $G_2H(j\omega)$, we solve the relationship (P5.11.2):

$$|G_2H(j\omega)| = \frac{1000}{\omega^2\sqrt{1+(0.1\omega)^2}} \qquad (P5.11.4)$$

$$\angle G_2H(j\omega) = -\left(180° + \tan^{-1}\frac{0.1\omega}{1}\right) = -180° - \tan^{-1}(0.1\omega) \qquad (P5.11.5)$$

We sketch the Nyquist plot, after computing the magnitude and the phase of $G_2H(j\omega)$ for intermediate values of ω:

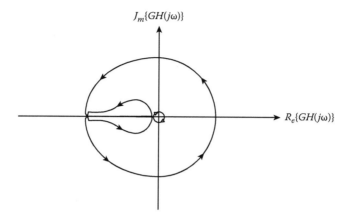

From the Nyquist plot we observe that $N = 2$, $P = 0$. Thus,

$$Z = N + P = 2 \neq -P \tag{P5.11.6}$$

From the relationship (P5.11.6), we conclude that the system is **unstable**.

b. The loop transfer function of the total system is

$$GH(s) = G_1(s) \cdot \frac{G_2(s)}{1 + G_2(s)H(s)} = \frac{10^4}{0.1s^3 + s^2 + 10^3} \tag{P5.11.7}$$

$$(P5.11.7) \Rightarrow GH(j\omega) = \frac{10^4}{10^3 - \omega^2 - j0.1\omega^3} \tag{P5.11.8}$$

Magnitude Computation:

$$|GH(j\omega)| = \frac{10^4}{\sqrt{(10^3 - \omega^2)^2 - (0.1\omega^3)^2}} \tag{P5.11.9}$$

Phase Computation:

$$\sphericalangle GH(j\omega) = -\left(-\tan^{-1}\frac{0.1\omega^3}{10^3 - \omega^2}\right) = \tan^{-1}\frac{0.1\omega^3}{10^3 - \omega^2} \tag{P5.11.10}$$

$$\left.\begin{aligned}
&\lim_{\omega \to +\infty} GH(j\omega) = 0^{\sphericalangle(-270°)} \\[4pt]
&\lim_{\omega \to 0^+} GH(j\omega) = 100^{\sphericalangle(0°)} \\[4pt]
&\lim_{\omega \to -\infty} GH(j\omega) = 0^{\sphericalangle(270°)} \\[4pt]
&\lim_{\omega \to 0^-} GH(j\omega) = 100^{\sphericalangle(0°)}
\end{aligned}\right\} \tag{P5.11.11}$$

We draw the Nyquist plot, after finding the magnitude and phase of $GH(j\omega)$ for intermediate values of ω:

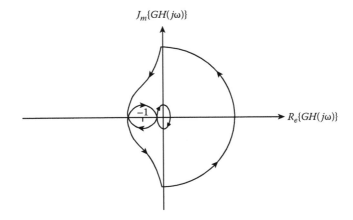

From the Nyquist plot we observe that $N = 0$, $P = 2$; hence,

$$Z = N + P = 2 \neq -P \qquad \text{(P5.11.12)}$$

From the relationship (P5.11.12), we conclude that the system is **unstable**.

5.12 Consider two systems with the following loop transfer functions:

a. $G(s)H(s) = \dfrac{200}{s(s+5)(s+10)}$

b. $G(s)H(s) = \dfrac{1+4s}{s^2(s+1)(1+2s)}$

Sketch the Nyquist plots for the two systems and draw conclusions about their stability.

Solution

a.

$$GH(s) = \dfrac{200}{s(s+5)(s+10)} \qquad \text{(P5.12.1)}$$

$$(\text{P5.12.1}) \Rightarrow GH(j\omega) = \dfrac{200}{j\omega\,(j\omega+5)(j\omega+10)} \qquad \text{(P5.12.2)}$$

$$(\text{P5.12.2}) \Rightarrow |GH(j\omega)| = \dfrac{200}{\omega\sqrt{25+\omega^2}\,\sqrt{100+\omega^2}} \qquad \text{(P5.12.3)}$$

$$(\text{P5.12.2}) \Rightarrow \sphericalangle\big(GH(j\omega)\big) = -90° - \tan^{-1}\dfrac{\omega}{5} - \tan^{-1}\dfrac{\omega}{10} \qquad \text{(P5.12.4)}$$

We plot the following table of values:

ω	$|GH(j\omega)|$	$<(GH(j\omega))$
0	∞	$-90°$
1	3.9	$-107°$
5	0.506	$-161.6°$
7.07	0.267	$-180°$
10	0.126	$-198.4°$
100	$1.98 \cdot 10^{-4}$	$-261.4°$
∞	0	$-270°$

We sketch the Nyquist plot based on the set of values of the previous table:

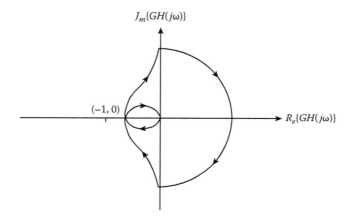

For $\omega = 7.07\,\text{rad/s}$, the angle is $-180°$ and the amplitude is $0.267 < 1$. The point $-1 + j0$ is at the left-hand side of the plot; hence, the system is **stable**.

b.
$$GH(s) = \frac{1+4s}{s^2(s+1)(1+2s)} \tag{P5.12.5}$$

$$(P5.12.5) \Rightarrow GH(j\omega) = \frac{1+4j\omega}{-\omega^2(1+j\omega)(1+2j\omega)} \tag{P5.12.6}$$

$$(P5.12.6) \Rightarrow |GH(j\omega)| = \frac{\sqrt{1+16\omega^2}}{\omega^2\sqrt{1+\omega^2}\sqrt{1+4\omega^2}} \tag{P5.12.7}$$

$$(P5.12.6) \Rightarrow \sphericalangle(GH(j\omega)) = \tan^{-1}(4\omega) - 180° - \tan^{-1}\omega - \tan^{-1}(2\omega) \tag{P5.12.8}$$

We plot the following table of values:

ω	$\lvert GH(j\omega)\rvert$	$\angle(GH(j\omega))$
0	∞	$-180°$
0.25	53.5	$-175.6°$
0.5	5.7	$-188.2°$
1	1.3	$-212.4°$
1.25	0.75	$-220.8°$
2	0.22	$-236.5°$
4	0.03	-252.5
∞	0	$-270°$

We sketch approximately the Nyquist plot based on the previous set of values:

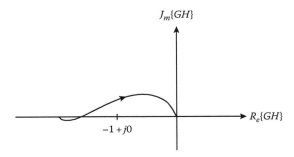

For $\omega = 0.38\,\text{rad/s}$, the angle is approximately $-180°$ and the amplitude is $8.55 > 1$. The system is **unstable**, since the point $-1 + j0$ is at the left-hand side of the Nyquist plot, as ω increases.

6

Root-Locus Analysis

6.1 Introduction

Root-locus analysis is a graphical method for examining how the roots of a system change under variation of a certain system parameter, commonly the gain of a feedback system.

Consider the loop transfer function

$$G(s)H(s) = K \frac{P(s)}{Q(s)} \tag{6.1}$$

where $P(s)$ and $Q(s)$ are polynomials of the complex variable s.

The closed-loop transfer function that describes the dynamic behavior of the system is

$$\frac{Y(s)}{X(s)} = \frac{G(s)}{1+G(s)H(s)} \overset{(6.1)}{=} \frac{G(s)Q(s)}{Q(s)+KP(s)} \tag{6.2}$$

The roots of the characteristic equation are the **poles** of the closed-loop system. They can be computed by the relationship

$$Q(s) + KP(s) = 0 \tag{6.3}$$

where K is the gain of the system.

The locations of the poles of the transfer function in the complex s-plane influence the transient response of the system and determine its stability. From relationship (6.3), we observe that every change in the value of the constant K results in the displacement of the poles in the complex plane.

The **root-locus diagram** is a method for representing the poles of the closed-loop system on the s-plane, in relation to a system parameter (usually the gain K).

From the root-locus diagram we obtain information about the stability and the overall behavior of the system.

The characteristic equation of the closed-loop system is

$$1 + G(s)H(s) = 0 \tag{6.4}$$

or

$$G(s)H(s) = -1 \tag{6.5}$$

$$(6.5) \Rightarrow |G(s)H(s)| = 1 \tag{6.6}$$

and

$$\sphericalangle(G(s)H(s)) = (2\rho + 1)\pi, \quad \rho = 0, \pm 1, \pm 2, \dots \tag{6.7}$$

Suppose that the open-loop transfer function is

$$G(s)H(s) = K \frac{(s+z_1)(s+z_2)\cdots(s+z_m)}{(s+p_1)(s+p_2)\cdots(s+p_n)} \tag{6.8}$$

Then relationships (6.6) and (6.7) become

$$|K| \frac{\prod_{j=1}^{m} |s+z_j|}{\prod_{i=1}^{n} |s+p_i|} = 1, \quad -\infty < K < \infty \tag{6.9}$$

and

$$\sum_{j=1}^{m} \sphericalangle(s+z_j) - \sum_{i=1}^{n} \sphericalangle(s+p_i) = \begin{cases} (2\rho+1)\pi, & K \geq 0 \\ 2\rho\pi, & K < 0 \end{cases}, \quad \rho = 0, \pm 1, \pm 2, \dots \tag{6.10}$$

The relationships (6.9) and (6.10) provide the **magnitude-phase condition** for the root locus. Once the root locus is drawn, the value of K for a specific point that corresponds to the root s_1 can be determined from Equation 6.9. The root locus that fulfills the relationships (6.9) and (6.10) for $K \in (-\infty, 0)$ is called **complementary** root locus.

6.2 Designing a Root-Locus Diagram

In this section, we introduce a 10-step procedure for drawing the root-locus diagram of a control system:

STEP 1: Branches start at the open-loop poles. The poles of $G(s)H(s)$ are called points of departure of the roots locus (RL).

STEP 2: Branches end at the open-loop zeros or at infinity. These points are called points of arrival of the RL.

STEP 3: The number of branches of the locus is equal to $\max(n, m)$, where m is the number of zeros and n is the number of the poles of $G(s)H(s)$.

STEP 4: The root locus is symmetric to the real axis (horizontal axis).

STEP 5: The intersection of the lines with the real axis can be found as

$$\sigma_\alpha = \frac{\sum_{i=1}^{n} (p_i) - \sum_{j=1}^{m} (z_j)}{n - m} \tag{6.11}$$

where

$\sum_{i=1}^{n} (p_i)$ is the algebraic sum of the values of the poles of $G(s)H(s)$

$\sum_{j=1}^{m} (z_j)$ is the algebraic sum of the values of the zeros of $G(s)H(s)$

STEP 6: For large values of s, RL tends asymptotically to the lines that form the following angles with the real axis:

$$\sphericalangle\varphi_\alpha = \frac{(2\rho + 1)\pi}{n - m}, \qquad \begin{array}{c} \rho = 0, 1, \ldots, |n - m| - 1 \\ K \geq 0 \end{array} \tag{6.12}$$

or

$$\sphericalangle\varphi_\alpha = \frac{2\rho\pi}{n - m}, \qquad \begin{array}{c} \rho = 0, 1, \ldots, |n - m| - 1 \\ K \leq 0 \end{array} \tag{6.13}$$

STEP 7: Part of the real axis can be a segment of the RL if

- For $K \geq 0$, the number of the poles and zeros that are at the right side of the segment is odd
- For $K \leq 0$, the number of the poles and zeros that are at the right side of the segment is even

STEP 8: The departure and arrival points are called **breakaway points** of the RL and can be found in two ways:

First way:

$$(6.3) \Rightarrow K = -\frac{Q(s)}{P(s)} \tag{6.14}$$

$$\frac{dK}{ds} = 0 \overset{(6.14)}{\Rightarrow} -\frac{Q'(s)P(s) - P'(s)Q(s)}{P^2(s)} = 0 \Rightarrow$$

$$Q'(s)P(s) - P'(s)Q(s) = 0 \tag{6.15}$$

Every root of the Equation 6.15 is accepted as a breakaway point if it satisfies the condition (6.4) for any real value of K.

Second way:

If the poles and zeros of $G(s)H(s)$ are real numbers, instead of (6.15) we can solve the following equation:

$$\sum_{i=1}^{n} \frac{1}{s - p_i} = \sum_{j=1}^{m} \frac{1}{s - z_j} \tag{6.16}$$

STEP 9: The angles of departure of the RL from a complex pole or the angles of arrival at a complex zero can be found as

$$\sphericalangle \varphi_d = (2\rho + 1)\pi - \left(\sum_{i=1}^{n} \varphi_{p_i} - \sum_{j=1}^{m} \varphi_{z_j} \right) \tag{6.17}$$

where

$\displaystyle\sum_{i=1}^{n} \varphi_{p_i}$ is the algebraic sum of the angles formed by the poles and the relevant complex pole (or zero)

$\displaystyle\sum_{j=1}^{m} \varphi_{z_j}$ is the algebraic sum of the angles formed by the zeros and the relevant complex pole (or zero)

STEP 10: The intersections of the root locus and the imaginary axis (vertical axis) are the points $\pm j\omega_c$, where the system from stable becomes unstable. They can be computed with the use of Routh's stability criterion.

6.3 Design of a Control System with the Use of the Root Locus

When designing a control system, we seek to adjust the time response and the frequency response to the technical requirements of the system. In doing so, we need to redistribute and add new poles or zeros in the open-loop transfer function $G(s)H(s)$ of the system. For this purpose, we can introduce controllers to the system, as follows:

1. By connecting a controller **in series** with the control units of the system
2. By connecting a controller as a **feedback loop** in the system
3. By connecting a controller **in parallel** to one or more control units of the system

The following controllers add poles and zeros to the loop transfer function in the s-plane.

6.3.1 Phase-Lead Controller

A phase-lead circuit is depicted in the below figure.

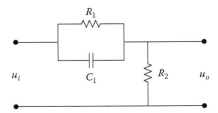

The transfer function of the circuit is

$$G(s) = \frac{V_o(s)}{V_i(s)} = \frac{R_2}{R_1 + R_2} \cdot \frac{sR_1C_1 + 1}{(R_1R_2C_1/(R_1 + R_2)) + 1} \tag{6.18}$$

It can be written as

$$G(s) = K\frac{(sT_1 + 1)}{(sT_2 + 1)} = \frac{1 + aTs}{a(1 + Ts)} \tag{6.19}$$

where

$$\left.\begin{array}{l} K = \dfrac{R_2}{R_1 + R_2}, \quad a = \dfrac{R_1 + R_2}{R_2} \\[2ex] T_1 = R_1C_1 \\[2ex] T_2 = \dfrac{R_1R_2C_1}{R_1 + R_2} = KT_1 = T \\[2ex] T_1 > T_2 \end{array}\right\} \tag{6.20}$$

The phase-lead controller is usually used to provide a sufficient phase margin for a system. With the use of phase-lead controller, we achieve

- Reduction of the rise time T_s
- Stability
- Increase of the critical gain K_c and of the critical frequency of oscillation ω_c

6.3.2 Phase-Lag Controller

A phase-lag circuit is shown in the following figure.

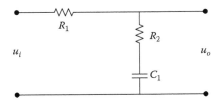

The transfer function is

$$G(s) = \frac{V_o(s)}{V_i(s)} = \frac{sR_2C_1 + 1}{s(R_1 + R_2)C_1 + 1} \qquad (6.21)$$

It can be written as

$$G(s) = \frac{(sT_1 + 1)}{(sT_2 + 1)} = \frac{1}{a}\frac{s + z}{s + p} \qquad (6.22)$$

where

$$\left.\begin{aligned}
&T_1 = R_2C_1 \\
&T_2 = (R_1 + R_2)C_1 \\
&T_1 < T_2 \\
&a = \frac{R_1 + R_2}{R_2}, \quad z = 1/T_1, \quad p = 1/T_2
\end{aligned}\right\} \qquad (6.23)$$

With the use of the phase-lag controller, we can reform the RL and determine the desired root locus in order to increase, for instance, its relative stability.

The effects of phase-lag compensation result in

- An increase of the rise time T_s
- An increase of the total static gain of the system
- The reduction of the steady state error e_{ss}

6.3.3 Lead-Lag Controller

The circuit of a lead-lag controller is shown in the following figure.

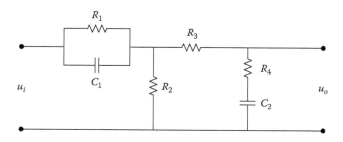

The transfer function of the system is

$$G(s) = \frac{V_o(s)}{V_i(s)}$$

$$= \frac{R_2(sR_4C_2+1)(sR_1C_1+1)}{R_1R_2(R_3+R_4)C_1C_2\left[s+\dfrac{1}{(R_3+R_4)C_2}\right]\left[s+\dfrac{R_1R_2+R_1R_3+R_1R_4+R_2R_3+R_2R_4}{R_1R_2(R_3+R_4)C_1}\right]} \qquad (6.24)$$

or

$$G(s) = K\frac{(sT_1+1)(sT_2+1)}{(sT_3+1)(sT_4+1)} \qquad (6.25)$$

where

$$\left.\begin{aligned}
K &= \frac{R_2(R_3+R_4)}{R_1R_2+R_1R_3+R_1R_4+R_2R_3+R_2R_4} \\[6pt]
T_1 &= R_1C_1 \\[6pt]
T_2 &= R_4C_2 \\[6pt]
T_3 &= (R_3+R_4)C_2 \\[6pt]
T_4 &= \frac{R_1R_2(R_3+R_4)C_1}{R_1R_2+R_1R_3+R_1R_4+R_2R_3+R_2R_4}
\end{aligned}\right\} \qquad (6.26)$$

Lead-lag controllers combine the advantages of the two controllers, but one has to be careful in the design in order to exploit the properties of the controller for different parts of the time response.

We describe here the system compensation process with the use of root loci and controller configurations:

1. The system requirements are associated to the desired dominant roots.
2. The root locus of the system is drawn and we examine if the needed roots belong to the root locus.
3. We choose the most suitable controller and determine its transfer function.
4. The new pole is determined so that the angle condition is satisfied. This means that the angle of the location of the desired root must be 180° and thus the root belongs to the new root locus of the system that includes the controller.
5. The total gain of the system is computed for the desired root. We can also calculate the error constant of the compensated system.
6. If the error is not acceptable, then we repeat the design process.

Formulas

TABLE F6.1

Steps for Designing the Root-Locus Diagram

S/N	Formulas for Designing a Root-Locus Diagram	Remarks
1	$G(s)H(s) = K \dfrac{(s+z_1)(s+z_2)\cdots(s+z_m)}{(s+p_1)(s+p_2)\cdots(s+p_n)}$	Open-loop transfer function
2	$\left\lvert K \right\rvert \dfrac{\prod_{i=1}^{m}\lvert s+z_i \rvert}{\prod_{j=1}^{n}\lvert s+p_j \rvert} = 1 \quad -\infty < K < \infty$	Magnitude condition for the points of the root locus
3	$\sum_{i=1}^{m} \sphericalangle(s+z_i) - \sum_{j=1}^{n} \sphericalangle(s+p_j) = \begin{cases} (2\rho+1)\pi, & K>0 \\ 2\rho\pi, & K<0 \end{cases}$ $\rho = \pm 1, \pm 2, \ldots$	Phase condition for the points of the root locus
4	$l = \max(m,n)$	Number of branches of the root locus
5(a)	$\sphericalangle \varphi_\alpha = \dfrac{(2\rho+1)\pi}{n-m}, \quad \begin{cases} \rho = 0,1,\ldots,\lvert n-m \rvert -1 \\ K \geq 0 \end{cases}$	Angles of asymptotes with the real axis for $K \geq 0$
5(b)	$\sphericalangle \varphi_\alpha = \dfrac{2\rho\pi}{n-m}, \quad \begin{cases} \rho = 0,1,\ldots,\lvert n-m \rvert -1 \\ K \leq 0 \end{cases}$	Angles of asymptotes with the imaginary axis for $K \leq 0$
6	$\sigma_\alpha = \dfrac{\sum_{i=1}^{n} p_i - \sum_{j=1}^{m} z_j}{n-m}$	Intersection of asymptotes with the real axis
7(a)	$\begin{cases} \dfrac{dK}{ds} = 0 \Rightarrow s_{b_i} \\ 1 + G(s_b)H(s_b) = 0 \quad \text{for } K \in R \end{cases}$	Computation of the breakaway points s_b (first way)
7(b)	$\sum_{i=1}^{n} \dfrac{1}{s-p_i} = \sum_{j=1}^{m} \dfrac{1}{s-z_j} p_i, \quad z_j \in R$	Computation of the breakaway points s_b (second way)
8	$\sphericalangle \varphi_d = (2\rho+1)\pi - \left(\sum_{i=1}^{n} \varphi_{p_i} - \sum_{j=1}^{m} \varphi_{z_j} \right)$	Angles of departure of the RL from complex poles or angles of arrival to complex zeros
9	$s_c = \pm j\omega_c$	Intersection points of the RL with the imaginary axis

TABLE F6.2

Compensation Circuits

S/N	Controller	Transfer Function
1	Phase lead	$\dfrac{V_o(s)}{V_i(s)} = \dfrac{R_2}{R_1 + R_2} \cdot \dfrac{sR_1C_1 + 1}{(R_1R_2C_1/(R_1 + R_2)) + 1}$

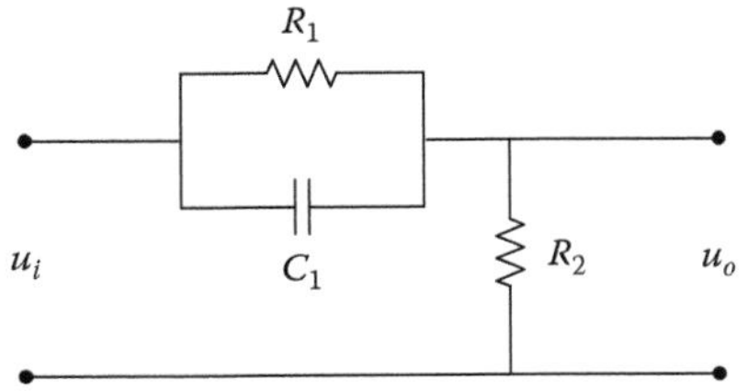

or

$$\dfrac{V_o(s)}{V_i(s)} = K\dfrac{(sT_1 + 1)}{(sT_2 + 1)} \quad T_1 > T_2$$

| 2 | Phase lag | $\dfrac{V_o(s)}{V_i(s)} = \dfrac{sR_2C_1 + 1}{s(R_1 + R_2)C_1 + 1}$ |

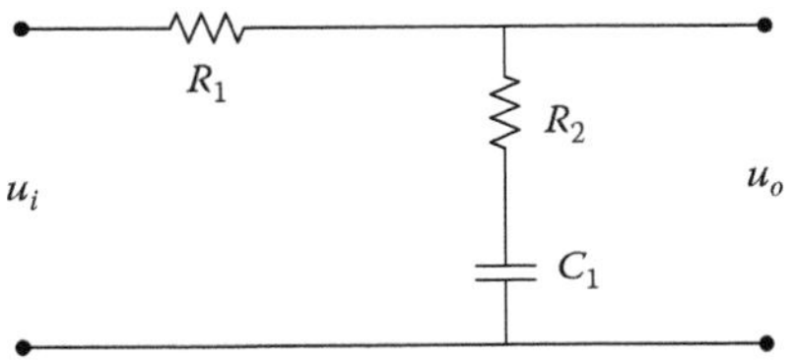

or

$$\dfrac{V_o(s)}{V_i(s)} = \dfrac{(sT_1 + 1)}{(sT_2 + 1)} \quad T_1 < T_2$$

| 3 | Lead-lag | $\dfrac{V_o(s)}{V_i(s)} = K\dfrac{(sT_1 + 1)(sT_2 + 1)}{(sT_3 + 1)(sT_4 + 1)}$ |

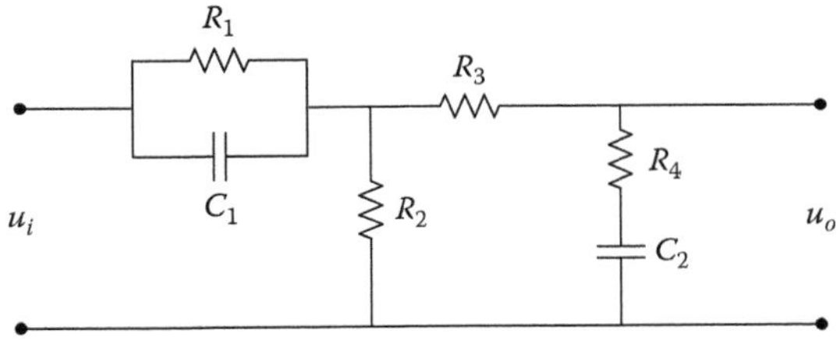

$$K = \dfrac{R_2(R_3 + R_4)}{R_1R_2 + R_1R_3 + R_1R_4 + R_2R_3 + R_2R_4}$$

$T_1 = R_1C_1$

$T_2 = R_4C_2$

$T_3 = (R_3 + R_4)C_2$

$T_4 = KT_1$

$T_1 > T_4$ (lead)

$T_2 < T_3$ (lag)

Problems

6.1 The following figure depicts with a bold line a segment of a root locus of the characteristic equation of a system with open-loop transfer function

$$GH(s) = \dfrac{K\left(s + (10/3)\right)}{s(s + 3)(s + 6)}, \quad K > 0$$

   a. Plot the rest of the straight-line segments of the locus, which are on the real axis.

   b. Mark the direction of the locus for every segment.

   c. Find the abscissa of point A.

d. What is the value of K at the point A?

e. Find the asymptotes of the locus.

f. Discuss the stability of the system.

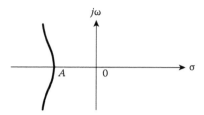

Solution

The open-loop transfer function is

$$GH(s) = \frac{K(s+(10/3))}{s(s+3)(s+6)} \tag{P6.1.1}$$

1. The poles are $p_1 = 0$, $p_2 = -3$, $p_3 = -6$ $(n = 3)$.
2. The zeros are $z_1 = -10/3$ $(m = 1)$.
3. The number of separate branches of the locus is $\max(3, 1) = 3$.
4. The intersection of the asymptotes is

$$\sigma_\alpha = \frac{\sum_{i=1}^{3} p_i - \sum_{j=1}^{1} z_j}{n-m} = \frac{-9-(-10/3)}{3-1} = -\frac{17}{6} = -2.83 \tag{P6.1.2}$$

Thus, the abscissa of point A is –2.83.

5. The angle of the asymptotes is

$$\left. \begin{aligned} \sphericalangle\varphi_\alpha = \frac{(2\rho+1)\pi}{n-m} \\ \rho = 0,1 \end{aligned} \right\} \Rightarrow \left. \begin{aligned} \hat{\varphi}_{\alpha_1} = 90° \quad (\rho = 0) \\ \hat{\varphi}_{\alpha_2} = 270° \quad (\rho = 1) \end{aligned} \right\} \tag{P6.1.3}$$

6. The segments of the real axis, which can be segments of the RL for $K > 0$, are the segment from 0 to –3 and the segment from –10/3 to –6.

7. The breakaway points of the RL are the roots of Equation P6.1.5. The characteristic equation of the system is

$$1 + \frac{K(s+(10/3))}{s(s+3)(s+6)} = 0 \Rightarrow s(s+3)(s+6) + K(s+(10/3)) = 0 \Rightarrow$$

$$K = -\frac{s(s+3)(s+6)}{s+(10/3)} = -\frac{s^3+9s^2+185}{s+(10/3)} \tag{P6.1.4}$$

and

$$\frac{dK}{ds} = 0 \overset{(P6.1.4)}{\Rightarrow} 2s^3 + 19s^2 + 60s + 60 = 0 \tag{P6.1.5}$$

The roots of Equation P6.1.5 are $s_1 = -2$, $s_{2,3} = -3.75 \pm j0.9682$. Apparently the root $s_1 = -2$ is also a breakaway point of the RL, since from relationship (P6.1.4) it follows that $K = 6 \in R$. Hence,

$$s_b = -2 \tag{P6.1.6}$$

8. We continue with Routh's tabulation in order to find the intersections of RL with the imaginary axis.

$$\text{C.E.:} \quad s^3 + 9s^2 + s(K + 18) + \frac{10}{3}K = 0 \tag{P6.1.7}$$

Routh's tabulation is

$$
\begin{array}{c|cc}
s^3 & 1 & K+18 \\
s^2 & 9 & \dfrac{10}{3}K \\
s^1 & b & \\
s^0 & \dfrac{10}{3}K &
\end{array}
$$

where

$$b = \frac{9(K+18) - (10/3)K}{9} = 8.63K + 18 \tag{P6.1.8}$$

In order to find the critical value of K that ensures the stability of the system from row s^1, we have

$$b = 0 \overset{(8)}{\Rightarrow} K = -\frac{18}{8.63} = -2.08 < 0$$

The critical value of K is negative; thus, there are no intersections of the RL with the imaginary axis.

9. We now plot the RL for $K > 0$.

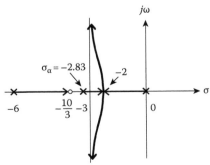

6.2 For the system shown in the following figure, design the RL diagram of the characteristic equation for $K > 0$, and find out the values of K for which the system is stable.

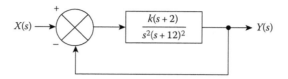

Solution

The open-loop transfer function is

$$GH(s) = \frac{K(s+2)}{s^2(s+12)^2} \qquad (P6.2.1)$$

1. The poles are $p_1 = p_2 = 0$, $p_3 = p_4 = -2$ ($n = 4$).
2. The zeros are $z_1 = -2$ ($m = 1$).
3. The number of separate branches of RL is $\max(4, 1) = 4$.
4. The intersection of the asymptotes is

$$\sigma_\alpha = \frac{\sum_{i=1}^{4} p_i - \sum_{j=1}^{1} z_j}{n - m} = \frac{-24 - (-2)}{4 - 1} = -\frac{22}{3} = -7.33 \qquad (P6.2.2)$$

5. The angles of the asymptotes are

$$\left. \begin{aligned} \sphericalangle \varphi_\alpha &= \frac{(2\rho + 1)\pi}{n - m} \\ \rho &= 0, 1, 2 \end{aligned} \right\} \Rightarrow \begin{aligned} \hat{\varphi}_{\alpha_1} &= \frac{180°}{3} = 60° & \text{for } (\rho = 0) \\ \hat{\varphi}_{\alpha_2} &= \frac{3 \cdot 180°}{3} = 180° & \text{for } (\rho = 1) \\ \hat{\varphi}_{\alpha_3} &= \frac{5 \cdot 180°}{3} = 300° & \text{for } (\rho = 2) \end{aligned} \qquad (P6.2.3)$$

6. The segments of the real axis, which can also be segments of RL for $K > 0$, are
   between $(-2, -12]$ and $[-12, -\infty)$.
7. The breakaway points of the RL are roots of Equation P6.2.6. The characteristic
   equation of the system is

$$1 + \frac{K(s+2)}{s^2(s+12)^2} = 0 \Rightarrow s^2(s+12)^2 + K(s+2) = 0 \Rightarrow \qquad (P6.2.4)$$

$$s^4 + 24s^3 + 144s^2 + Ks + 2K = 0$$

Solving for K, we get

$$K = -\frac{s^2(s+12)^2}{s+2} \tag{P6.2.5}$$

$$\frac{dK}{ds} = 0 \overset{(P6.2.5)}{\Rightarrow} 3s^4 + 56s^3 + 288s^2 - 576s = 0 \tag{P6.2.6}$$

An apparent root of Equation P6.2.6 is $s = 0$, which is acceptable as a departure point of the branches from the real axis; as for $s = 0$, from the relationship (P6.2.5), we get $K = 0$.

8. The intersections of RL with the imaginary axis can be found as follows:
We proceed with Routh's tabulation for the characteristic equation of relationship (P6.2.7):

$$s^4 + 24s^3 + 144s^2 + Ks + 2K = 0 \tag{P6.2.7}$$

Routh's tabulation is

s^4	1	144	$2K$
s^3	24	K	
s^2	$\dfrac{3456-K}{24}$	$2K$	
s^1	$\dfrac{K(3456-K)-1152K}{3456-K}$		
s^0	$2K$		

From row s^1, we get

$$\frac{K(3456-K)-1152K}{3456-K} = 0 \Rightarrow 2304K - K^2 = 0 \Rightarrow K(2304-K) = 0 \tag{P6.2.8}$$

$$\Rightarrow K_c = 2304$$

From the auxiliary equation of row s^2, we have

$$\frac{3456-K_c}{24}s^2 + 2K_c \Rightarrow \omega_c = 9.8\,\text{rad/s} \tag{P6.2.9}$$

9. By using the previous findings, we now plot the RL of the system for $K > 0$.

We observe that for $0 < K < 2304$ the system is stable.

6.3 The loop transfer function of a system is

$$G(s)H(s) = \frac{K}{s(s^2 + 4s + 8)}, \quad K > 0$$

a. Find the asymptotes and the angles of departure of the RL.
b. Compute the critical value of K so that the closed-loop system is stable, and find the intersections of RL with the imaginary axis.
c. Plot the RL of the characteristic equation of the system.

Solution

a. The loop transfer function is

$$G(s)H(s) = \frac{K}{s(s^2 + 4s + 8)} \tag{P6.3.1}$$

The intersection of the asymptotes is

$$\sigma_\alpha = \frac{\sum_{i=1}^{3} p_i - \sum_{j=0}^{} z_j}{n - m} = \frac{(-2 - 2j) + (-2 + 2j) + 0}{3} = -\frac{4}{3} \tag{P6.3.2}$$

The angles of the asymptotes for $K > 0$ are

$$\left. \begin{array}{l} \sphericalangle\varphi_\alpha = \dfrac{(2\rho + 1)\pi}{n - m} \\[2mm] \rho = 0, 1, 2 \end{array} \right\} \Rightarrow \left. \begin{array}{l} \hat{\varphi}_{\alpha_1} = \dfrac{180°}{3} = 60° \quad \text{for } (\rho = 0) \\[2mm] \hat{\varphi}_{\alpha_2} = \dfrac{3 \cdot 180°}{3} = 180° \quad \text{for } (\rho = 1) \\[2mm] \hat{\varphi}_{\alpha_3} = \dfrac{5 \cdot 180°}{3} = 300° \quad \text{for } (\rho = 2) \end{array} \right\} \tag{P6.3.3}$$

The angle of departure from the complex pole $(-2 + 2j)$ is given by:

$$\sphericalangle\varphi_d = 180° + \sphericalangle GH'(s) \tag{P6.3.4}$$

where

$$\sphericalangle\left(GH'(s)\right) = \sphericalangle\left(\left.\frac{K}{s(s+2+2j)}\right|_{s=-2+2j}\right) = \sphericalangle\left(\frac{K}{4j(-2+2j)}\right) \Rightarrow$$

$$\sphericalangle\left(GH'(s)\right) = 0° - 90° - \tan^{-1}\left(\frac{2}{-2}\right) = 0° - 90° - 135° = -225° \tag{P6.3.5}$$

Thus,

$$\sphericalangle\varphi_d = 180° - 225° = -45° \tag{P6.3.6}$$

Another way of computing the angle of departure φ_d is shown as follows:

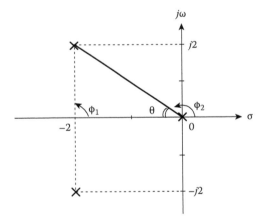

From relationship (6.17), we have

$$\sphericalangle\varphi_d = (2\rho+1)\pi - \left(\sum_{i=1}^{2}\varphi_{p_i} - \sum_{j=0}\varphi_{z_j}\right) = 180° - (\hat{\varphi}_1 + \hat{\varphi}_2) \tag{P6.3.7}$$

But

$$\sphericalangle\varphi_1 = 90° \tag{P6.3.8}$$

$$\sphericalangle\varphi_2 = 180° - \hat{\theta} = 180° - \tan^{-1}\frac{2}{2} = 180° - 45° = 135° \tag{P6.3.9}$$

$$\overset{(P6.3.8)}{\underset{(P6.3.9)}{(P6.3.7) \Rightarrow}} \sphericalangle\varphi_d = 180° - (90° + 135°) = 180° - 225° = -45° \tag{P6.3.10}$$

As the RL is symmetrical with respect to the real axis, the angle of departure from the complex pole $(-2 - 2j)$ is $+45°$. The angle of departure from the pole to zero is $180°$.

b. The characteristic equation is

$$1 + G(s)H(s) = 0 \overset{\text{(P6.3.1)}}{\Rightarrow} s(s^2 + 4s + 8) + K = 0 \Rightarrow$$

$$s^3 + 4s^2 + 8s + K = 0 \tag{P6.3.11}$$

Routh's tabulation is

$$
\begin{array}{c|cc}
s^3 & 1 & 8 \\
s^2 & 4 & K \\
s^1 & 8 - \dfrac{K}{4} & \\
s^0 & K &
\end{array}
$$

The closed-loop system is stable if

$$\left.\begin{array}{c} 8 - \dfrac{K}{4} > 0 \\[2mm] K > 0 \end{array}\right\} \Rightarrow 0 < K < 32 \tag{P6.3.12}$$

Given that $K > 0$, for $K = K_c = 32$, the RL intersects the imaginary axis. We find the intersection $\pm j\omega_c$ from the auxiliary equation of row s^2:

$$4s^2 + K_c = 0 \Rightarrow s^2 = -\frac{32}{4} \Rightarrow s = \pm j2\sqrt{2} \Rightarrow$$

$$\omega_c = 2\sqrt{2}\ \text{rad/s} \tag{P6.3.13}$$

c. Based on the previous computations, we plot the root-locus diagram of the system.

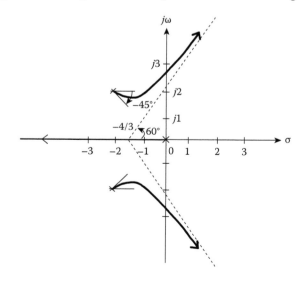

6.4 Given the following loop transfer function

$$GH(s) = \frac{K}{s(s+1)(s+3)(s+4)}$$

sketch the RL of the characteristic equation for $K > 0$ and for $K < 0$.

Solution

The open-loop transfer function is

$$GH(s) = \frac{K}{s(s+1)(s+3)(s+4)} \tag{P6.4.1}$$

1. The poles are $p_1 = 0$, $p_2 = -1$, $p_3 = -3$, $p_4 = -4$ ($n = 4$).
2. There are no zeros.
3. There are $4 = \max(4,0)$ separate loci.
4. The intersection of the asymptotes is

$$\sigma_\alpha = \frac{\sum_{i=1}^{4} p_i}{n-m} = \frac{0+(-1)+(-3)+(-4)}{3} = -\frac{8}{4} = -2 \tag{P6.4.2}$$

5. The angles of the asymptotes are

$$\sphericalangle\varphi_\alpha = \begin{cases} \dfrac{(2\rho+1)\pi}{n-m} & \text{for } K > 0 \tag{P6.4.3} \\[2mm] \dfrac{(2\rho)\pi}{n-m} & \text{for } K < 0 \tag{P6.4.4} \end{cases}$$

$$\rho = 0,1,2,\dots,n-m-1 \Rightarrow \rho = 0,1,2,3$$

Hence, for $K > 0$, we have

$$\left.\begin{aligned} \hat{\varphi}_{a_1} &= \frac{180°}{4} = 45° \\[2mm] \hat{\varphi}_{a_2} &= \frac{3\cdot 180°}{4} = 135° \\[2mm] \hat{\varphi}_{a_3} &= \frac{5\cdot 180°}{4} = 225° \\[2mm] \hat{\varphi}_{a_4} &= \frac{7\cdot 180°}{4} = 315° \end{aligned}\right\} \tag{P6.4.5}$$

While for $K < 0$, we have

$$
\left.\begin{aligned}
\hat{\phi}_{a_1} &= \frac{0 \cdot 180°}{4} = 0° \\[2mm]
\hat{\phi}_{a_2} &= \frac{2 \cdot 180°}{4} = 90° \\[2mm]
\hat{\phi}_{a_3} &= \frac{4 \cdot 180°}{4} = 180° \\[2mm]
\hat{\phi}_{a_4} &= \frac{6 \cdot 180°}{4} = 270°
\end{aligned}\right\}
\tag{P6.4.6}
$$

6. The segments of the real axis that belong to the RL for $K > 0$ are those between 0 and −1 and between −3 and −4.

   For $K < 0$, the RL includes the segment that begins from 0 and tends to $+\infty$, the segment between −1 to −3, and the segment from −4 that tends to $-\infty$.

7. In order to find the breakaway points, we have

$$
\text{C.E.:} \quad 1 + GH(s) = 0 \overset{\text{(P6.4.1)}}{\Rightarrow} K = -s(s+1)(s+3)(s+4) \tag{P6.4.7}
$$

$$
\frac{dK}{ds} = 0 \overset{\text{(P6.4.7)}}{\Rightarrow} 2s^3 + 12s^2 + 19s + 6 = 0 \tag{P6.4.8}
$$

By solving (8), we get the roots

$$
s_1 = -2, \quad s_2 \simeq -0.42, \quad \text{and} \quad s_3 \simeq -3.58.
$$

The root $s_{b_1} = -2$ is the breakaway point for $K < 0$, and the roots $s_{b_2} = -0.42$ and $s_{b_3} = -3.58$ are the breakaway points for $K > 0$.

8. In order to find the intersections of RL with the imaginary axis, we proceed with Routh's tabulation.

   The characteristic equation is

$$
s^4 + 8s^3 + 19s^2 + 12s + K = 0 \tag{P6.4.9}
$$

Routh's tabulation is

$$
\begin{array}{c|ccc}
s^4 & 1 & 19 & K \\
s^3 & 8 & 12 & \\
s^2 & 17.5 & K & \\
s^1 & \dfrac{210 - 8K}{17.5} & & \\
s^0 & K & &
\end{array}
$$

From row s^1, it follows that

$$\frac{210-8K}{17.5} = 0 \Rightarrow K_c = \frac{210}{8} = 26.25 \qquad \text{(P6.4.10)}$$

The intersections are determined with the use of the auxiliary equation of row s^2:

$$17.5s^2 + K_c = 0 \Rightarrow 17.5s^2 + 26.25 = 0 \Rightarrow$$
$$s = \pm 1.2j \qquad \text{(P6.4.11)}$$

9. Based on the previous, we now plot the RL diagram.
   For $K > 0$, the RL is:

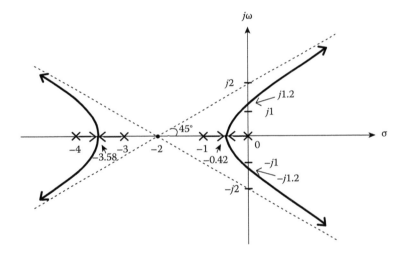

For $K < 0$, the RL is shown in:

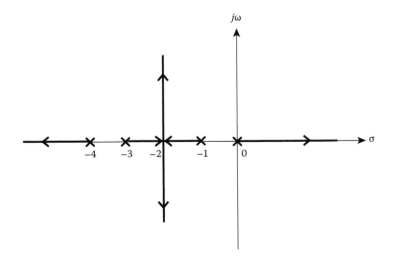

6.5 The loop transfer function of a system is

$$GH(s) = \frac{K(s+1)}{s(s-1)(s^2+4s+16)}, \quad K > 0$$

Plot the RL of the characteristic equation of the system.

Solution

The open-loop transfer function is

$$G(s)H(s) = \frac{K(s+1)}{s(s-1)(s^2+4s+10)} \tag{P6.5.1}$$

1. The poles are $p_1 = 0$, $p_2 = 1$, $p_3 = -2 + j3.465$, $p_4 = -2 - j3.465$ and $(n = 4)$.
2. The zeros are $z_1 = -1$ $(m = 1)$.
3. There are $4 = \max(4, 1)$ separate loci.
4. The intersection of the asymptotes is

$$\sigma_\alpha = \frac{1+(-2-j3.465)+(-2+j3.465)-(-1)}{3} = -\frac{2}{3} \tag{P6.5.2}$$

5. The angles of the asymptotes are

$$\left.\begin{aligned}\sphericalangle\varphi_\alpha &= \frac{(2\rho+1)\pi}{n-m} \\[4pt] \rho &= 0,1,2\end{aligned}\right\} \Rightarrow \begin{aligned} \hat{\varphi}_{\alpha_1} &= \frac{180°}{3} = 60° & (\rho=0) \\[4pt] \hat{\varphi}_{\alpha_2} &= \frac{3\cdot180°}{3} = 180° & (\rho=1) \\[4pt] \hat{\varphi}_{\alpha_3} &= \frac{5\cdot180°}{3} = 300° & (\rho=2) \end{aligned}\right\} \tag{P6.5.3}$$

6. In order to find the angles of departure from the complex poles, we proceed as follows: For the complex pole $(-2 + j3.465)$, we have

$$\sphericalangle\varphi_d = 180° - (\Sigma\varphi_p - \Sigma\varphi_z) =$$

$$180° - \left(\tan^{-1}\frac{3.465}{2} + 180° - \tan^{-1}\frac{3.465}{3} + 90° - 180° + \tan^{-1}\frac{3.465}{1}\right) \Rightarrow$$

$$\varphi_d = 180° - (180° - 60° + 180° - 49.1° + 90° - 180° + 73.9°) \simeq -54.8°$$

Hence,

$$\hat{\varphi}_{d(-2+j3.465)} = -54.8° \tag{P6.5.4}$$

The angle of departure from the complex pole $(-2 - j3.465)$, due to symmetry, is

$$\hat{\varphi}_{d(-2-j3.465)} = 54.8° \tag{P6.5.5}$$

A second way of finding the angle $\hat{\varphi}_d$ is

$$\hat{\varphi}_{d(-2+j3.465)} = \sphericalangle(s_1 + 2 - j3.465)$$

$$= -(2\rho + 1)\pi + \sphericalangle(s_1 + 1) - \sphericalangle(s_1) - \sphericalangle(s_1 - 1) - \sphericalangle(s_1 + 2 + j3.465) \Rightarrow$$

$$\hat{\varphi}_{d(-2+j3.465)} = -\pi + \sphericalangle(-2 + j3.465 + 1) - \sphericalangle(-2 + j3.465) - \sphericalangle(-3 + j3.5)$$

$$-\sphericalangle(-2 + j3.465 + 2 + j3.465) \simeq -54.8°$$

7. In order to find the breakaway points, we have

$$\text{C.E.:} \quad 1 + G(s)H(s) = 0 \overset{\text{(P6.5.1)}}{\Rightarrow} K = -\frac{s(s-1)(s^2 + 4s + 10)}{(s+1)} \tag{P6.5.6}$$

$$\frac{dK}{ds} = 0 \overset{\text{(P6.5.6)}}{\Rightarrow} 3s^4 + 10s^3 + 21s^2 + 24s - 16 = 0 \tag{P6.5.7}$$

By solving (P6.5.7), we get the roots

$$\left.\begin{aligned} s_{1,2} &= -0.7595 \pm j2.1637 \\ s_3 &= -2.2627 \\ s_4 &= 0.4483 \end{aligned}\right\} \tag{P6.5.8}$$

The complex roots $s_{1,2}$ are rejected. The roots s_3 and s_4 are accepted as breakaway points, because they give real values of K, that is, 35.48 and 2.048, respectively.

8. In order to find the intersections of RL with the imaginary axis, we proceed with Routh's tabulation.

The characteristic equation is

$$s^4 + 3s^3 + 12s^2 + (K-16)s + K = 0 \tag{P6.5.9}$$

Routh's tabulation is

s^4	1	12	K
s^3	3	$K-16$	
s^2	$\dfrac{52-K}{3}$	K	
s^1	b		
s^0	K		

where

$$b = \frac{((52-K)/3)(K-16)-3K}{(52-K)/3} \qquad \text{(P6.5.10)}$$

We now compute the critical values of K for stability.
From row s^1 it suffices that $b = 0$.

$$(\text{P6.5.10}) \Rightarrow K^2 - 59K + 832 = 0 \Rightarrow \begin{cases} \to K_1 = 35.7 \\ \to K_2 = 23.3 \end{cases}$$

Thus, we get two values for critical stability:

$$K_{c_1} = 35.7 \quad \text{and} \quad K_{c_2} = 23.3 \qquad \text{(P6.5.11)}$$

The intersections of RL with the imaginary axis are found with the use of the auxiliary equation of row s^2:

$$\frac{52 - K_{cr}}{3} s^2 + K_{cr} = 0 \qquad \text{(P6.5.12)}$$

From (P6.5.12) for $K_{cr_1} = 35.7$, we get $j\omega_{cr_1} = j2.56$, and for $K_{cr_2} = 23.3$, we have $j\omega_{cr_2} = j1.56$.

9. Based on the previous queries, we plot the RL of the characteristic equation of the system.

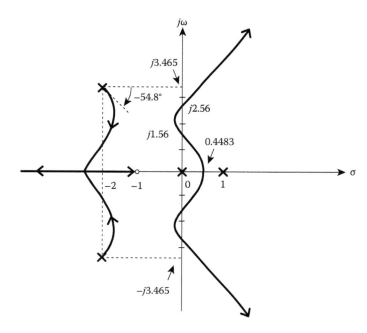

6.6 The loop transfer function of a system is given by

$$G(s)H(s) = \frac{K}{s(s+1)(s^2+4s+8)}, \quad K > 0$$

a. Find the asymptotes and the angles of departure of the RL.
b. Find the breakaway points (if there are any) of the RL. Take into account that one root of the equation $4s^3 + 15s^2 + 24s + 8 = 0$ is $s = -1.6549 + j1.3432$.
c. Find the critical value of K so that the system is stable.
d. Plot the RL.

Solution

The open-loop transfer function is

$$GH(s) = \frac{K}{s(s+1)(s^2+4s+8)}, \quad K > 0 \tag{P6.6.1}$$

a. The intersection of the asymptotes is

$$\sigma_\alpha = \frac{\sum_{i=1}^{n} p_i - \sum_{j=1}^{m} z_j}{n-m} = \frac{0 + (-1) + (-2+j2) + (-2-j2)}{4-0} = -\frac{5}{4} \tag{P6.6.2}$$

The angles of the asymptotes for $K > 0$ are

$$\left.\begin{array}{l} \sphericalangle\varphi_\alpha = \dfrac{(2\rho+1)\pi}{n-m} \\[2mm] \rho = 0,1,2,\ldots,n-m-1 \end{array}\right\} \Rightarrow \left.\begin{array}{ll} \hat{\varphi}_{\alpha_1} = \dfrac{180°}{4} = 45° & (\rho = 0) \\[2mm] \hat{\varphi}_{\alpha_2} = \dfrac{3\cdot180°}{4} = 135° & (\rho = 1) \\[2mm] \hat{\varphi}_{\alpha_3} = \dfrac{5\cdot180°}{4} = 225° & (\rho = 2) \\[2mm] \hat{\varphi}_{\alpha_4} = \dfrac{7\cdot180°}{4} = 315° & (\rho = 3) \end{array}\right] \tag{P6.6.3}$$

The angle of departure from the complex pole $(-2 + j2)$ is

$$\sphericalangle\varphi_d = 180° + \sphericalangle GH'(s) \tag{P6.6.4}$$

where

$$GH'(s) = \sphericalangle\frac{K}{s(s+1)(s+2+j2)} = 0° - 90° - \tan^{-1}\left(\frac{2}{-2}\right) - \tan^{-1}\left(\frac{2}{-1}\right)$$

$$= 0° - 90° - 135° - 116.37° \simeq -341.57°$$

Hence,

$$\sphericalangle\varphi_d \overset{(P6.6.4)}{=} 180° - 341.57° = -161.57° \tag{P6.6.5}$$

The angle of departure from the complex conjugate pole $(-2 - j2)$ is, due to symmetry of RL, 161.57°. The angle of departure from the pole to 0 is 180° and from the pole to −1 is 0°.

b. The breakaway points satisfy the following relationship:

$$\sum_{i=1}^{n}\frac{1}{(s_b + p_i)} = \sum_{j=1}^{m}\frac{1}{(s_b + z_j)} \tag{P6.6.6}$$

$$(P6.6.6) \Rightarrow \frac{1}{s_b} + \frac{1}{s_b + 1} + \frac{1}{s_b + 2 - j2} + \frac{1}{s_b + 2 + j2} = 0 \Rightarrow$$

$$4s_b^3 + 15s_b^2 + 24s_b + 8 = 0 \tag{P6.6.7}$$

The roots of Equation P6.6.7 are $s_1 = -0.4402$ and $s_{2,3} = -1.6549 \pm j1.3432$. The breakaway point is the root $s_1 = s_b = -0.4402$ for which we get a real value of K.

c. The characteristic equation is

$$s^4 + 5s^3 + 12s^2 + 8s + K = 0 \tag{P6.6.8}$$

Routh's tabulation is

$$
\begin{array}{c|ccc}
s^4 & 1 & 12 & K \\
s^3 & 5 & 8 & \\
s^2 & 10.4 & K & \\
s^1 & b & & \\
s^0 & K & &
\end{array}
$$

where

$$b = \frac{10.4 \cdot 8 - 5K}{10.4} \tag{P6.6.9}$$

The closed-loop system is stable for $b > 0$ and $K > 0$:

$$b > 0 \Rightarrow 8 - \frac{5K}{10.4} > 0 \Rightarrow K < 16.64 \tag{P6.6.10}$$

Thus,

$$0 < K < 16.64 \qquad\qquad (P6.6.11)$$

For $K_c = 16.64$, the roots are on the imaginary axis.

By substituting the intersections with the imaginary axis in the auxiliary equation of row s^2, we get

$$10.4s^2 + K_c = 0 \Rightarrow \pm j\omega_c = \pm j1.2649 \qquad\qquad (P6.6.12)$$

d. The number of separate loci is $\max(n, m) = \max(4, 0) = 4$. These depart from the poles of the open-loop transfer function, $p_1 = 0$, $p_2 = -1$, and $p_{3,4} = -2 \pm j2$. The number of branches that approach infinity is $n - m = 4$, as $K \to \infty$.

Based on the previous queries we now plot the RL of the characteristic equation of the system:

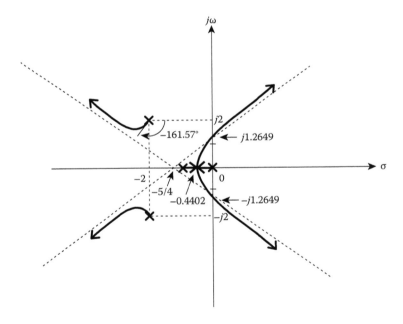

6.7 Sketch the RL for the systems with the following loop transfer functions:

a. $G(s)H(s) = \dfrac{K(1+10s)}{s^2(1+5s)(s+1)}, \quad K > 0$

b. $G(s)H(s) = \dfrac{K(s+10)}{s(s+6)(s+40)^2}, \quad K > 0$

c. $G(s)H(s) = \dfrac{K(s+6)}{s(s+2)(s^2+20s+200)}, \quad K > 0$

Solution

a. The loop transfer function of the system is

$$G(s)H(s) = \frac{K(1+10s)}{s^2(1+5s)(s+1)} \tag{P6.7.1}$$

$$(P6.7.1) \Rightarrow G(s)H(s) = \frac{2K(s+0.1)}{s^2(s+0.2)(s+1)} \tag{P6.7.2}$$

1. The poles are $p_1 = p_2 = 0$, $p_3 = -0.2$, $p_4 = -1$ ($n = 4$).
2. The zeros are $z = -0.1$ ($m = 1$).
3. The number of separate loci is $4 = \max(n, m) = \max(4, 1)$.
4. The intersection of the asymptotes is

$$\sigma_\alpha = \frac{\sum_{i=1}^{4} p_i - \sum_{j=1}^{1} z_j}{n-m} = \frac{-1.2-(-0.1)}{4-1} = -0.366 \tag{P6.7.3}$$

5. The angles of the asymptotes are

$$\left. \begin{array}{l} \sphericalangle\varphi_\alpha = \dfrac{(2\rho+1)\pi}{n-m} \\[2mm] \rho = 0,1,2 \end{array} \right\} \Rightarrow \begin{array}{l} \hat\varphi_{\alpha_1} = \dfrac{180°}{3} = 60° \quad (\rho = 0) \\[2mm] \hat\varphi_{\alpha_2} = \dfrac{3\cdot180°}{3} = 180° \quad (\rho = 1) \\[2mm] \hat\varphi_{\alpha_3} = \dfrac{5\cdot180°}{3} = 300° \quad (\rho = 2) \end{array} \right\} \tag{P6.7.4}$$

6. The breakaway points of RL are roots of Equation P6.7.6.
    The characteristic equation of the system is

$$1+G(s)H(s) = 0 \Rightarrow 1 + \frac{K(1+10s)}{s^2(1+5s)(s+1)} = 0 \Rightarrow$$

$$K = -\frac{s^4+1.2s^3+0.2s^2}{2(s+0.1)} \tag{P6.7.5}$$

$$\frac{dK}{ds} = 0 \stackrel{(P6.7.5)}{\Rightarrow} 3s^4+2.8s^3+0.56s^2+0.04s = 0 \tag{P6.7.6}$$

The roots of Equation P6.7.6 are $s_1 = 0$, $s_2 = 0.6912$, $s_{3,4} = -0.1211 \pm j0.068$.
The only one accepted is $s_1 = 0$, because it is a breakaway point and it gives $K = 0$. The roots $s_{3,4}$ are complex. The root $s_2 = 0.6912$ gives $K = -0.26 < 0$, and it is also at the right side of an even number of poles and zeros. Thus, $s_b = s_1 = 0$.

7. The intersections of RL with the imaginary axis can be found as follows:

$$\text{C.E.:} \quad s^4+1.2s^3+0.2s^2+2Ks+0.2K = 0 \tag{P6.7.7}$$

Routh's tabulation is

$$
\begin{array}{c|ccc}
s^4 & 1 & 0.2 & 0.2K \\
s^3 & 1.2 & 2K & \\
s^2 & \dfrac{0.24-2K}{1.2} & 0.2K & \\
s^1 & b & & \\
s^0 & 0.2K & &
\end{array}
$$

where

$$b = \frac{\big((0.24-2K)/1.2\big)\cdot 2K - 0.24K}{(0.24-2K)/1.2} \tag{P6.7.8}$$

From row s^1, we have

$$b = 0 \Rightarrow 0.48K - 4K^2 = 0 \Rightarrow K(0.192 - 4K) = 0 \Rightarrow$$

$$\left.\begin{array}{l} K_{1c} = 0 \\ K_{2c} = 0.048 \end{array}\right\} \tag{P6.7.9}$$

For the intersections of the RL with the imaginary axis, from the auxiliary equation of s^2, we have

$$\frac{0.24 - 2K_c}{1.2}s^2 + 0.2K_c = 0 \tag{P6.7.10}$$

$$\left.\begin{array}{l} \text{For} \quad K_c = K_{c_1} = 0 \Rightarrow \pm j\omega_{c_1} = \pm j0 \\ \text{while for} \quad K_c = K_{c_2} = 0.048 \Rightarrow \pm j\omega_{c_2} = \pm j0.283 \end{array}\right\} \tag{P6.7.11}$$

8. Based on the previous queries, we now plot the root-locus diagram:

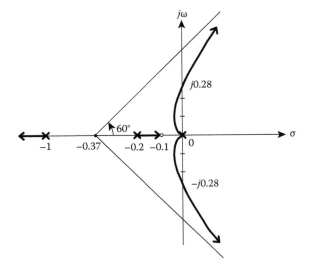

b. The loop transfer function is

$$G(s)H(s) = \frac{K(s+10)}{s(s+6)(s+40)^2} \qquad \text{(P6.7.12)}$$

1. The poles are $p_1 = 0$, $p_2 = -6$, $p_3 = p_4 = -40$ ($n = 4$).
2. The zeros are $z = -10$ ($m = 1$).
3. The number of separate loci is $4 = \max(n, m) = \max(4, 1)$.
4. The intersection of the asymptotes is

$$\sigma_\alpha = \frac{\sum_{i=1}^{4} p_i - \sum_{j=1}^{1} z_j}{n-m} = \frac{-6+(-40)+(-40)-(-10)}{4-1} = -\frac{76}{3} \qquad \text{(P6.7.13)}$$

5. The angles of the asymptotes are

$$\left. \begin{aligned} \sphericalangle\varphi_\alpha = \frac{(2\rho+1)\pi}{n-m} \\ \rho = 0,1,2 \end{aligned} \right] \Rightarrow \left. \begin{aligned} \hat{\varphi}_{\alpha_1} &= \frac{180°}{3} = 60° \quad (\rho=0) \\ \hat{\varphi}_{\alpha_2} &= \frac{3\cdot180°}{3} = 180° \quad (\rho=1) \\ \hat{\varphi}_{\alpha_3} &= \frac{5\cdot180°}{3} = 300° \quad (\rho=2) \end{aligned} \right\} \qquad \text{(P6.7.14)}$$

6. The breakaway points with the real axis are roots of Equation P6.7.16.
   The characteristic equation is

$$1+G(s)H(s) = 0 \overset{\text{(P6.7.12)}}{\Rightarrow} s(s+6)(s+40)^2 + K(s+10) = 0 \Rightarrow$$

$$K = -\frac{s(s+6)(s+40)^2}{s+10} \qquad \text{(P6.7.15)}$$

$$\frac{dK}{ds} = 0 \overset{\text{(P6.7.15)}}{\Rightarrow} 3s^4 + 212s^3 + 4,660s^2 + 41,600s + 96,000 = 0 \qquad \text{(P6.7.16)}$$

The roots of (P6.7.16) are $s_1 = -40$, $s_2 = -3.4298$, $s_{3,4} = -13.6184 \pm j6.9129$.
   Roots $s_{3,4}$ are rejected as breakaway points because they are complex. We accept $s_1 = -40$ and $s_2 = -3.4298$, which correspond to real values of K and belong to acceptable segments of the real axis, that is,

$$\left. \begin{aligned} s_{b_1} &= -3.43 \\ s_{b_2} &= -40 \end{aligned} \right\} \qquad \text{(P6.7.17)}$$

7. The intersections of RL with the imaginary axis are $\pm j\omega_c$. Therefore,

C.E.: $1+G(s)H(s) = 0 \Rightarrow s^4 + 86s^3 + 2080s^2 + (9600+K)s + 10K = 0$ (P6.7.18)

Routh's tabulation is

$$
\begin{array}{c|ccc}
s^4 & 1 & 2,080 & 10K \\
s^3 & 86 & 9,600+K & \\
s^2 & \dfrac{169,280-K}{86} & 10K & \\
s^1 & b & & \\
s^0 & 10K & &
\end{array}
$$

where

$$
b = \frac{(169,280-K)/86\cdot(9,600+K)-860K}{(169,280-K)/86} = \frac{-K^2+85,720K+1,625,088,000}{169,280-K} \qquad \text{(P6.7.19)}
$$

From s^1, we have

$$
b = 0 \Rightarrow -K^2+85,720K+1,625,088,000 = 0 \Rightarrow
\begin{cases}
K_1 = -15,979.335 \\
K_2 = 101,699.3
\end{cases}
$$

We reject the negative value of K; hence, the critical value of K for the stability of the closed-loop system is

$$
K_c = 101,699.3 \qquad \text{(P6.7.20)}
$$

From row s^2 by substituting $K = K_c$ to the auxiliary equation, we get the intersections

$$
\frac{169,280-K_c}{86}s^2+9,600+K_c = 0 \Rightarrow \pm j\omega_c = \pm j35.97 \qquad \text{(P6.7.21)}
$$

8. We now plot the RL of the characteristic equation of the system:

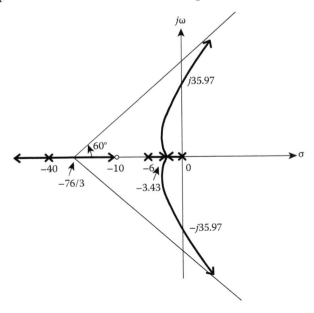

c. The loop transfer function is

$$G(s)H(s) = \frac{K(s+6)}{s(s+2)(s^2+20s+200)} \tag{P6.7.22}$$

1. The poles are $p_1 = 0$, $p_2 = -2$, $p_{3,4} = -10 \pm j10$ ($n = 4$).
2. The zeros are $z = -6$ ($m = 1$).
3. The number of separate loci is $\max(n, m) = \max(4, 1) = 4$.
4. The intersection of the asymptotes is

$$\sigma_\alpha = \frac{\sum_{i=1}^{4} p_i - \sum_{j=1}^{1} z_j}{n-m} = \frac{-2+(-10+j10)+(-10-j10)-(-6)}{4-1} = -\frac{16}{3} \tag{P6.7.23}$$

5. The angles of the asymptotes are

$$\left.\begin{array}{c} \sphericalangle\varphi_\alpha = \dfrac{(2\rho+1)\pi}{n-m} \\[2mm] \rho = 0,1,2 \end{array}\right\} \Rightarrow \left.\begin{array}{c} \hat\varphi_{\alpha_1} = \dfrac{180°}{3} = 60° \qquad (\rho = 0) \\[2mm] \hat\varphi_{\alpha_2} = \dfrac{3\cdot180°}{3} = 180° \quad (\rho = 1) \\[2mm] \hat\varphi_{\alpha_3} = \dfrac{5\cdot180°}{3} = 300° \quad (\rho = 2) \end{array}\right\} \tag{P6.7.24}$$

6. The breakaway points are roots of Equation P6.7.26:

$$\text{C.E.:} \quad 1+G(s)H(s) = 0 \overset{(\text{P6.7.22})}{\Rightarrow} s(s+2)(s^2+20s+200)+K(s+6) = 0 \Rightarrow$$

$$K = -\frac{s(s+2)(s^2+20s+200)}{s+6} \tag{P6.7.25}$$

$$\frac{dK}{ds} = 0 \overset{(\text{P6.7.25})}{\Rightarrow} 3s^4+68s^3+636s^2+2880s+2400 = 0 \tag{P6.7.26}$$

The roots of Equation P6.7.26 are $s_1 = -1.0512$, $s_2 = -10.4863$, $s_{3,4} = -5.5646 \pm j6.4505$. The roots $s_{3,4}$ are rejected as they are negative. The roots $s_1 = -1.0512$ και $s_2 = -10.4863$ are accepted as breakaway points, because they belong to acceptable segments of the real axis, and they give positive values of K. Therefore,

$$\left.\begin{array}{c} s_{b_1} = -1.0512 \\[2mm] s_{b_2} = -10.4863 \end{array}\right\} \tag{P6.7.27}$$

7. The intersections of RL with the imaginary axis are $\pm j\omega_c$. We have

$$\text{C.E.:} \quad s^4+22s^3+240s^2+(400+K)s+6K = 0 \tag{P6.7.28}$$

We proceed with Routh's tabulation:

$$
\begin{array}{c|ccc}
s^4 & 1 & 240 & 6K \\
s^3 & 22 & 400+K & \\
s^2 & \dfrac{4480-K}{22} & 6K & \\
s^1 & b & & \\
s^0 & 6K & &
\end{array}
$$

where

$$
b = -\frac{-K^2 + 1,576K + 1,952,000}{4,480 - K} \tag{P6.7.29}
$$

We assume that the term of row s^1 is 0 and we find K_c as

$$
b = 0 \Rightarrow -K^2 + 1,576K + 1,952,000 = 0 \Rightarrow
\begin{cases}
\to K_1 = -816.04 \\
\to K_2 = 2,392.04
\end{cases}
$$

The negative value is rejected; thus,

$$
K_c = 2392.04 \tag{P6.7.30}
$$

Substituting K by K_c in row s^2, we get

$$
\frac{4480 - K_c}{22} \cdot s^2 + 6K_c = 0 \Rightarrow \pm j\omega_c = \pm j11.26 \tag{P6.7.31}
$$

8. We now compute the angle of departure φ_d from the complex poles. For the complex pole $-10 + j10$ it is

$$
\sphericalangle \varphi_d = 180^\circ - \left(\sum \varphi_p - \sum \varphi_z \right) \tag{P6.7.32}
$$

where

$$
\sum \varphi_p = \left(180^\circ - \tan^{-1} \frac{10}{10} \right) + \left(180^\circ - \tan^{-1} \frac{10}{8} \right) + 90^\circ = 353.66^\circ \tag{P6.7.33}
$$

and

$$
\sum \varphi_z = 180^\circ - \tan^{-1} \frac{10}{4} = 111.8^\circ \tag{P6.7.34}
$$

Hence,

$$\sphericalangle \varphi_d = 180° - (353.66° - 111.8°) = -61.8° \qquad \text{(P6.7.35)}$$

Due to symmetry, the angle of departure φ_d from the complex conjugate pole $-10 - j10$ is $61.8°$.

9. We finally plot the RL of the characteristic equation of the system:

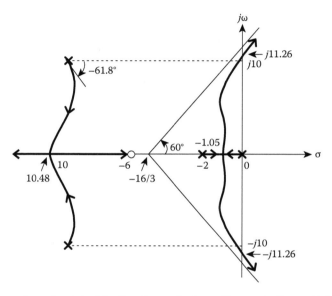

6.8 Consider the system depicted in the figure below. Compute the suitable compensation configuration that results in the following characteristics of the time response: maximum percent overshoot $M_p = 1.25$ and peak time $t_p = 0.5\,\text{s}$.

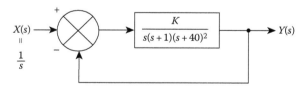

Solution

The loop transfer function is

$$G(s)H(s) = \frac{K}{s(s+1)(s+40)^2}, \quad K > 0 \qquad \text{(P6.8.1)}$$

Based on the characteristics for the time response for $M_p = 1.25$ and $t_p = 0.5\,\text{s}$ we compute the dominant roots of the system as follows:

$$s_{1,2} = -J\omega_n \pm j\omega_n\sqrt{1-J^2} \qquad \text{(P6.8.2)}$$

but

$$M_p = 1 + e^{-J\pi/\sqrt{1-J^2}} = 1.25 \Rightarrow J \simeq 0.4 < 1 \qquad \text{(P6.8.3)}$$

$$t_p = \frac{\pi}{\omega_n \sqrt{1-J^2}} = 0.5 \Rightarrow \omega_n = 6.86 \, \text{rad/s} \qquad \text{(P6.8.4)}$$

$$\text{(P6.8.2)} \underset{\text{(P6.8.4)}}{\overset{\text{(P6.8.3)}}{\Rightarrow}} s_{1,2} = -2.74 \pm j6.29 \qquad \text{(P6.8.5)}$$

We continue to design the RL diagram of the system's characteristic equation. From this we will examine the location of the dominant roots of Equation P6.8.5.

The poles are $p_1 = 0$, $p_2 = -1$, $p_3 = p_4 = -40$ ($n = 4$).

The number of separate loci is $4 = \max(n, m) = \max(4, 0)$.

The intersection of the asymptotes is

$$\sigma_\alpha = \frac{\sum_{i=1}^4 p_i}{4} = \frac{(-1) + (-40) + (-40)}{4} = -20.25 \qquad \text{(P6.8.6)}$$

The angles of the asymptotes are

$$\left. \begin{array}{l} \sphericalangle \hat{\varphi}_\alpha = \dfrac{(2\rho + 1)\pi}{n - m} \\[2mm] \rho = 0, 1, 2, 3 \end{array} \right\} \Rightarrow \left. \begin{array}{ll} \hat{\varphi}_{\alpha_1} = \dfrac{180°}{4} = 45° & (\rho = 0) \\[2mm] \hat{\varphi}_{\alpha_2} = \dfrac{3 \cdot 180°}{4} = 135° & (\rho = 1) \\[2mm] \hat{\varphi}_{\alpha_3} = \dfrac{5 \cdot 180°}{4} = 225° & (\rho = 2) \\[2mm] \hat{\varphi}_{\alpha_4} = \dfrac{7 \cdot 180°}{4} = 315° & (\rho = 3) \end{array} \right\} \qquad \text{(P6.8.7)}$$

From the characteristic equation, we have

$$1 + G(s)H(s) = 0 \Rightarrow s(s+1)(s+40)^2 + K = 0 \Rightarrow$$

$$K = -s(s+1)(s+40)^2 \qquad \text{(P6.8.8)}$$

$$\frac{dK}{ds} = 0 \overset{(8)}{\Rightarrow} s^3 + 60.75s^2 + 840s + 400 = 0 \qquad \text{(P6.8.9)}$$

The roots of Equation P6.8.9 are $s_1 = -0.4973$, $s_2 = -20.2563$, and $s_3 = -40$. The root s_2 is rejected since it is located on the right side of an even number of poles and zeros. Thus,

$$\left. \begin{array}{l} s_{b_1} = -0.4973 \\[2mm] s_{b_2} = -40 \end{array} \right\} \qquad \text{(P6.8.10)}$$

We proceed with Routh's tabulation in order to find the intersections of the branches of RL with the imaginary axis.

The characteristic equation is

$$\text{C.E.:} \quad s^4 + 81s^3 + 1680s^2 + 1600s + K = 0 \tag{P6.8.11}$$

Routh's tabulation is

$$
\begin{array}{c|ccc}
s^4 & 1 & 1680 & K \\
s^3 & 81 & 1600 & \\
s^2 & 1660.25 & K & \\
s^1 & b & & \\
s^0 & K & &
\end{array}
$$

where

$$b = \frac{1660.25 \cdot 1600 - 81 \cdot K}{1660.25} = 1600 - 0.049K \tag{P6.8.12}$$

From the row s^1 by setting b equal to zero we get

$$K_c = 3279 \tag{P6.8.13}$$

We compute the intersections $\pm j\omega_c$ with the imaginary axis, from row s^2:

$$1660.25s^2 + K_c = 0 \Rightarrow \pm j\omega_c = \pm j4.44 \tag{P6.8.14}$$

The RL of the system is plotted:

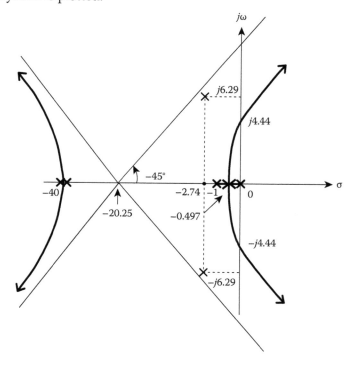

We observe that the dominant roots $-2.74 \pm j6.29$ are not roots of the RL that we plotted. Consequently, we must connect in series to the system a phase-lead compensator with a transfer function such that the pole at -1 is eliminated and thus the dominant roots of Equation P6.8.5 become roots of the locus.

The transfer function of the compensator is

$$G_D(s) = \frac{s+1}{s+s_x} \tag{P6.8.15}$$

We compute s_x from the angles condition for the complex root $s_1 = -2.74 + j6.29$. We have

$$\sphericalangle(G(s_1)H(s_1)) = (2\rho+1)\pi = 180° \tag{P6.8.16}$$

The graphic computation of s_x is shown next

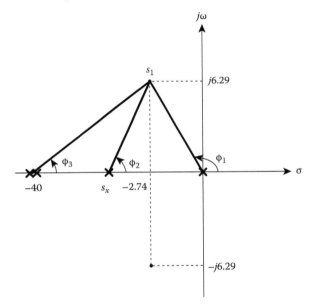

It is

$$\varphi_1 + \varphi_2 + 2\varphi_3 = 0 \Rightarrow$$

$$\left(180 - \tan^{-1}\frac{6.29}{2.74}\right) + \left(\tan^{-1}\frac{6.29}{s_x - 2.74}\right) + \left(2\tan^{-1}\frac{6.29}{40 - 2.74}\right) = 180° \Rightarrow \tag{P6.8.17}$$

$$\tan^{-1}\frac{6.29}{s_x - 2.74} = 47.3° \Rightarrow \frac{6.29}{s_x - 2.74} = \tan(47.3°) \Rightarrow$$

$$s_x = 8.6$$

Thus, the compensation controller's transfer function is

$$G_D(s) = \frac{s+1}{s+8.6} \tag{P6.8.18}$$

The circuit of a phase-lead controller is

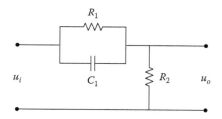

Thus,

$$G_D(s) = \frac{R_2}{R_1 + R_2} \cdot \frac{sR_1C_1 + 1}{s(R_1R_2C_1)/(R_1 + R_2) + 1} \tag{P6.8.19}$$

From (P6.8.18) and (P6.8.19), we obtain

$$\left.\begin{aligned} R_1C_1 &= 1 \\ \frac{R_2}{R_1 + R_2} &= 0.116 \end{aligned}\right\} \tag{P6.8.20}$$

For $C_1 = 1\,\mu F$, we get $R_1 = 1\,M\Omega$ and $R_2 = 131, 22\,K\Omega$.
The new system is shown below

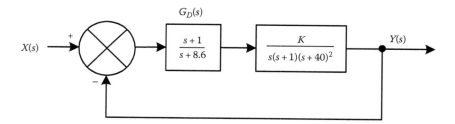

6.9 The block diagram of a position control system is shown in the following Figure (a)

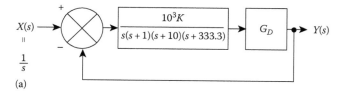

(a)

a. Plot the RL for the system if $G_D(s) = 1$.
b. Assume that G_D is replaced by the compensation controller depicted in Figure (b). Plot the RL again and discuss the influence of the compensator in the behavior of the system.

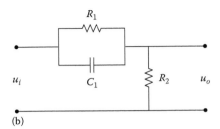

(b)

$$R_1 = 1\,M\Omega, \quad R_2 = 100\,K\Omega, \quad C_1 = 1\,\mu F, \quad G_D(s) = \frac{V_o(s)}{V_i(s)}$$

Solution

a. The loop transfer function of the system for $G_D(s) = 1$ is

$$G(s)H(s) = \frac{10^3 K}{s(s+1)(s+10)(s+333.3)} \tag{P6.9.1}$$

1. The poles are $p_1 = 0$, $p_2 = -1$, $p_3 = -10$, $p_4 = -333.3$ ($n = 4$).
2. There are no zeros ($m = 0$).
3. The number of separate loci is $\max(n, m) = \max(4, 0) = 4$.
4. The intersection of the asymptotes is

$$\sigma_\alpha = \frac{\sum_{i=1}^{4} p_i}{4} = \frac{-1 + (-10) + (-333.3)}{4} = -86.075 \tag{P6.9.2}$$

5. The angles of the asymptotes are

$$\left. \sphericalangle \varphi_\alpha = \frac{(2\rho + 1)\pi}{n - m} \atop \rho = 0, 1, 2, 3 \right\} \Rightarrow \left. \begin{array}{ll} \hat{\varphi}_{\alpha_1} = \dfrac{180°}{4} = 45° & (\rho = 0) \\[2mm] \hat{\varphi}_{\alpha_2} = \dfrac{3 \cdot 180°}{4} = 135° & (\rho = 1) \\[2mm] \hat{\varphi}_{\alpha_3} = \dfrac{5 \cdot 180°}{4} = 225° & (\rho = 2) \\[2mm] \hat{\varphi}_{\alpha_4} = \dfrac{7 \cdot 180°}{4} = 315° & (\rho = 3) \end{array} \right\} \tag{P6.9.3}$$

6. The possible breakaway points are roots of Equation P6.9.5
   The characteristic equation is

$$1+G(s)H(s)=0 \overset{(P6.9.1)}{\Rightarrow} s(s+1)(s+10)(s+333.3)+10^3K=0 \Rightarrow$$

$$K=\frac{s(s+1)(s+10)(s+333.3)}{10^3} \qquad (P6.9.4)$$

$$\frac{dK}{ds}=0 \overset{(P6.9.4)}{\Rightarrow} 4s^3+1032.9s^2+7352.6s+3333=0 \qquad (P6.9.5)$$

The roots of Equation P6.9.5 are $s_1=-0.4865$, $s_2=-6.8261$, $s_3=-250.9124$.
   The roots s_1 and s_3 are accepted as breakaway points, because they are on branches of the real axis which are on the right side of an odd number of poles and zeros. Therefore,

$$\left.\begin{array}{l} s_{b_1}=-0.4865 \\[2mm] s_{b_2}=-250.9124 \end{array}\right\} \qquad (P6.9.6)$$

7. The intersections of RL with the imaginary axis are at $\pm j\omega_c$.

$$\text{C.E.:} \quad s^4+344.3s^3+3676.3s^2+3333s+10^3K=0 \qquad (P6.9.7)$$

We proceed with Routh's tabulation:

$$\begin{array}{c|ccc}
s^4 & 1 & 3676.3 & 10^3K \\
s^3 & 344.3 & 3333 & \\
s^2 & 3666.6 & 10^3K & \\
s^1 & b & & \\
s^0 & 10^3K & &
\end{array}$$

where

$$b=3333-93.9K \qquad (P6.9.8)$$

The critical value of K for stability is computed by row s^1 as

$$3666.6s^2+10^3K_c=0 \Rightarrow \pm j\omega_c=\pm j3.1 \qquad (P6.9.10)$$

From the equation of row s^2 we find the intersections $\pm j\omega_c$ of the RL with the imaginary axis:

$$3666.6s^2+10^3K_c=0 \Rightarrow \pm j\omega_c=\pm j3.1 \qquad (P6.9.10)$$

8. We plot the RL of the characteristic equation of the system.

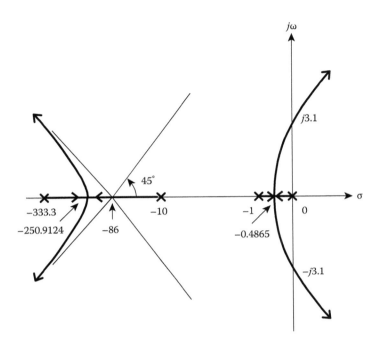

b. The transfer function of the phase-lead controller shown in Figure (a) is

$$G_D(s) = \frac{R_2}{R_1 + R_2} \cdot \frac{sR_1C_1 + 1}{s(R_1R_2C_1)/(R_1 + R_2) + 1}$$ (P6.9.11)

By substituting the given values, we get

$$G_D(s) = \frac{s+1}{s+11}$$ (P6.9.12)

The loop transfer function after the controller is connected in series becomes

$$(G(s)H(s))' = \frac{10^3 K}{s(s+1)(s+10)(s+333.3)} \cdot \frac{s+1}{s+11} \Rightarrow$$ (P6.9.13)

$$(G(s)H(s))' = \frac{10^3 K}{s(s+10)(s+11)(s+333.3)}$$

We follow the usual procedure for drawing the RL of the roots of the characteristic equation of the system:

1. The poles are $p_1 = 0$, $p_2 = -10$, $p_3 = -11$, $p_4 = -333.3$ ($n = 4$).
2. There are no zeros ($m = 0$).
3. The number of separate loci is $\max(n, m) = \max(4, 0) = 4$.

4. The intersection of the asymptotes is

$$\sigma_\alpha = \frac{\sum_{i=1}^{4} p_i}{4} = \frac{-10 + (-11) + (-333.3)}{4} = -88.6 \qquad \text{(P6.9.14)}$$

5. The angles of the asymptotes are

$$\left. \begin{aligned} \sphericalangle \varphi_\alpha &= \frac{(2\rho+1)\pi}{n-m} \\ \rho &= 0,1,2,3 \end{aligned} \right\} \Rightarrow \left. \begin{aligned} \hat{\varphi}_{\alpha_1} &= \frac{180°}{4} = 45° \quad (\rho = 0) \\ \hat{\varphi}_{\alpha_2} &= \frac{3 \cdot 180°}{4} = 135° \quad (\rho = 1) \\ \hat{\varphi}_{\alpha_3} &= \frac{5 \cdot 180°}{4} = 225° \quad (\rho = 2) \\ \hat{\varphi}_{\alpha_4} &= \frac{7 \cdot 180°}{4} = 315° \quad (\rho = 3) \end{aligned} \right\} \qquad \text{(P6.9.15)}$$

6. The possible breakaway points with the real axis are roots of Equation P6.9.17.

$$1 + (G(s)H(s))' = 0 \overset{\text{(P6.9.13)}}{\Rightarrow} s(s+10)(s+11)(s+333.3) + 10^3 K = 0 \Rightarrow$$

$$K = -\frac{s(s+10)(s+11)(s+333.3)}{10^3} \qquad \text{(P6.9.16)}$$

$$\frac{dK}{ds} = 0 \overset{\text{(P6.9.16)}}{\Rightarrow} 4s^3 + 1{,}062.9s^2 + 14{,}218.6s + 36{,}663 = 0 \qquad \text{(P6.9.17)}$$

The roots of Equation P6.9.17 are $s_1 = -3.4637$, $s_2 = -10.5115$, $s_3 = -251.7499$. The breakaway points are

$$\left. \begin{aligned} s_{b_1} &= -3.4637 \\ s_{b_2} &= -251.7499 \end{aligned} \right\} \qquad \text{(P6.9.18)}$$

7. The intersections of RL with the imaginary axis are computed as follows

$$\text{C.E.:} \quad s^4 + 354.3s^3 + 7{,}109.3s^2 + 36{,}663s + 10^3 K = 0 \qquad \text{(P6.9.19)}$$

We proceed with Routh's tabulation:

s^4	1	7,109.3	$10^3 K$
s^3	354.3	36,663	
s^2	7,005.8	$10^3 K$	
s^1	b		
s^0	$10^3 K$		

where

$$b = \frac{7,005.8 \cdot 36,663 - 354.3 \cdot 10^3 K}{7,005.8} = 36,663 - 50.572K \qquad \text{(P6.9.20)}$$

From row s^1, we have

$$b = 0 \overset{\text{(P6.9.20)}}{\Rightarrow} K_c = 724.96 \qquad \text{(P6.9.21)}$$

From row s^2, we get

$$7005.8s^2 + 10^3 K_c = 0 \Rightarrow \pm j\omega_c = \pm j10.17 \qquad \text{(P6.9.22)}$$

8. We plot the RL of the characteristic equation of the system:

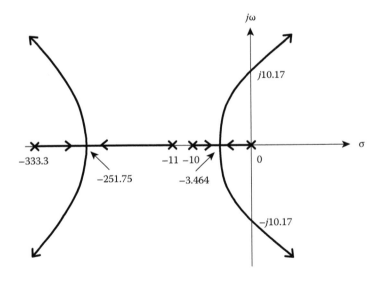

We observe that the compensator increased the relative stability of the closed-loop system, since it increased the limits of K for stability, that is, $0 < K < 724.96$.

6.10 For the system depicted in the following figure

   a. Plot the RL.

   b. Derive the suitable compensation circuit that will result in a time response with the following characteristics: maximum percent overshoot $M_p = 1.15$ and peak time $t_p = 0.03\,\text{s}$. Take into account that $R_1 = 1\,\text{M}\Omega$, $R_2 = 222\,\text{K}\Omega$, and $C_1 = C_2 = 1\,\mu\text{F}$.

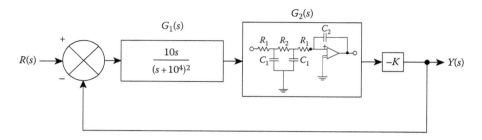

Solution

The loop transfer function of the system is

$$G(s)H(s) = G_1(s) \cdot \underbrace{\left(-\frac{Z_o(s)}{Z_i(s)} \right) \cdot (-K)}_{G_2(s)} = G_1(s) \cdot \frac{Z_o(s)}{Z_i(s)} \cdot K \qquad \text{(P6.10.1)}$$

However,

$$\frac{Z_o(s)}{Z_i(s)} = \frac{1}{C_2 s(2R_1 + R_2)\left(1 + (R_1 R_2 C_1/(2R_1 + R_2))s\right)(1 + R_1 C_1 s)} \qquad \text{(P6.10.2)}$$

We have

$$\left. \begin{aligned} R_1 C_1 &= 10^6 \cdot 10^{-6} = 1 \\ \frac{R_1 R_2 C_1}{2R_1 + R_2} &\simeq 0.1 \end{aligned} \right\} \qquad \text{(P6.10.3)}$$

$$\overset{\text{(P6.10.3)}}{(2) \Rightarrow} \frac{Z_o(s)}{Z_i(s)} = \frac{1}{10^{-6} s(2 \cdot 10^6 + 222 \cdot 10^3)(1 + 0.1s)(1 + s)} \Rightarrow$$

$$\frac{Z_o(s)}{Z_i(s)} = \frac{4.5}{s(s + 10)(s + 1)} \qquad \text{(P6.10.4)}$$

Hence, the loop transfer function is

$$G(s)H(s) \overset{\text{(P6.10.1)}}{=} \frac{10s}{(s + 10^4)^2} \cdot \frac{4.5}{s(s + 10)(s + 1)} \cdot K \Rightarrow$$

$$G(s)H(s) = \frac{45K}{(s + 1)(s + 10)(s + 10^4)^2} \qquad \text{(P6.10.5)}$$

We continue as usual with the design procedure of the RL diagram:

1. The poles are $p_1 = -1$, $p_2 = -10$, $p_3 = p_4 = -10^4$ ($n = 4$).
2. There are no zeros ($m = 0$).
3. The number of separate branches is $\max(n, m) = \max(4, 0) = 4$.

4. The intersection of the asymptotes is

$$\sigma_\alpha = \frac{\sum_{i=1}^{4} p_i}{4} = \frac{-1+(-10)+(-10^4)+(-10^4)}{4} = -5002.8 \qquad \text{(P6.10.6)}$$

5. The angles of the asymptotes are

$$\left. \begin{array}{c} \sphericalangle\varphi_\alpha = \dfrac{(2\rho+1)\pi}{n-m} \\[2mm] \rho = 0,1,2,3 \end{array} \right\} \Rightarrow \left. \begin{array}{l} \hat\varphi_{\alpha_1} = \dfrac{180°}{4} = 45° \qquad (\rho = 0) \\[2mm] \hat\varphi_{\alpha_2} = \dfrac{3\cdot 180°}{4} = 135° \quad (\rho = 1) \\[2mm] \hat\varphi_{\alpha_3} = \dfrac{5\cdot 180°}{4} = 225° \quad (\rho = 2) \\[2mm] \hat\varphi_{\alpha_4} = \dfrac{7\cdot 180°}{4} = 315° \quad (\rho = 3) \end{array} \right\} \qquad \text{(P6.10.7)}$$

6. The possible breakaway points are roots of Equation P6.10.9:

$$1+G(s)H(s) = 0 \overset{\text{(P6.10.5)}}{\Rightarrow} (s+1)(s+10)(s+10^4)^2 + 45K = 0 \Rightarrow$$

$$K = -\frac{(s+1)(s+10)(s+10^4)^2}{45} \qquad \text{(P6.10.8)}$$

$$\frac{dK}{ds} = 0 \overset{\text{(P6.10.8)}}{\Rightarrow} 4s^3 + 60,033s^2 + 2.0044\cdot 10^8 s + 1.1002\cdot 10^9 = 0 \qquad \text{(P6.10.9)}$$

The roots of Equation P6.10.9 are $s_1 \simeq -5$, $s_2 = -5003$, $s_3 = -10^4$. From these, we reject the root $s_2 = -5003$ because it belongs to an unacceptable branch of the real axis.

Thus, the breakaway points are

$$\left. \begin{array}{l} s_{b_1} = -5 \\[2mm] s_{b_2} = -10^4 \end{array} \right\} \qquad \text{(P6.10.10)}$$

7. The intersections of RL with the imaginary axis are computed as follows:

$$\text{C.E.:} \quad s^4 + 20,011s^3 + 1.0022\cdot 10^8 s^2 + 1.1002\cdot 10^9 s + 10^9 + 45K = 0 \qquad \text{(P6.10.11)}$$

Routh's tabulation is

$$\begin{array}{c|ccc} s^4 & 1 & 1.0022\cdot 10^8 & 10^9 + 45K \\ s^3 & 20,011 & 1.1002\cdot 10^9 & \\ s^2 & 1.00165\cdot 10^8 & 10^9 + 45K & \\ s^1 & b & & \\ s^0 & 10^9 + 45K & & \end{array}$$

where

$$b = \frac{1.1018 \cdot 10^{17} - 900,495K}{1.00165 \cdot 10^8} \tag{P6.10.12}$$

Computation of K_c,

$$b = 0 \Rightarrow K_c = 1.22 \cdot 10^{11} \tag{P6.10.13}$$

Computation of $\pm j\omega_c$,

$$1.00165 \cdot 10^8 \cdot s^2 + 10^9 + 45K_c = 0 \Rightarrow \pm j\omega_c = \pm j234.5 \tag{P6.10.14}$$

8. We plot the RL of the characteristic equation of the system:

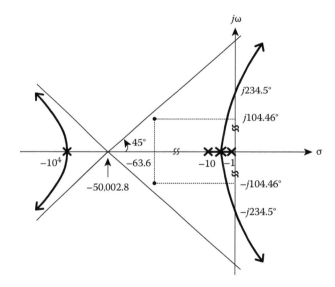

From the time-response characteristics M_p and t_p, we compute the dominant roots of the system:

$$M_p = 1 + e^{-J\pi/\sqrt{1-J^2}} = 1.15 \Rightarrow J \simeq 0.52 < 1 \tag{P6.10.15}$$

$$t_p = \frac{\pi}{\omega_n\sqrt{1-J^2}} = 0.03 \Rightarrow \omega_n = 122.3\,\text{rad/s} \tag{P6.10.16}$$

The dominant roots are

$$s_{1,2} = -J\omega_n \pm j\omega_n\sqrt{1-J^2} = -63.6 \pm j104.46 \tag{P6.10.17}$$

These roots must be also roots of the RL, so we have to connect with the system in series a phase-lead controller. Its transfer function is

$$G_D(s) = \frac{s+10}{s+s_x} \tag{P6.10.18}$$

Computation of s_x

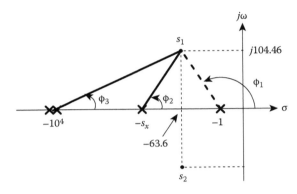

From the condition for the angles, for s_1, we have

$$\sphericalangle\big(G(s_1)H(s_1)\big) = 180° \implies \hat{\phi}_1 + \hat{\phi}_2 + 2\hat{\phi}_3 = 180° \tag{P6.10.19}$$

$$(P6.10.19) \implies \left(180° - \tan^{-1}\frac{104.46}{62.6}\right) + \left(\tan^{-1}\frac{104.46}{s_x-63.6}\right) + \left(2\tan^{-1}\frac{104.46}{10^4-63.6}\right) = 180° \implies$$

$$\tan^{-1}\frac{104.46}{s_x-63.6} = 57.9° \implies \frac{104.46}{s_x-63.6} = 1.594 \implies \tag{P6.10.20}$$

$$s_x = 129.138$$

The phase-lead controller's transfer function is

$$G_D(s) = \frac{R_2}{R_1+R_2} \cdot \frac{sR_1C_1+1}{s(R_1R_2C_1/(R_1+R_2))+1} \tag{P6.10.21}$$

By combining (P6.10.18), (P6.10.20), and (P6.10.21), it follows that

$$\left.\begin{aligned} R_1C_1 &= \frac{1}{10} \\[2mm] \frac{R_1R_2C_1}{R_1+R_2} &= \frac{1}{129.138} \end{aligned}\right\} \tag{P6.10.22}$$

We choose $C_1 = 0.1\,\mu F$; hence, $R_1 = 1\,M\Omega$ and $R_2 \simeq 84\,K\Omega$.

6.11 The next figure shows the movement of a robot arm, where $\theta_c(t)$ is the desired angle of the movement of the arm and $\theta_L(t)$ is the real angle of its movement. Plot the RL of the system for $k_D/k_p = 0.05$, $k_D/k_p = 0.5$, and for $k_D/k_p = 5$ and discuss the stability of the system for each case.

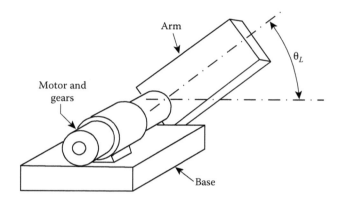

Solution

In order to find the loop transfer function $G(s)H(s)$ of the system we must plot its block diagram. The Laplace-transformed mathematical model of the system is

$$\Theta_c(s) - \Theta_L(s) = \Theta_e(s) \tag{P6.11.1}$$

$$E(s) = (k_p + sk_D)\Theta_e(s) \tag{P6.11.2}$$

$$E_a(s) = nE(s) \tag{P6.11.3}$$

$$E_a(s) - k_m\Omega_m(s) = (R_m + sL_m)I_a(s) \tag{P6.11.4}$$

$$T(s) = k_T I_a(s) \tag{P6.11.5}$$

$$T(s) = (Js + B)\Omega_m(s) \tag{P6.11.6}$$

$$\Omega_m(s) = s\Theta_m(s) \tag{P6.11.7}$$

$$\Theta_L(s) = \frac{1}{n}\Theta_m(s) \tag{P6.11.8}$$

The block diagram is

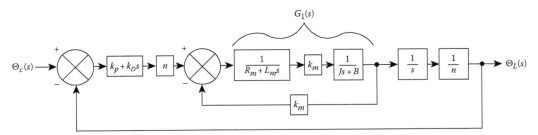

$G_1(s)$

The loop transfer function is

$$G(s)H(s) = (k_p + k_D s) \cdot n \cdot G_1(s) \cdot \frac{1}{s} \cdot n \qquad \text{(P6.11.9)}$$

$$G_1(s) = \frac{k_T/(R_m + L_m s)(Js + B)}{1 + (k_m k_T/(R_m + L_m s)(Js + B))} = \frac{k_T}{(R_m + L_m s)(Js + B) + k_m k_T} \qquad \text{(P6.11.10)}$$

$$(\text{P6.11.9}) \overset{(\text{P6.11.10})}{\Rightarrow} G(s)H(s) = \frac{(k_p + k_D s) \cdot k_T}{s[(R_m + L_m s)(Js + B) + k_m k_T]} \qquad \text{(P6.11.11)}$$

By substituting the values in (P6.11.11), we get

$$G(s)H(s) = \frac{9.5(k_p + k_D s)}{s(s+1)(s+10)} \qquad \text{(P6.11.12)}$$

i. For $k_D/k_p = 0.05$, it is

$$G(s)H(s) = \frac{9.5k_p(0.05s + 1)}{s(s+1)(s+10)} \qquad \text{(P6.11.13)}$$

We now proceed with the plot of the RL of the characteristic equation of the system:

1. The poles are $p_1 = 0$, $p_2 = -1$, $p_3 = -10$ ($n = 3$).
2. The zeros are $z_1 = -20$ ($m = 1$).
3. The number of separate loci is $3 = \max(n, m) = \max(3, 1)$.
4. The intersection of the asymptotes is

$$\sigma_\alpha = \frac{\sum_{i=1}^{3} p_i - \sum_{j=1}^{1} z_j}{n - m} = \frac{-1 + (-10) - (-20)}{3 - 1} = 4.5 \qquad \text{(P6.11.14)}$$

5. The angles of the asymptotes are

$$
\left.\begin{aligned}
\sphericalangle\varphi_\alpha &= \frac{(2\rho+1)\pi}{n-m} \\
\rho &= 0,1
\end{aligned}\right\} \Rightarrow
\left.\begin{aligned}
\hat\varphi_{\alpha_1} &= \frac{180°}{2} = 90° \quad (\rho=0) \\
\hat\varphi_{\alpha_2} &= \frac{3\cdot 180°}{2} = 270° \quad (\rho=1)
\end{aligned}\right\}
\tag{P6.11.15}
$$

6. The breakaway points are computed as follows:

$$
\text{C.E.:} \quad 1+G(s)H(s) = 0 \overset{(P6.11.13)}{\Rightarrow} s(s+1)(s+10)+9.5k_p(0.05s+1)=0 \Rightarrow
$$
$$
k_p = -\frac{s(s+1)(s+10)}{9.5(0.05s+1)}
\tag{P6.11.16}
$$

$$
\frac{dk_p}{ds} = 0 \overset{(P6.11.16)}{\Rightarrow} 0.95s^3 + 33.725s^2 + 209s + 95 = 0
\tag{P6.11.17}
$$

The roots of Equation P6.11.17 are $s_1 = -0.4933$, $s_2 = -7.3232$, $s_3 = -27.6835$.
   Only the root $s_1 = -0.4933$ is accepted, because it belongs to a branch of the real axis, in the right side of which there is an odd number of poles and zeros. Therefore,

$$
s_b = -0.4933
\tag{P6.11.18}
$$

7. The intersections of RL with the imaginary axis are $\pm j\omega_c$. We have

$$
\text{C.E.:} \quad s^3 + 11s^2 + (0.475k_p+10)s + 9.5k_p = 0
\tag{P6.11.19}
$$

We proceed with Routh's tabulation:

$$
\begin{array}{c|cc}
s^3 & 1 & 0.475k_p+10 \\
s^2 & 11 & 9.5k_p \\
s^1 & b & \\
s^0 & 9.5k_p &
\end{array}
$$

where

$$
b = \frac{11\cdot(0.475k_p+10)-9.5k_p}{11}
\tag{P6.11.20}
$$

From row s^1, we have

$$b = 0 \Rightarrow k_{p_c} = 25.7 \qquad \text{(P6.11.21)}$$

From row s^2, we get

$$11s^2 + 9.5k_{p_c} = 0 \Rightarrow \pm j\omega_c = \pm j4.7 \qquad \text{(P6.11.22)}$$

8. We finally plot the RL of the characteristic equation of the system.

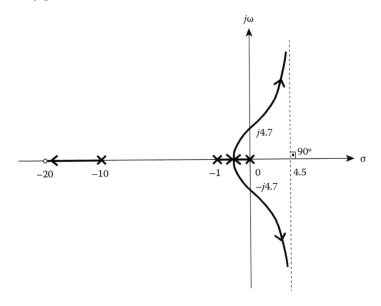

ii. For $k_D/k_p = 0.5$, we have

$$G(s)H(s) = \frac{9.5k_p(0.5s + 1)}{s(s+1)(s+10)} \qquad \text{(P6.11.23)}$$

We continue with the design procedure for the RL of the characteristic equation of the system:

1. The poles are $p_1 = 0$, $p_2 = -1$, $p_3 = -10$ $(n = 3)$.
2. The zeros are $z_1 = -2$ $(m = 1)$.
3. The number of separate loci is $\max(n, m) = \max(3, 1) = 3$.
4. The intersection of the asymptotes is

$$\sigma_\alpha = \frac{\sum_{i=1}^{3} p_i - \sum_{j=1}^{1} z_j}{n - m} = \frac{-1 + (-10) - (-2)}{3 - 1} = -4.5 \qquad \text{(P6.11.24)}$$

5. The angles of the asymptotes are

$$\sphericalangle \varphi_\alpha = \frac{(2\rho+1)\pi}{n-m} \Biggr\} \atop \rho = 0,1 \Rightarrow \begin{array}{l} \hat{\varphi}_{\alpha_1} = \dfrac{180°}{2} = 90° \quad (\rho=0) \\[4mm] \hat{\varphi}_{\alpha_2} = \dfrac{3 \cdot 180°}{2} = 270° \quad (\rho=1) \end{array} \Biggr\} \tag{P6.11.25}$$

6. The breakaway points are computed as follows:

$$\text{C.E.:} \quad 1+G(s)H(s) = 0 \overset{(P6.11.23)}{\Rightarrow} s(s+1)(s+10)+9.5k_p(0.5s+1) = 0 \Rightarrow$$
$$k_p = -\frac{s(s+1)(s+10)}{9.5(0.5s+1)} \tag{P6.11.26}$$

$$\frac{dk_p}{ds} = 0 \overset{(P6.11.26)}{\Rightarrow} 9.5s^3 + 80.75s^2 + 209s + 95 = 0 \tag{P6.11.27}$$

The roots of Equation P6.11.27 are $s_1 = -0.5727$, $s_2 = -3.9636 + j1.3226$, $s_3 = -3.9636 - j1.3226$.

Only the root s_1 is accepted, because it belongs to a branch of the real axis. The roots $s_{2,3}$ are rejected because they are complex. Therefore,

$$s_b = -0.5727 \tag{P6.11.28}$$

7. The intersections $\pm j\omega_c$ of RL with the imaginary axis are computed as follows:

$$\text{C.E.:} \quad s^3 + 11s^2 + (4.75k_p + 10)s + 9.5k_p = 0 \tag{P6.11.29}$$

We proceed with Routh's tabulation:

$$\begin{array}{c|cc} s^3 & 1 & 4.75k_p+10 \\ s^2 & 11 & 9.5k_p \\ s^1 & b & \\ s^0 & 9.5k_p & \end{array}$$

where

$$b = \frac{11 \cdot (4.75k_p+10) - 9.5k_p}{11} \tag{P6.11.30}$$

From row s^1, we have

$$b = 0 \Rightarrow k_{p_c} = -2.57 \tag{P6.11.31}$$

The value of k_p is negative; therefore, there are no intersections with the imaginary axis.

8. We finally plot the RL of the characteristic equation of the system:

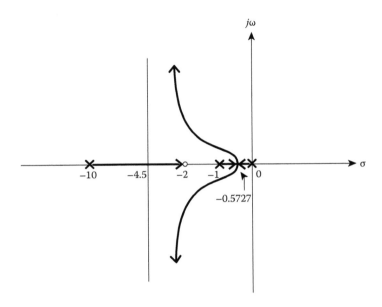

iii. For $k_D/k_p = 5$, we have

$$G(s)H(s) = \frac{9.5k_p(5s+1)}{s(s+1)(s+10)} \qquad \text{(P6.11.32)}$$

We continue with the design procedure for the RL of the characteristic equation of the system:

1. The poles are $p_1 = 0$, $p_2 = -1$, $p_3 = -10$ ($n = 3$).
2. The zeros are $z_1 = -0.2$ ($m = 1$).
3. The number of separate loci is $\max(n, m) = \max(3, 1) = 3$.
4. The intersection of the asymptotes is

$$\sigma_\alpha = \frac{\sum_{i=1}^{3} p_i - \sum_{j=1}^{1} z_j}{n - m} = \frac{-1 + (-10) - (-0.2)}{3 - 1} = -5.4 \qquad \text{(P6.11.33)}$$

5. The angles of the asymptotes are

$$\left. \begin{aligned} \sphericalangle\varphi_\alpha &= \frac{(2\rho+1)\pi}{n-m} \\ \rho &= 0,1 \end{aligned} \right\} \Rightarrow \left. \begin{aligned} \hat{\varphi}_{\alpha_1} &= \frac{180°}{2} = 90° \quad (\rho = 0) \\ \hat{\varphi}_{\alpha_2} &= \frac{3 \cdot 180°}{2} = 270° \quad (\rho = 1) \end{aligned} \right\} \qquad \text{(P6.11.34)}$$

6. The breakaway point is

$$s_b = -5.4287 \qquad\qquad (P6.11.35)$$

7. There are no intersections of the RL with the imaginary axis.
8. We plot the new RL of the characteristic equation of the system:

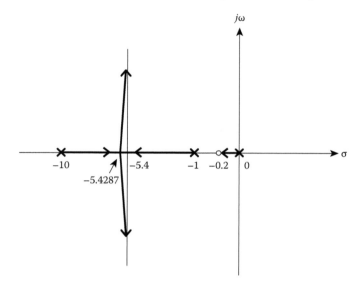

We observe that as the value of the ratio k_D/k_p increases, the relative stability of the closed-loop system increases as well.

7

Frequency Response: Bode Diagrams

7.1 Steady-State Response for Sinusoid Input Signal

Frequency response is the steady-state response of a linear time-invariant (lti) system to a sinusoid input signal of constant amplitude and variable frequency. The time response of an lti system to a sinusoid input signal is also a sinusoid signal, with the frequency of the input signal but different amplitude and phase angle.

The frequency response of a system can be found from its **transfer function**, if the complex variable s is replaced by the imaginary variable $j\omega$, where ω is the frequency in rad/s. The new transfer function $G(j\omega)$ is a complex function of a real variable, which has **magnitude** and **phase**. The magnitude-phase diagrams provide important information for analyzing and designing a control system.

Suppose that the system shown in the figure below is linear and time invariant:

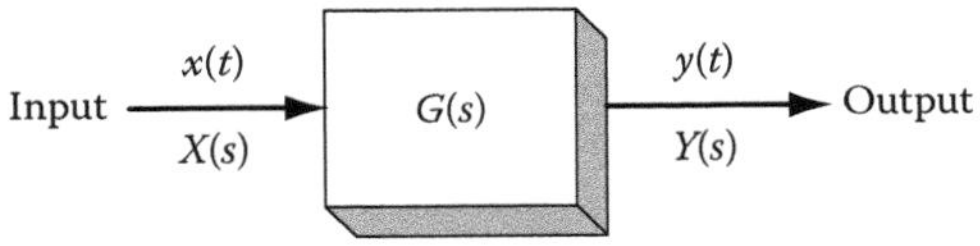

Input $\xrightarrow{\begin{array}{c} x(t) \\ X(s) \end{array}}$ $\boxed{G(s)}$ $\xrightarrow{\begin{array}{c} y(t) \\ Y(s) \end{array}}$ Output

We have

$$G(s) = \frac{Y(s)}{X(s)} \tag{7.1}$$

We assume that the input is

$$x(t) = A\sin\omega t \Rightarrow X(s) = \frac{A\omega}{s^2 + \omega^2} \tag{7.2}$$

The output $y(t)$ is given by

$$y(t) = A\,|G(j\omega)|\sin(\omega t + \varphi) \tag{7.3}$$

where

$$|G(j\omega)| = \left|\frac{Y(j\omega)}{X(j\omega)}\right| \tag{7.4}$$

and

$$\varphi = \sphericalangle(G(j\omega)) = \sphericalangle\left(\frac{Y(j\omega)}{X(j\omega)}\right) \tag{7.5}$$

From relationship (7.3), we conclude that if the input of a stable lti system is a sinusoidal signal, then the steady-state output is also sinusoidal with amplitude equal to $A \cdot |G(j\omega)|$ and a phase that differs by $\varphi = \sphericalangle(G(j\omega))$ from the phase of the input signal.

7.2 Frequency Response Characteristics

Figure 7.1a and b illustrates a typical form of the magnitude and phase of the transfer function versus the frequency ω.

The characteristics of the frequency response, as it can be observed in Figure 7.1, are as follows:

1. **Maximum value of the frequency response** M_p is the maximum value of the magnitude of the frequency response. The value of resonant peak indicates the relative stability of the system. As the value of M_p increases, the percent overshoot of the step response increases as well.

2. **Bandwidth** *Bw* is the frequency ω_b for which the magnitude of the frequency response is reduced by 3 dB in relation to the value that corresponds to low frequencies. Bandwidth is a measure of the ability of the system to reproduce the input signal. For ω_b, we have

$$|G(j\omega_b)| = 0.707 \tag{7.6}$$

By increasing the bandwidth we reduce the rise time of the step response.

3. **Resonant frequency** ω_r is the frequency for which $|G(j\omega_r)| = M_p$. If the damping ratio approaches zero, resonant frequency coincides with the natural oscillation frequency of the system.

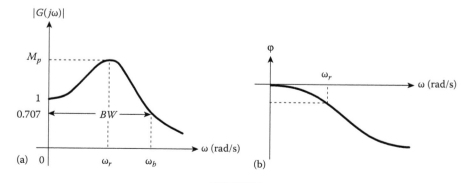

FIGURE 7.1
(a) Magnitude plot of a transfer function. (b) Phase plot of a transfer function.

7.3 Time Response and Frequency Response

7.3.1 First-Order Systems

The transfer function of a first-order system is

$$G(s) = \frac{K}{Ts+1} \tag{7.7}$$

In the sinusoidal steady state, the transfer function $G(j\omega)$ is

$$G(j\omega) = G(s)\big|_{s=j\omega} = \frac{K}{j\omega T+1} = \frac{K}{\sqrt{\omega^2 T^2+1}} e^{-j\tan^{-1}(\omega T)} \tag{7.8}$$

From the graphs of $|G(j\omega)|$ and $\sphericalangle G(j\omega)$, shown in figures (a) and (b) respectively, we have

$$\left.\begin{array}{l} M_p = K \\[4pt] \omega_r = 0 \\[4pt] \omega_b = \dfrac{1}{T} = Bw \end{array}\right\} \tag{7.9}$$

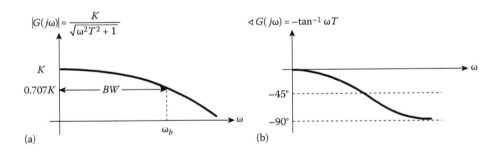

(a)

(b)

7.3.2 Second-Order Systems

The transfer function of a second-order system is

$$G(s) = \frac{\omega_n^2}{s^2 + 2J\omega_n s + \omega_n^2} \tag{7.10}$$

In the sinusoidal steady state $G(j\omega)$ is

$$G(j\omega) = G(s)\big|_{s=j\omega} = \frac{\omega_n^2}{(j\omega)^2 + 2J\omega_n(j\omega) + \omega_n^2} \tag{7.11}$$

The magnitude $|G(j\omega)|$ is

$$|G(j\omega)| = \frac{1}{\sqrt{(1-(\omega/\omega_n)^2)^2 + (2J(\omega/\omega_n))^2}}$$ (7.12)

The phase $\varphi = \sphericalangle G(j\omega)$ is

$$\varphi = \sphericalangle G(j\omega) = -\tan^{-1}\frac{2J(\omega/\omega_n)}{1-(\omega/\omega_n)^2}$$ (7.13)

In order to determine the resonant frequency ω_r, we differentiate $|G(j\omega)|$ with respect to ω/ω_n and set the result equal to zero. Hence,

$$\omega_r = \omega_n\sqrt{1-2J^2}, \quad J < 0.707$$ (7.14)

The maximum value of the frequency response M_p is

$$M_p = \frac{1}{2J\sqrt{1-J^2}}, \quad J < 0.707$$ (7.15)

For $J \geq 0.707$, it holds that $M_p = 1$.

The bandwidth $BW = \omega_b$ is found by setting at (7.12) $|G(j\omega)|$ equal to zero:

$$BW = \omega_b = \omega_n\sqrt{1-2J^2 + \sqrt{4J^4 - 4J^2 + 2}}$$ (7.16)

The graph of $|G(j\omega)|$ versus the normalized frequency ω/ω_n in relation to the damping ratio J is depicted in Figure 7.2 while the graph of the ratio BW/ω_n versus the damping ratio J is illustrated in Figure 7.3.

The maximum value of time response y_m for a unit-step input is

$$y_m = 1 + e^{-J\pi/\sqrt{1-J^2}}$$ (7.17)

Both y_m and M_p are functions of the damping ratio J.

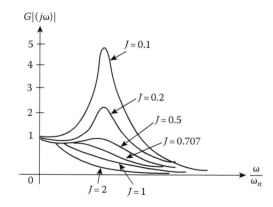

FIGURE 7.2
Graph of $|G(j\omega)|$ versus the normalized frequency ω/ω_n.

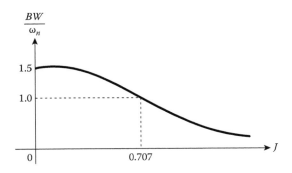

FIGURE 7.3
The ratio BW/ω_n versus the damping ratio J.

The damped natural frequency is

$$\omega_d = \omega_n \sqrt{1-J^2} \tag{7.18}$$

Both ω_d and ω_r are functions of ω_n and J. It holds that

$$\frac{\omega_r}{\omega_d} = \frac{\sqrt{1-2J^2}}{\sqrt{1-J^2}} \tag{7.19}$$

Recall that the time response of a second-order system is

$$y(t) = L^{-1}\{Y(s)\} = 1 - \frac{e^{-J\omega_n t}}{\sqrt{1-J^2}} \sin(\omega_d t + \varphi) \tag{7.20}$$

For a constant damping ratio, the larger the natural frequency, the faster the output reaches its steady state. The usual requirements are

- Relatively low values of M_p
- Relatively large values of bandwidth so that the time constant of the system $\tau = 1/\omega_n$ is small

7.4 Bode Diagrams

Bode diagrams or Bode plots consist of two curves. The first is the curve of $|G(j\omega)|$ in decibel (dB) as a function of ω and the other one is the phase curve $\varphi = \sphericalangle G(j\omega)$ as a function of ω.
   Bode plots provide information about the absolute and the relative stability of linear closed-loop systems.
   An important advantage of the frequency response analysis is that it makes possible to examine the stability of a closed-loop system from the frequency response of the open-loop system.

Suppose that the general form of a transfer function $G(j\omega)$ is

$$G(j\omega) = \frac{K(j\omega T_1 + 1)(j\omega T_2 + 1)^2}{(j\omega)^n (j\omega T_3 + 1)\left[(j\omega)^2 + 2J\omega_n(j\omega) + \omega_n^2\right]} \tag{7.21}$$

Bode diagrams, that is, magnitude and phase curves of $G(j\omega)$ versus the frequency ω, are plotted in semi-logarithmic scale.

The magnitude is expressed in decibel (dB). Thus,

$$20\log|G(j\omega)| = 20\log|K| + 20\log|j\omega T_1 + 1| + 40\log|j\omega T_2 + 1|$$

$$-20n\log|j\omega| - 20\log|j\omega T_3 + 1| - 20\log|(j\omega)^2 + 2J\omega_n(j\omega) + \omega_n^2| \tag{7.22}$$

or

$$20\log|G(j\omega)| = 20\log|K| + 20\log\sqrt{1+(\omega T_1)^2} + 40\log\sqrt{1+(\omega T_2)^2}$$

$$-20n\log\omega - 20\log\sqrt{1+(\omega T_3)^2} - 20\log\sqrt{(\omega_n^2 - \omega^2)^2 + (2J\omega_n\omega)^2} \tag{7.23}$$

The phase is expressed in degrees. Thus,

$$\varphi = \sphericalangle G(j\omega) = \sphericalangle(j\omega T_1 + 1) + \sphericalangle(j\omega T_2 + 1)^2 - \sphericalangle(j\omega)^n - \sphericalangle(j\omega T_3 + 1)$$

$$-\sphericalangle\left((j\omega)^2 + 2J\omega_n(j\omega) + \omega_n^2\right) \tag{7.24}$$

or

$$\varphi = \sphericalangle G(j\omega) = \tan^{-1}\omega T_1 + 2\tan^{-1}\omega T_2 - n\cdot 90° - \tan^{-1}\omega T_3 - \tan^{-1}\frac{2J\omega_n\omega}{\omega_n^2 - \omega^2} \tag{7.25}$$

We now introduce the procedure of plotting the individual terms given in relationships (7.23) and (7.25). The magnitude is denoted by π and the phase is denoted by φ.

1. **Real Constant K**

$$\Pi = 20\log K \tag{7.26}$$

$$\varphi = \begin{cases} 0°, & K > 0 \\ 180°, & K < 0 \end{cases} \tag{7.27}$$

2. **Integration and differentiation terms**

$$\Pi = \pm 20n\log\omega \tag{7.28}$$

$$\varphi = \pm 90°n \tag{7.29}$$

Relationship (7.28) represents a set of straight lines in semi-logarithmic scale with a slope of $\pm 20n$ dB per decade.

3. **Poles and zeros of the form** $(j\omega T + 1)^{\pm n}$

$$\Pi = \pm 20n \log |j\omega T + 1| = \pm 20n \log \sqrt{1 + (\omega T)^2} \qquad (7.30)$$

$$\varphi = \pm n \tan^{-1} \omega T \qquad (7.31)$$

The angular frequency $\omega = 1/T$ is called **corner frequency.**
We have

$$\Pi \simeq \begin{cases} 0, & \omega \ll \dfrac{1}{T} \\[2mm] \pm 3n, & \omega = \dfrac{1}{T} \\[2mm] \pm 20n \log \omega T, & \omega \gg \dfrac{1}{T} \end{cases} \qquad (7.32)$$

$$\varphi = \begin{cases} 0, & \omega = 0 \\ \pm 45°n, & \omega T = 1 \\ \pm 90°n, & \omega \to \infty \end{cases} \qquad (7.33)$$

4. **Terms of the form** $\left[\left(j\dfrac{\omega}{\omega_n} \right)^2 + 2J\left(j\dfrac{\omega}{\omega_n} \right) + 1 \right]^{\pm n}$

$$\Pi = \pm 20n \log \sqrt{\left(1 - \left(\dfrac{\omega}{\omega_n} \right)^2 \right)^2 + 4J^2 \left(\dfrac{\omega}{\omega_n} \right)^2} \qquad (7.34)$$

$$\varphi = \pm n \tan^{-1} \dfrac{2J(\omega/\omega_n)}{1 - (\omega/\omega_n)^2} \qquad (7.35)$$

We have

$$\Pi \simeq \begin{cases} 0, & \dfrac{\omega}{\omega_n} \ll 1 \\[3mm] \pm 40n \log\left(\dfrac{\omega}{\omega_n} \right), & \dfrac{\omega}{\omega_n} \gg 1 \\[3mm] \pm 20n \log\left| 1 + \left(\dfrac{\omega}{\omega_n} \right)^2 \right|, & J = 1 \\[3mm] \pm 20n \log\left| 1 - \left(\dfrac{\omega}{\omega_n} \right)^2 \right|, & J = 0 \end{cases} \qquad (7.36)$$

and

$$\varphi = \begin{cases} 0, & \dfrac{\omega}{\omega_n} = 0 \\[2ex] \pm 90°\,n, & \dfrac{\omega}{\omega_n} = 1 \\[2ex] \pm 180°\,n, & \dfrac{\omega}{\omega_n} \to \infty \end{cases} \tag{7.37}$$

7.5 Bode Stability Criterion

A system is unstable if its gain is greater than unity (or 0 dB), at a frequency for which the phase is 180°.

A Bode diagram provides two measures of the relative stability of the system. These are the gain margin K_g and the phase margin φ_a.

7.5.1 Gain Margin K_g

Gain margin K_g is the quantity that results from the following relationship:

$$K_g(\text{dB}) = -20\log|G(j\omega_c)H(j\omega_c)| \tag{7.38}$$

where ω_c is the frequency for which

$$\varphi(\omega_c) = -180° \tag{7.39}$$

A closed-loop system is **stable** if

$$K_g > 0 \tag{7.40}$$

7.5.2 Phase Margin φ_a

Phase margin is the quantity that results from the relationship

$$\varphi_a = 180° + \varphi \tag{7.41}$$

where
   φ is the angle of $G(j\omega_1)H(j\omega_1)$
   ω_1 is the frequency for which

$$|G(j\omega_1)H(j\omega_1)| = 1 \tag{7.42}$$

A closed-loop system is stable if

$$\varphi_a > 0 \tag{7.43}$$

Remarks:

- Bode diagrams are often used with an open-loop transfer function $G(j\omega)H(j\omega)$ for examining the stability of a closed-loop system.
- It is usually an easy task to experimentally compute the frequency response of a system. This is important as sometimes it is not possible to find the unknown transfer function of a system or of one of its components.
- The plot of approximate logarithmic curves of a Bode diagram is facilitated by a method of asymptotic approximations. The curve can be easily sketched if we apply the proper corrections to the asymptotic lines.

  In particular, every term of the form $(1 + j\omega)^{\pm n}$ must be corrected by $\pm 3n$ dB at the corner frequency $\omega_o = 1/T$ rad/s and by $\pm n$ dB at $\omega = 2/T$ and at $\omega = 1/2T$.
- A negative phase angle is called **phase lag**, while a positive phase angle is called **phase lead**.
- A transfer function is called a **minimum-phase transfer function** if it has no poles or zeros at the right-half s-plane. If it has poles or zeros in the right-half s-plane, it is called nonminimum-phase transfer function. For minimum-phase systems, the phase angle for $\omega \to \infty$ is $-90° \times (n - m)$, where n is the degree of the numerator's polynomial, and m is the degree of the denominator's polynomial of the transfer function. The slope of the magnitude curve tends to $-20 \times (n - m)$ dB/dec, as the frequency ω approaches infinity.

Formulas

TABLE F7.1

Frequency Response of First- and Second-Order Systems

	Transfer Function	Frequency Response	Characteristics
First-order system	$G(s) = \dfrac{K}{Ts+1}$	$G(j\omega) = \dfrac{K}{j\omega T + 1}$	$M_p = K$
			$\omega_r = 0$
		$\|G(j\omega)\| = K$	$\omega_b = \dfrac{1}{T} = BW$
		$\angle G(j\omega) = -\tan^{-1}\omega T$	
Second-order system	$G(s) = \dfrac{\omega_n^2}{s^2 + 2J\omega_n s + \omega_n^2}$	$G(j\omega) = \dfrac{\omega_n^2}{(j\omega)^2 + 2J\omega_n(j\omega) + \omega_n^2}$	$M_p = \dfrac{1}{2J\sqrt{1-J^2}}, \quad J < 0.707$
		$\|G(j\omega)\| = \dfrac{1}{\sqrt{(1-(\omega/\omega_n)^2)^2 + (2J(\omega/\omega_n))^2}}$	$M_p = 1, \quad J \ge 0.707$
			$\omega_r = \omega_n\sqrt{1-2J^2}, \quad J < 0.707$
		$\angle G(j\omega) = -\tan^{-1}\dfrac{2J(\omega/\omega_n)}{1-(\omega/\omega_n)^2}$	$BW = \omega_b$
			$\quad = \omega_n\sqrt{1-2J^2 + \sqrt{4J^4 - 4J^2 + 2}}$
			$\dfrac{\omega_r}{\omega_d} = \dfrac{\sqrt{1-2J^2}}{\sqrt{1-J^2}}$
In both cases		$\|G(j\omega_b)\| = 0.707$	
		$\|G(j\omega_r)\| = M_p$	

TABLE F7.2

Bode Diagrams

S/N	Various Terms	Diagram
1	Real constant K	Π (dB)

$$\Pi = 20 \log K$$

$$\varphi = \begin{cases} 0^\circ, & K > 0 \\ 180^\circ, & K < 0 \end{cases}$$

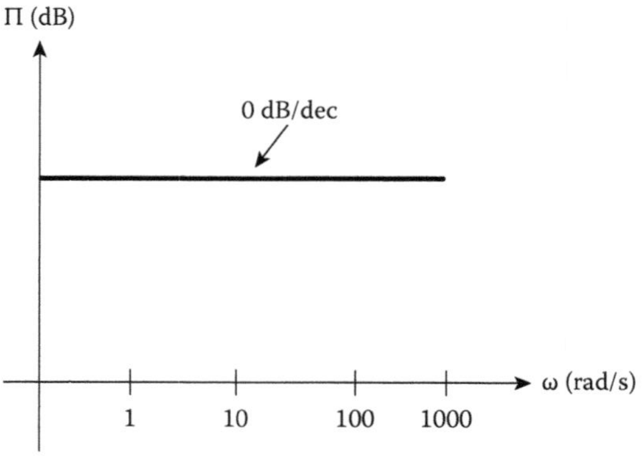

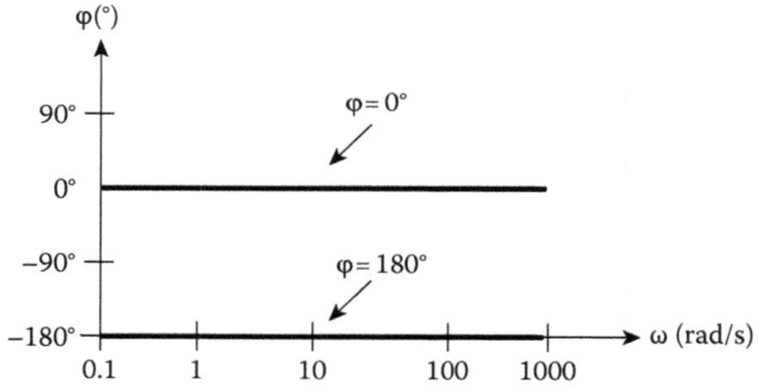

| 2 | Term of the form $(j\omega)^{\pm n}$ | 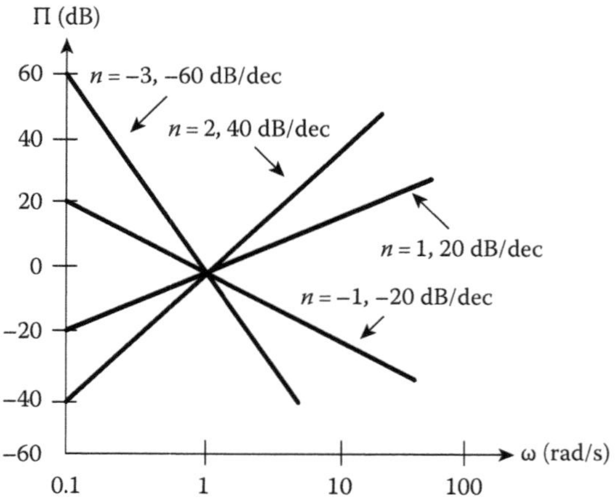 Π (dB) |

TABLE F7.2 (continued)

Bode Diagrams

S/N	Various Terms	Diagram
	$\Pi = \pm 20n \log \omega$ $\varphi = \pm 90° n$	$\varphi(°)$ $n = 2$ $180°$ $90°$ $0°$ $n = -1$ $-90°$ $-180°$ $n = -3$ $-270°$ ω (rad/s) 0.1 1 10
3	Term of the form $(j\omega T + 1)^{\pm n}$	Π (dB) 20 dB/dec $n = 1$ 0.1 ω (T) $n = -1$ 1 10 -20 dB/dec
	$\Pi = \pm 20n \log \sqrt{1 + (\omega T)^2}$ or $\Pi = \begin{cases} 0, & \omega \ll 1/T \\ \pm 3n, & \omega = 1/T \\ \pm 20n \log \omega T, & \omega \gg 1/T \end{cases}$ $\varphi = \pm n \tan^{-1} \omega T$ or $\varphi = \begin{cases} 0, & \omega = 0 \\ \pm 45° n, & \omega T = 1 \\ \pm 90° n, & \omega \to \infty \end{cases}$	$\varphi(°)$ $90°$ $n = 1$ $0°$ ω (T) $n = -1$ $-90°$

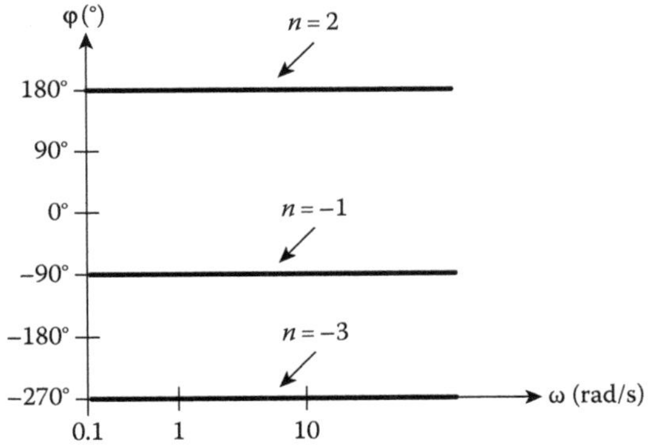

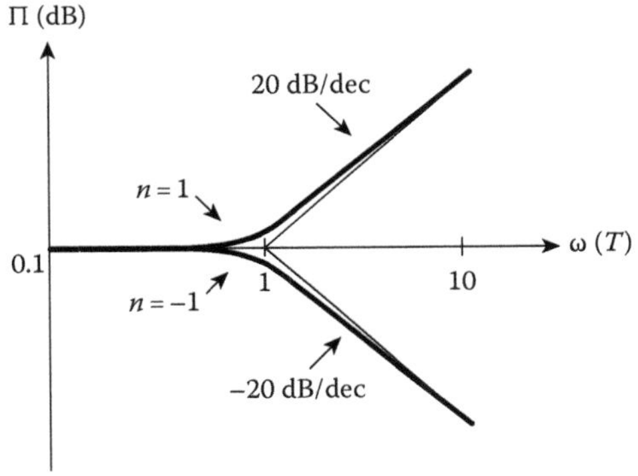

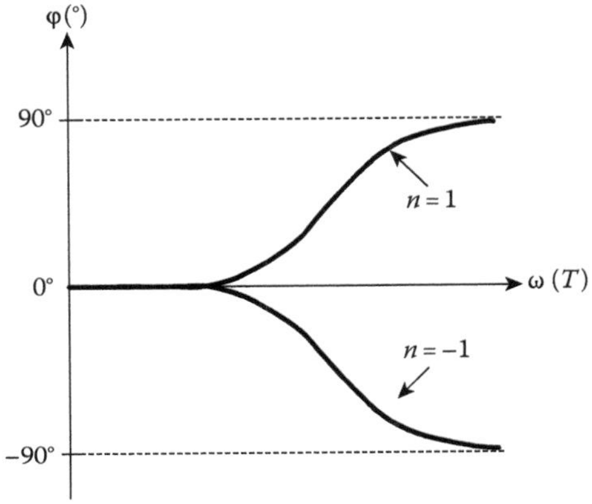

(continued)

TABLE F7.2 (continued)

Bode Diagrams

S/N	Various Terms	Diagram
4	Term of the form $$\left[\left(j\frac{\omega}{\omega_n}\right)+2J\left(j\frac{\omega}{\omega_n}\right)^2+1\right]^{\pm n}$$	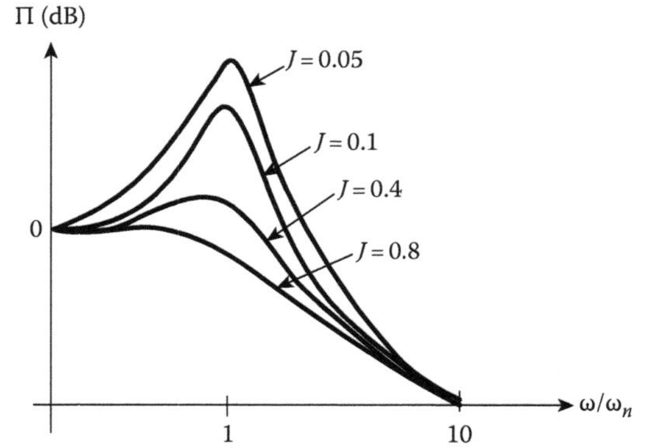

$$\Pi = \pm 20n\log\sqrt{\left(1-\left(\frac{\omega}{\omega_n}\right)^2\right)^2+4J^2\left(\frac{\omega}{\omega_n}\right)^2}$$

or

$$\Pi = \begin{cases} 0, & \frac{\omega}{\omega_n}\ll 1 \\ \pm 40nl\log\frac{\omega}{\omega_n}, & \frac{\omega}{\omega_n}\gg 1 \\ \pm 20n\log\left|1+\left(\frac{\omega}{\omega_n}\right)^2\right|, & J=1 \\ \pm 20n\log\left|1-\left(\frac{\omega}{\omega_n}\right)^2\right|, & J=0 \end{cases}$$

$$\varphi = \pm 90°n$$

or

$$\varphi = \begin{cases} 0, & \frac{\omega}{\omega_n}=0 \\ \pm 90°n, & \frac{\omega}{\omega_n}=1 \\ \pm 180°n, & \omega\to\infty \end{cases}$$

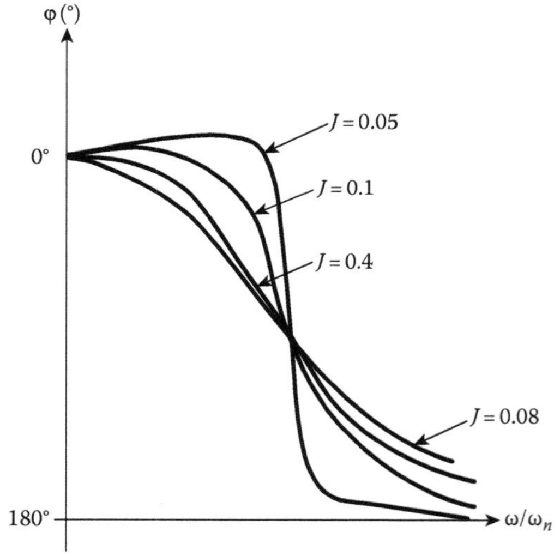

TABLE F7.3

Gain and Phase Margin

S/N	Name	Formula-Remarks
1	Gain margin (K_g)	$K_g(\text{dB}) = -20\log\lvert G(j\omega_c)H(j\omega_c)\rvert$
		$\varphi(\omega_c) = -180°$
		For $K_g > 0 \Rightarrow$ The closed-loop system is stable
2	Phase margin (φ_α)	$\varphi_\alpha = 180° + \varphi$
		$\varphi : \lvert G(j\omega_1)H(j\omega_1)\rvert = 1$
		For $\varphi_\alpha > 0 \Rightarrow$ The closed-loop system is stable

Problems

7.1 The electric circuit shown in the figure has an input voltage of the form $e_i(t) = E \sin \omega t$. Find the steady-state current that passes through the resistor R.

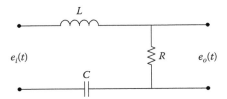

Solution

We apply Kirchhoff's law for voltage to the circuit. We have

$$L\frac{di(t)}{dt} + Ri(t) + \frac{1}{C}\int_0^t i(t)\,dt = e_i(t) \tag{P7.1.1}$$

The transfer function between $I(s)$ and $E_i(s)$ is

$$\frac{I(s)}{E_i(s)} = \frac{1}{Ls + R + (1/Cs)} = G(s) \tag{P7.1.2}$$

For an input $e_i(t) = E \sin \omega t$, the steady-state current $i_{ss}(t)$ is

$$i_{ss}(t) = E\lvert G(j\omega)\rvert \sin(\omega t + \sphericalangle G(j\omega)) \tag{P7.1.3}$$

where

$$G(j\omega) \overset{(P7.1.2)}{=} \frac{1}{Lj\omega + R - j(1/C\omega)} \tag{P7.1.4}$$

Hence,

$$|G(j\omega)|^{(P7.1.4)} = \frac{1}{\sqrt{R^2 + (L\omega - (1/C\omega))^2}} \tag{P7.1.5}$$

and

$$\angle G(j\omega)^{(P7.1.4)} = -\tan^{-1}\left(\frac{L\omega - (1/C\omega)}{R}\right) \tag{P7.1.6}$$

Therefore, the steady-state current that passes through the resistor R is

$$i_{ss}(t) = \frac{E}{\sqrt{R^2 + (L\omega - (1/C\omega))^2}} \sin\left[\omega t - \tan^{-1}\left(\frac{L\omega - (1/C\omega)}{R}\right)\right] \tag{P7.1.7}$$

7.2 The block diagram of a feedback system is depicted in Figure (a). Figures (b) and (c) illustrate the frequency responses of the stages $G_1(s)$ and $G_2(s)$, respectively. The feedback loop transfer function is

$$H(s) = \frac{1}{4}$$

a. Find the transfer function of the system $F(s) = \dfrac{C(s)}{R(s)}$.

b. Compute the natural frequency ω_n and the damping ratio J.

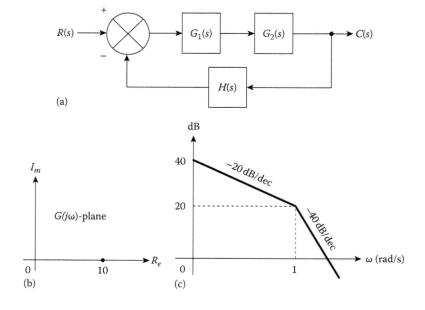

(a)

(b)

(c)

Solution

a. From the system block diagram, we obtain the closed-loop transfer function of the system:

$$F(s) = \frac{C(s)}{R(s)} = \frac{G_1(s)G_2(s)}{1 + H(s)G_1(s)G_2(s)} \tag{P7.2.1}$$

From Figure (b), the transfer function $G_1(s)$ is computed as

$$G_1(j\omega) = 10 + j0 \Rightarrow G_1(s) = 10 \tag{P7.2.2}$$

From Figure (c), the transfer function $G_2(s)$ is computed as

$$G_2(j\omega) = \frac{K}{j\omega(j(\omega/1) + 1)} \tag{P7.2.3}$$

where

$$20 \log K = 20 \Rightarrow \log K = 1 \Rightarrow K = 10 \tag{P7.2.4}$$

$$(\text{P7.2.3}), (\text{P7.2.4}) \Rightarrow G_2(s) = \frac{10}{s(s+1)} \tag{P7.2.5}$$

By substituting to (P7.2.1), we get

$$F(s) = \frac{C(s)}{R(s)} = \frac{10 \cdot (10/s(s+1))}{1 + (1/4) \cdot 10 \cdot (10/s(s+1))} = \frac{100}{s^2 + s + 25} \tag{P7.2.6}$$

b. The characteristic equation of the system is

$$s^2 + s + 25 \equiv s^2 + 2J\omega_n s + \omega_n^2 \tag{P7.2.7}$$

Thus,

$$\omega_n = \sqrt{25} \Rightarrow \omega_n = 5\,\text{rad/s} \tag{P7.2.8}$$

and

$$2J\omega_n = 1 \Rightarrow J = \frac{1}{2\omega_n} \Rightarrow J = 0.1 < 1 \tag{P7.2.9}$$

7.3 Consider the system shown in the following figure.

a. Sketch the Bode plot of the open-loop transfer function for $K = 1$.

b. Indicate the gain and phase margins. Is the closed-loop system stable?

c. For which value of K is the phase margin equal to $45°$?

d. For which value of K is the gain margin equal to $16\,\text{dB}$?

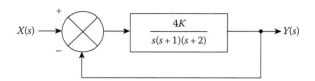

Solution

a. The transfer function of the open-loop system is

$$G(s)H(s) = \frac{4K}{s(s+1)(s+2)} \tag{P7.3.1}$$

For $s = j\omega$ from relationship (P7.3.1), and for $K = 1$, we have

$$G(j\omega)H(j\omega) = \frac{4}{j\omega(j\omega+1)(j\omega+2)} \tag{P7.3.2}$$

or

$$G(j\omega)H(j\omega) = \frac{2}{j\omega\left(j\underbrace{\frac{\omega}{1}}_{\omega_{01}}+1\right)\left(j\underbrace{\frac{\omega}{2}}_{\omega_{02}}+1\right)} \tag{P7.3.3}$$

There is a pole at $s = 0$ and there are two real poles, one at $s = -1$, with a corner frequency $\omega_{01} = 1\,\text{rad/s}$, and another at $s = -2$, with corner frequency $\omega_{02} = 2\,\text{rad/s}$. The magnitude of $G(j\omega)H(j\omega)$ in dB is computed as follows:

$$(P7.3.3) \Rightarrow |G(j\omega)H(j\omega)|_{\text{dB}} = 20\log|G(j\omega)H(j\omega)| = 20\log\left|\frac{2}{j\omega(j(\omega/1)+1)(j(\omega/2)+1)}\right|$$

$$= 20\log 2 - 20\log\left|j\omega(j\omega+1)\left(j\frac{\omega}{2}+1\right)\right|$$

$$= 20\log 2 - 20\left[\log|j\omega| + \log|j\omega+1| + \log\left|j\left(\frac{\omega}{2}\right)+1\right|\right]$$

$$= 20\log 2 - 20\log\omega - 20\log\sqrt{1+\omega^2} - 20\log\sqrt{1+\left(\frac{\omega}{2}\right)^2} \tag{P7.3.4}$$

The phase of $G(j\omega)H(j\omega)$ in degrees is computed as follows:

$$(P7.3.3) \Rightarrow \sphericalangle G(j\omega)H(j\omega) = \sphericalangle\left(\frac{2}{j\omega(j(\omega/1)+1)(j(\omega/2)+1)}\right)$$

$$= -\tan^{-1}\frac{\omega}{0} - \tan^{-1}\frac{\omega}{1} - \tan^{-1}\frac{\omega}{2} = -90° - \tan^{-1}\omega - \tan^{-1}\frac{\omega}{2} \tag{P7.3.5}$$

We proceed by plotting the **magnitude curve** with the use of asymptotes:

1. The term $-20 \log \omega$, that is, the term $(j\omega)^{-1}$, is a straight line with a slope of $-20\,\text{dB/s}$.

2. The term $-20 \log \sqrt{1+\omega^2}$, that is, the term $(j\omega + 1)^{-1}$, is activated at the corner frequency $\omega_{01} = 1\,\text{rad/s}$ and contributes to the slope by $-20\,\text{dB/dec}$. Consequently, the total slope is given by $(-20\,\text{dB/dec}) + (-20\,\text{db/dec}) = -40\,\text{dB/dec}$.

3. The term $-20 \log \sqrt{1+(\omega/2)^2}$, that is, the term $(j(\omega/2) + 1)^{-1}$, is activated at the corner frequency $\omega_{02} = 2\,\text{rad/s}$ and contributes to the slope by $-20\,\text{dB/dec}$. The curve's slope is now given $(-40\,\text{dB/dec}) + (-20\,\text{dB/dec}) = -60\,\text{dB/dec}$.

4. The approximate magnitude curve for $\omega_{01} = 1\,\text{rad/s}$ begins at $26\,\text{dB}$ as from (P7.3.4) $20 \log 2 - 20 \log 0.1 = 6 - (-20) = 26\,\text{dB}$.

We plot the exact phase curve after constructing the following table of values, based on the relationship (P7.3.5):

ω	$-90°$	$-\tan^{-1}\omega$	$-\tan^{-1}\dfrac{\omega}{2}$	$G(j\omega)H(j\omega)$
0.1	$-90°$	$-5.7°$	$-2.9°$	$-98.6°$
0.2	$-90°$	$-11.3°$	$-5.7°$	$-107°$
0.5	$-90°$	$-26.6°$	$-14°$	$-130.6°$
1	$-90°$	$-45°$	$-26.6°$	$-161.6°$
1.5	$-90°$	$-56.3°$	$-36.9°$	$-183.2°$
2	$-90°$	$-63.4°$	$-45°$	$-198.4°$
5	$-90°$	$-78.7°$	$-68.2°$	$-236.9°$
10	$-90°$	$-84.3°$	$-78.7°$	$-253°$
20	$-90°$	$-87.1°$	$-84.3°$	$-261.4°$

The following sketch represents the magnitude (approximate) and phase (exact) curves in semi-logarithmic scale.

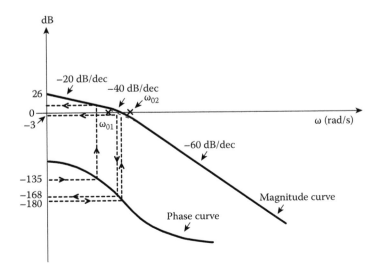

b. From the magnitude curve of the Bode diagram we observe that for the frequency $\omega_c \cong 1.15\,\text{rad/s}$ we have $20 \log |G(j\omega_1)H(j\omega_1)| = 0$. For this frequency the phase is approximately $-168°$. Thus, the phase margin is

$$\varphi_\alpha = 180° - 168° = 12° > 0 \qquad\qquad (P7.3.6)$$

For the frequency $\omega_c \cong 1.4\,\text{rad/s}$, the phase is $-180°$. The gain at this frequency is $-3\,\text{dB}$. Hence, the gain margin is

$$K_g = -20 \log |G(j\omega_c)H(j\omega_c)| = -|G(j\omega_c)H(j\omega_c)|_{\text{dB}} \simeq 3\,\text{dB} > 0 \qquad (P7.3.7)$$

Since $\varphi_\alpha > 0$ and $K_g > 0$ the system is stable.

c. The phase margin will be $45°$, if the phase at frequency ω_c is $-180° + 45° = -135°$. The frequency that corresponds to this phase is $\omega_1 = 0.6\,\text{rad/s}$, for which $20 \log |G(j\omega_1')H(j\omega_1')| = 5\,\text{dB}$.

Therefore, we must insert an additional gain to the system, as shown in the following figure:

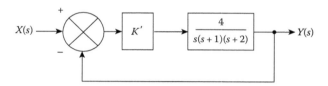

It must hold that

$$20 \log |K'G(j\omega_1')H(j\omega_1')| = 0 \Rightarrow 20 \log K' + 20 \log |G(j\omega_1')H(j\omega_1')| = 0 \Rightarrow$$

$$20 \log K' = -5 \Rightarrow \log K' = -0.25 \Rightarrow K' = 0.56 \qquad\qquad (P7.3.8)$$

The total gain of the system is

$$K' \cdot 4K \overset{(K=1)}{=} 0.56 \cdot 4 = 2.24 \qquad\qquad (P7.3.9)$$

d. The gain margin will become $16\,\text{dB}$, if we add a gain K'' to the system, so that for $\omega_1 = 1.4\,\text{rad/s}$ we have

$$|K''G(j\omega_c)H(j\omega_c)|_{\text{dB}} = -16\,\text{dB} \Rightarrow$$

$$20 \log K'' = -16\,\text{dB} - 20 \log |G(j\omega_c)H(j\omega_c)| = -16\,\text{dB} + 3\,\text{dB} = -13\,\text{dB} \Rightarrow$$

$$\log K'' = -\frac{13}{20} \Rightarrow K'' = 0.224 \qquad\qquad (P7.3.10)$$

7.4 Consider the system depicted in the following figure:

a. Draw the Bode plot of the open-loop transfer function for $K = 1$.

b. Determine graphically the value of the gain constant so that the phase margin is 30°. Is the system stable?

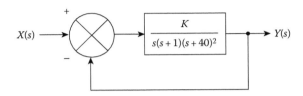

Solution

a. The open-loop transfer function of the system shown is

$$G(s)H(s) = \frac{K}{s(s+1)(s+40)^2} \qquad \text{(P7.4.1)}$$

From relationship (P7.4.1) and for $s = j\omega$ and $K = 1$, we have

$$G(j\omega)H(j\omega) = \frac{1}{j\omega(j\omega+1)(j\omega+40)^2} \qquad \text{(P7.4.2)}$$

or

$$G(j\omega)H(j\omega) = \frac{6.25 \cdot 10^{-4}}{j\omega(j(\omega/1)+1)(j(\omega/40)+1)^2} \qquad \text{(P7.4.3)}$$

There is a pole at the point $s = 0$, a pole at the point $s = -1$ with a corner frequency $\omega_{01} = 1\,\text{rad/s}$, and a double pole at the point $s = -40$ with a corner frequency $\omega_{02} = 40\,\text{rad/s}$.

The magnitude in dB of $G(j\omega)H(j\omega)$ is computed as

$$\text{(P7.4.3)} \Rightarrow |G(j\omega)H(j\omega)|_{dB} = 20\log|G(j\omega)H(j\omega)| = 20\log\left|\frac{6.25 \cdot 10^{-4}}{j\omega(j(\omega/1)+1)(j(\omega/40)+1)^2}\right|$$

$$= 20\log 6.25 \cdot 10^{-4} - 20\log\omega - 20\log\sqrt{1+\omega^2} - 40\log\sqrt{1+\left(\frac{\omega}{40}\right)^2} \qquad \text{(P7.4.4)}$$

The phase (in degrees) of $G(j\omega)H(j\omega)$ is given by

$$\text{(P7.4.3)} \Rightarrow \sphericalangle G(j\omega)H(j\omega) = \sphericalangle\left(\frac{6.25 \cdot 10^{-4}}{j\omega(j(\omega/1)+1)(j(\omega/40)+1)^2}\right)$$

$$= -90° - \tan^{-1}\omega - 2\tan^{-1}\frac{\omega}{40} \qquad \text{(P7.4.5)}$$

We plot the magnitude curve with the use of asymptotes:

1. The term $20 \log 6.25 \cdot 10^{-4} \simeq -64\,\text{dB}$ is a straight line with a slope of $0\,\text{dB/dec}$.

2. The term $-20 \log \omega$ is a straight line with a slope of $-20\,\text{dB/dec}$.

3. The term $-20 \log \sqrt{1 + \omega^2}$ is activated at the corner frequency $\omega_{01} = 1\,\text{rad/s}$ and contributes with a slope of $-20\,\text{dB/dec}$. Hence, the total slope is $(-20\,\text{dB/dec}) + (-20\,\text{dB/dec}) = -40\,\text{dB/dec}$.

4. The term $-40 \log \sqrt{1 + (\omega/40)^2}$ is activated at the corner frequency $\omega_{02} = 40\,\text{rad/s}$ and contributes with a slope of $-40\,\text{dB/dec}$.

The slope of the magnitude curve is $(-40\,\text{dB/dec}) + (-40\,\text{dB/dec}) = -80\,\text{dB/dec}$.

We plot the exact phase curve by constructing the following table of values, based on relationship (P7.4.5):

ω	$-90°$	$-\tan^{-1}\omega$	$-2\tan^{-1}\dfrac{\omega}{40}$	$G(j\omega)H(j\omega)$
0.1	$-90°$	$-5.7°$	$-0.3°$	$-96°$
0.5	$-90°$	$-26.6°$	$-1.4°$	$-118°$
1.0	$-90°$	$-45°$	$-2.9°$	$-137.9°$
5.0	$-90°$	$-78.7°$	$-14.3°$	$-183°$
10	$-90°$	$-84.3°$	$-28.1°$	$-202.4°$
20	$-90°$	$-87.1°$	$-53.1°$	$-230.2°$
40	$-90°$	$-88.6°$	$-90°$	$-268.6°$
100	$-90°$	$-89.4°$	$-136.4°$	$-315.8°$
1000	$-90°$	$-89.9°$	$-175.4°$	$-355.3°$

The following figure shows the magnitude curve (approximate and real) and phase (real) curve:

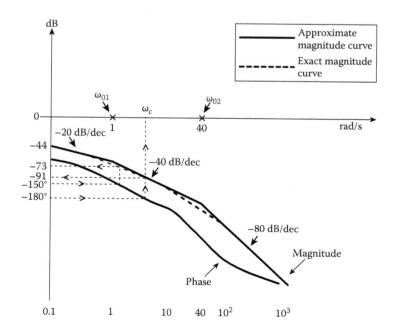

The magnitude curve, for $\omega = 0.1\,rad/s$, begins from $-44\,dB$ as from (P7.4.4) we have $20\log 6.25 \cdot 10^{-4} - 20\log \omega = -64 - 20\log(0.1) = -64 + 20 = -44\,dB$.

b. The phase margin is equal to $30°$, if

$$\varphi = \varphi_\alpha - 180° = 30° - 180° = -150° \qquad (P7.4.6)$$

From the Bode diagram, we see that a gain of $-73\,dB$ corresponds to a phase of $-150°$. Hence,

$$20\log|K| = 73 \Rightarrow K \simeq 4467 \qquad (P7.4.7)$$

The critical frequency of oscillations for a phase of $-180°$ is estimated from the Bode diagram. It is

$$\omega_c \simeq 4.4\,rad/s \qquad (P7.4.8)$$

The critical gain for phase $-180°$ is

$$K_c \simeq 91\,dB \Rightarrow K_c \simeq 35,481.3 \qquad (P7.4.9)$$

7.5 The transfer function of the open-loop system is

$$G(s)H(s) = \frac{100K(0.1s+1)}{s(0.001s+1)(s^2+2s+100)}$$

a. Plot the Bode diagrams of the open-loop transfer function for $K = 1$ and for $K = 10$.
b. Draw conclusions for the stability of the system.
c. Derive graphically the critical value of the gain constant K for stability and the critical frequency of oscillations.
d. What is the phase margin if the gain K is equal to $1/3$ of its critical value?

Solution

a. The open-loop transfer function of the system is

$$G(s)H(s) = \frac{100K(0.1s+1)}{s(0.001s+1)(s^2+2s+100)} \qquad (P7.5.1)$$

From the relationship (P7.5.1), for $s = j\omega$ and $K = 1$, we have

$$G(j\omega)H(j\omega) = \frac{100(0.1j\omega+1)}{j\omega(0.001j\omega+1)((j\omega)^2+2j\omega+100)} \qquad (P7.5.2)$$

or

$$G(j\omega)H(j\omega) = \frac{(j(\omega/10)+1)}{j\omega(j(\omega/1000)+1)((j(\omega/10))^2 + 2j(\omega/100)+1)} \qquad (P7.5.3)$$

From the relationship (P7.5.3), the corner frequency for the zero at $s = -10$ is $\omega_{01} = 10\,\text{rad/s}$. For the complex conjugate poles of the polynomial $s^2 + 2s + 100$, the corner frequency is $\omega_{02} = 10\,\text{rad/s}$. For the real pole at $s = -1000$ it is $\omega_{03} = 1000\,\text{rad/s}$. The magnitude of $G(j\omega)H(j\omega)$ in dB is computed as follows:

$$(P7.5.3) \Rightarrow 20\log|G(j\omega)H(j\omega)| = 20\log\left|\frac{(j(\omega/10)+1)}{j\omega(j(\omega/1000)+1)((j(\omega/10))^2 + 2j(\omega/100)+1)}\right|$$

$$= 20\log\left|j\frac{\omega}{10}+1\right| - 20\log|j\omega| - 20\log\left|j\frac{\omega}{1000}+1\right| - 20\log\left|\left(j\frac{\omega}{10}\right)^2 + 2j\frac{\omega}{100}+1\right| \Rightarrow$$

$$|G(j\omega)H(j\omega)|_{dB} = 20\log\sqrt{1+\left(\frac{\omega}{10}\right)^2} - 20\log\omega - 20\log\sqrt{1+\left(\frac{\omega}{1000}\right)^2}$$

$$-20\log\sqrt{\left(1-\left(\frac{\omega}{10}\right)^2\right)^2 + \left(\frac{2\omega}{100}\right)^2} \qquad (P7.5.4)$$

The phase of $G(j\omega)H(j\omega)$ in degrees is

$$(P7.5.3) \Rightarrow \sphericalangle G(j\omega)H(j\omega) = \sphericalangle\left(\frac{(j(\omega/10)+1)}{j\omega(j(\omega/1000)+1)((j(\omega/10))^2 + 2j(\omega/100)+1)}\right) \Rightarrow$$

$$\sphericalangle G(j\omega)H(j\omega) = \sphericalangle\left(j\frac{\omega}{10}+1\right) - \sphericalangle\left[j\omega\left(j\frac{\omega}{1000}+1\right)\left(\left(j\frac{\omega}{10}\right)^2 + 2j\frac{\omega}{100}+1\right)\right] \Rightarrow$$

$$\sphericalangle G(j\omega)H(j\omega) = \tan^{-1}\frac{\omega}{10} - 90° - \tan^{-1}\frac{\omega}{1000} - \tan^{-1}\frac{(2\omega/100)}{1-(\omega/10)^2} \qquad (P7.5.5)$$

The magnitude curve is plotted with the use of asymptotes:
1. For $\omega = 1\,\text{rad/s}$, from relationship (P7.5.4), we have

$$|G(j1)H(j1)|_{dB} \simeq -20\log 1 \simeq 0\,\text{dB}$$

For $\omega = 10$ rad/s, from relationship (P7.5.4), we have

$$|G(j10)H(j10)|_{dB} \simeq -20\log 10 = -20\,\text{dB}$$

Therefore, the approximate magnitude curve for $\omega = 1\,\text{rad/s}$ begins at $\omega = 1\,\text{rad/s}$ and has a slope of $-20\,\text{dB/s}$, which corresponds to the term $-20\log\omega$, that is, to the term $(j\omega)^{-1}$.

2. The term $20\log\sqrt{1+(\omega/10)^2}$, that is, the term $(1+j(\omega/10))^1$, is activated at the corner frequency $\omega_{01} = 10\,\text{rad/s}$ and contributes with a slope of $+20\,\text{dB/dec}$. At the same corner frequency $\omega_{02} = 10\,\text{rad/s}$, the term $-20\log\sqrt{(1-(\omega/10)^2)^2+(2\omega/100)^2}$, that is, $((j(\omega/10))^2 + 2j(\omega/100) + 1)^{-1}$, contributes with a slope of $-40\,\text{dB/dec}$.
   Hence, the total slope is $-20\,\text{dB/dec} + 20\,\text{dB/dec} + (-40\,\text{dB/dec}) = -40\,\text{dB/dec}$.

3. The term $-20\log\sqrt{1+(\omega/1000)^2}$, that is, $(1 + j(\omega/1000))^{-1}$, is activated at the corner frequency $\omega_{03} = 1000\,\text{rad/s}$ and contributes with a slope of $-40\,\text{dB/dec} + (-20\,\text{dB/dec}) = -60\,\text{dB/dec}$.
   We plot the exact phase curve after constructing the following table of values, based on the relationship (P7.5.5):

ω	$\left(1+J\dfrac{\omega}{40}\right)$	$(j\omega)^{-1}$	$\left(1+J\dfrac{\omega}{1000}\right)$	$\left(J\dfrac{\omega}{10}+2J\dfrac{\omega}{100}\right)^{-1}$	$G(j\omega)H(j\omega)$
1	5.7°	−90°	−0.1°	−1.2°	−85.6°
5	26.6°	−90°	−0.3°	−7.6°	−71.3°
10	45°	−90°	−0.6°	−45°	−90.6°
11	47.7°	−90°	−0.6°	−133.6°	−176.5°
15	56.3°	−90°	−0.9°	−166.5°	−201.1°
20	63.4°	−90°	−1.1°	−172.4°	−200.1°
50	78.7°	−90°	−2.9°	−177.6°	−191.8°
100	84.3°	−90°	−5.7°	−178.8°	−190.2°
500	88.9°	−90°	−26.6°	−179.8°	−207.5°
1000	89.4°	−90°	−45°	−179.9°	−225.5°
5000	89.9°	−90°	−78.7°	−179.9°	−258.7°

Remark:

For the argument of the term $((j(\omega/10))^2 + 2j(\omega/100) + 1)^{-1}$, it holds that

$$\sphericalangle\left(\left(j\frac{\omega}{10}\right)^2 + 2j\frac{\omega}{100} + 1\right)^{-1} = \begin{cases} -45°, & \omega = 10\,\text{rad/s} \\ -\tan^{-1}\dfrac{(2\omega/100)}{1-(\omega/10)^2}, & \omega < 10\,\text{rad/s} \\ -180° + \tan^{-1}\dfrac{(2\omega/100)}{(\omega/10)^2 - 1}, & \omega > 10\,\text{rad/s} \end{cases} \quad (\text{P7.5.6})$$

The following figure depicts the phase and magnitude plots of the loop transfer function of the system for $K = 1$ and for $K = 10$. For $K = 10$, the difference is that the magnitude curve is displaced by $20\,\text{dB}$ upward, because of the term $20\log K = 20\log 10 = 20\,\text{dB}$. The phase curve does not change.

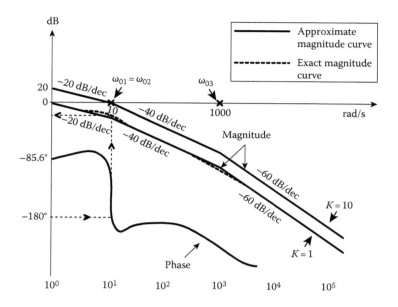

b. From the Bode diagram, we observe that for a phase of $G(j\omega)H(j\omega)$ equal to $-180°$, we have

$$K_c \simeq 20\,\text{dB} \Rightarrow K_c \simeq 10 \tag{P7.5.7}$$

The system is stable for

$$0 < K < 10 \tag{P7.5.8}$$

The phase margin is positive since

$$\varphi_\alpha \ (\text{for } K = 1) = 180° - (-85.6°) = 94.4° > 0 \tag{P7.5.9}$$

c. The critical value of K for stability is $K_c \simeq 10$, while the critical frequency of oscillations is

$$\omega_c \simeq 12\,\text{rad/s} \tag{P7.5.10}$$

d. For

$$K = \frac{1}{3}K_c \overset{(P7.5.7)}{\Rightarrow} K \simeq 3.33 = 10.5\,\text{dB} \tag{P7.5.11}$$

From the Bode diagram, we find that for $K = 10.5\,\text{dB}$ the phase is $\varphi \cong -76°$. Consequently, the phase margin is

$$\sphericalangle\varphi_\alpha = 180° - \left|-76°\right| = 180° - 76° = 140° > 0$$

We conclude that for $K = (1/3)K_c$ the relative stability of the system increases.

7.6 Plot the Bode diagrams for the systems with the following loop transfer functions:

a. $G(s)H(s) = \dfrac{(s+1)}{(s+2)(s+10)^2(s+100)}$

b. $G(s)H(s) = \dfrac{10}{s^2(1+0.5s)(1+0.1s)^2}$

Solution

a. The loop transfer function is

$$G(s)H(s) = \frac{s+1}{(s+2)(s+10)^2(s+100)} = \frac{5 \cdot 10^{-5}(s+1)}{((s/2)+1)((s/10)+1)^2((s/100)+1)} \qquad (P7.6.1)$$

From the relationship (P7.6.1) for $s = j\omega$ and for $K = 1$, we have

$$G(j\omega)H(j\omega) = \frac{5 \cdot 10^{-5}(j(\omega/1)+1)}{(j(\omega/2)+1)(j(\omega/10)+1)^2(j(\omega/100)+1)} \qquad (P7.6.2)$$

The magnitude of $G(j\omega)H(j\omega)$ in dB is

$$(P7.6.2) \Rightarrow |G(j\omega)H(j\omega)|_{dB} = 20\log|G(j\omega)H(j\omega)|$$

$$= 20\log\left|5 \cdot 10^{-5}\left(j\frac{\omega}{1}+1\right)\right| - 20\log\left|\left(j\frac{\omega}{2}+1\right)\left(j\frac{\omega}{10}+1\right)^2\left(j\frac{\omega}{100}+1\right)\right| \Rightarrow$$

$$|G(j\omega)H(j\omega)|_{dB} = 20\log 5 \cdot 10^{-5} + 20\log\sqrt{1+\left(\frac{\omega}{1}\right)^2} - 20\log\sqrt{1+\left(\frac{\omega}{2}\right)^2}$$

$$-40\log\sqrt{1+\left(\frac{\omega}{10}\right)^2} - 20\log\sqrt{1+\left(\frac{\omega}{100}\right)^2} \qquad (P7.6.3)$$

The phase of $G(j\omega)H(j\omega)$ in degrees is

$$(P7.6.2) \Rightarrow \sphericalangle G(j\omega)H(j\omega) = \sphericalangle\left(5 \cdot 10^{-5}\left(j\frac{\omega}{1}+1\right)\right)$$

$$-\sphericalangle\left(\left(j\frac{\omega}{2}+1\right)\left(j\frac{\omega}{10}+1\right)^2\left(j\frac{\omega}{100}+1\right)\right) \Rightarrow$$

$$\sphericalangle G(j\omega)H(j\omega) = \tan^{-1}\frac{\omega}{1} - \tan^{-1}\frac{\omega}{2} - 2\tan^{-1}\frac{\omega}{10} - \tan^{-1}\frac{\omega}{100} \qquad (P7.6.4)$$

The corner frequencies are $\omega_{01} = 1\,\text{rad/s}$ for the zero at $s = -1$, $\omega_{02} = 2\,\text{rad/s}$ for the pole at $s = -2$, $\omega_{03} = 10\,\text{rad/s}$ for the pole at $s = -10$, and $\omega_{04} = 100\,\text{rad/s}$ for the pole at $s = -100$. In order to design the magnitude curve we proceed as follows:

1. For $\omega = 0.1\,\text{rad/s}$, the curve begins from $20\log 5 \cdot 10^{-5} = -86\,\text{dB}$.

2. The term $20\log\sqrt{1+(\omega/1)^2}$ is activated at the corner frequency $\omega_{01} = 1\,\text{rad/s}$ and contributes to the slope by $+20\,\text{dB/dec}$. Hence, the magnitude curve for $\omega \geq 1\,\text{rad/s}$ has a slope of $20\,\text{dB/dec}$.

3. The term $-20\log\sqrt{1+(\omega/2)^2}$ is activated at the corner frequency $\omega_{02} = 2\,\text{rad/s}$ and contributes to the slope by $-20\,\text{dB/dec}$. Therefore, for $\omega \geq 2\,\text{rad/s}$, the slope of the magnitude curve is $+20\,\text{dB/dec} + (-20\,\text{dB/dec}) = 0\,\text{dB/dec}$.

4. The term $-40\log\sqrt{1+(\omega/10)^2}$ is activated at the corner frequency $\omega_{03} = 10\,\text{rad/s}$ and contributes to the slope by $-40\,\text{dB/dec}$. The total slope of the magnitude curve for $\omega \geq 10\,\text{rad/s}$ is $0\,\text{dB/dec} + (-40\,\text{dB/dec}) = -40\,\text{dB/dec}$.

5. The term $-20\log\sqrt{1+(\omega/10)^2}$ is activated at the corner frequency $\omega_{03} = 10\,\text{rad/s}$ and contributes to the slope by $-20\,\text{dB/dec}$. The total slope of the magnitude curve for $\omega \geq 100\,\text{rad/s}$ is $-40\,\text{dB/dec} + (-20\,\text{dB/dec}) = -60\,\text{dB/dec}$.

The exact phase curve is plotted by the use of the following table of values, based on the relationship (P7.6.4):

ω	$\tan^{-1}\dfrac{\omega}{1}$	$-\tan^{-1}\dfrac{\omega}{2}$	$-2\tan^{-1}\dfrac{\omega}{10}$	$-\tan^{-1}\dfrac{\omega}{100}$	$G(j\omega)H(j\omega)$
0.1	5.7°	−2.9°	−0.6°	−0.1°	2.1°
0.5	26.6°	−14°	−5.7°	−0.3°	6.6°
1	45°	−26.6°	−11.4°	−0.6°	6.4°
2	63.4°	−45°	−22.6°	−1.1°	−5.3°
5	78.7°	−68.2°	−53.1°	−2.9°	−45.5°
10	84.3°	−78.7°	−90°	−5.7°	−90.1°
50	88.9°	−87.7°	−157.4°	−26.6°	−182.8°
100	89.4°	−88.9°	−168.6°	−45°	−213.1°
500	89.9°	−89.8°	−177.7°	−78.7°	−256.3°

The following figure represents the Bode diagram (approximate and real magnitude-phase curves) in semi-logarithmic paper:

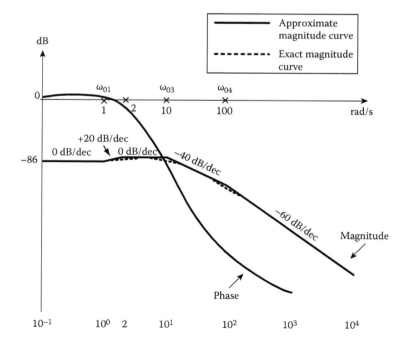

b. The loop transfer function is

$$G(s)H(s) = \frac{10}{s^2(1+0.5s)(1+0.1s)^2} = \frac{10}{s^2(1+(s/2))(1+(s/10))^2} \tag{P7.6.5}$$

For $s = j\omega$, from the relationship (P7.6.5), we have

$$G(j\omega)H(j\omega) = \frac{10}{(j\omega)^2(1+j(\omega/2))(1+j(\omega/10))^2} \tag{P7.6.6}$$

The magnitude of $G(j\omega)H(j\omega)$ in dB is

$$|G(j\omega)H(j\omega)|_{dB} = 20\log|G(j\omega)H(j\omega)|$$

$$= 20\log 10 - 20\log|(j\omega)^2| - 20\log\left|1+j\frac{\omega}{2}\right| - 20\log\left|\left(1+j\frac{\omega}{10}\right)^2\right|$$

$$= 20\log 10 - 40\log\omega - 20\log\sqrt{1+\left(\frac{\omega}{2}\right)^2} - 40\log\sqrt{1+\left(\frac{\omega}{10}\right)^2} \tag{P7.6.7}$$

The phase of $G(j\omega)H(j\omega)$ in degrees is

$$(\text{P7.6.6}) \Rightarrow \sphericalangle G(j\omega)H(j\omega) = \sphericalangle 10 - \sphericalangle(j\omega)^2 - \sphericalangle\left(1+j\frac{\omega}{2}\right) - \sphericalangle\left(1+j\frac{\omega}{10}\right)^2 \Rightarrow$$

$$\sphericalangle G(j\omega)H(j\omega) = -\underbrace{2\tan^{-1}\frac{\omega}{0}}_{-180^\circ} - \tan^{-1}\frac{\omega}{2} - 2\tan^{-1}\frac{\omega}{10} \tag{P7.6.8}$$

The corner frequencies are $\omega_{01} = 2\,\text{rad/s}$ for the pole at $s = -2$, and $\omega_{02} = 10\,\text{rad/s}$ for the pole at $s = -10$.

The magnitude curve is plotted with the use of asymptotes as follows:

1. For $\omega = 0.1\,\text{rad/s}$, it begins from $20\log 10 - 40\log(0.1) = 60\,\text{dB}$.
2. The term $-40\log\omega$ is a straight line with a slope of $-40\,\text{dB/dec}$.
3. The term $-20\log\sqrt{1+(\omega/2)^2}$ is activated at the corner frequency $\omega_{01} = 2\,\text{rad/s}$ and contributes with a slope of $-20\,\text{dB/dec}$. Therefore, the slope of the magnitude curve is $-40\,\text{dB/dec} + (-20\,\text{dB/dec}) = -60\,\text{dB/dec}$.
4. The term $-40\log\sqrt{1+(\omega/10)^2}$ is activated at the corner frequency $\omega_{02} = 10\,\text{rad/s}$ and contributes with a slope of $-40\,\text{dB/dec}$. Hence, the total slope of the magnitude curve for $\omega \geq 10\,\text{rad/s}$ is $-60\,\text{dB/dec} + (-40\,\text{dB/dec}) = -100\,\text{dB/dec}$.

For the phase curve, we construct the following table of values from the relationship (P7.6.8):

ω	$-180°$	$-\tan^{-1}\dfrac{\omega}{2}$	$-2\tan^{-1}\dfrac{\omega}{10}$	$G(j\omega)H(j\omega)$
0.1	$-180°$	$-2.7°$	$-1.1°$	$-183.8°$
0.5	$-180°$	$-14°$	$-5.7°$	$-199.7°$
1	$-180°$	$-26.6°$	$-11.4°$	$-218°$
2	$-180°$	$-45°$	$-22.6°$	$-247.6°$
10	$-180°$	$-78.7°$	$-90°$	$-348.7°$
50	$-180°$	$-87.7°$	$-157.4°$	$-425.1°$
100	$-180°$	$-88.8°$	$-168.6°$	$-437.4°$

The Bode diagram is

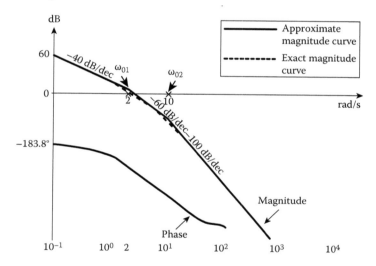

7.7 The figure below depicts the magnitude curves (asymptotic approach) of two minimum-phase transfer functions, $G_1(s)$ and $G_2(s)$:

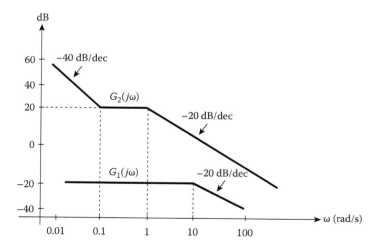

a. Sketch the approximate graph of the magnitude of the transfer function $G(s) = G_1(s)G_2(s)$.

b. Determine the analytic expression of the transfer function $G(s)$. If you derive that this is a second-order system, suppose that the damping ratio is $J = 0.5 < 1$.

c. Repeat the previous queries, supposing that the first slope of $G_2(j\omega)$ is $-20\,\text{dB/dec}$.

Solution

a. We know that

$$G(s) = G_1(s) \cdot G_2(s) \tag{P7.7.1}$$

Therefore,

$$20\log|G(j\omega)| = 20\log|G_1(j\omega) \cdot G_2(j\omega)| = 20\log|G_1(j\omega)| + 20\log|G_2(j\omega)| \tag{P7.7.2}$$

The graph of $G(s)$ is the result of the sum of the two given magnitude (in dB) graphs.

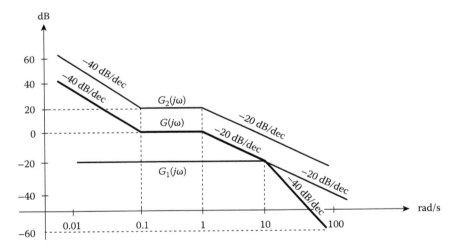

b. From the graph of $G(s)$, we have

$$G(j\omega) = \frac{K(1+(j\omega/0.1)^2 + 2Jj(\omega/0.1))}{(j\omega)^2(1+j(\omega/1))(1+j(\omega/10))} = \frac{K(1+(j\omega/0.1)^2 + 2(0.5)(j\omega/0.1))}{(j\omega)^2(1+j(\omega/1))(1+j(\omega/10))} \tag{P7.7.3}$$

In order to compute K, we assume that $\omega \simeq 0$. Hence,

$$G(j\omega) = \frac{K}{(j\omega)^2} = -\frac{K}{\omega^2} \tag{P7.7.4}$$

and

$$20\log|G(j\omega)| = 20\log\left|-\frac{K}{\omega^2}\right| = 20\log|K| - 20\log\omega^2 \tag{P7.7.5}$$

For magnitude of 0 dB, we have

$$0 = 20 \log |K| - 20 \log \omega^2 \tag{P7.7.6}$$

$$\Rightarrow |K| = K = \omega^2 \Rightarrow \omega = \sqrt{K} = 0.1 \tag{P7.7.7}$$

Thus,

$$G(s) = \frac{0.1^2(1 + (s^2/0.1^2) + (s/0.1))}{s^2(s+1)((s/10)+1)} \tag{P7.7.8}$$

c. The first slope of $G_2(j\omega)$ is −20 dB/dec instead of −40 dB/dec; thus, the magnitude curve of $G(s)$ has the following form:

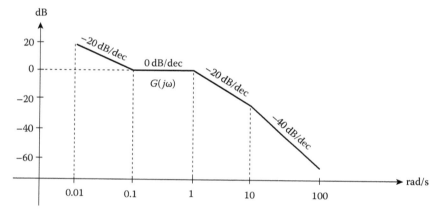

From the graph it follows that

$$G(j\omega) = \frac{K(1 + j(\omega/0.1))}{j\omega(1 + j(\omega/1))(1 + j(\omega/10))} \tag{P7.7.9}$$

For $\omega \simeq 0$, it is

$$G(j\omega) = \frac{K}{j\omega} \tag{P7.7.10}$$

Then

$$20 \log |G(j\omega)| = 20 \log \left| \frac{K}{j\omega} \right| = 20 \log |K| - 20 \log |j\omega| \tag{P7.7.11}$$

For magnitude of 0 dB, we have

$$0 = 20 \log |K| - 20 \log \omega \tag{P7.7.12}$$

$$\Rightarrow |K| = K = \omega = 0.1 \tag{P7.7.13}$$

Hence,

$$G(s) = \frac{0.1(1+(s/0.1))}{s(s+1)((s/10)+1)} \tag{P7.7.14}$$

7.8 The experimental graph of the approximate magnitude curve of a transfer function $G(s)$ is shown in the figure below. The transfer function $G(s)$ is minimum phase and has positive gain K. Compute the transfer function $G(s)$.

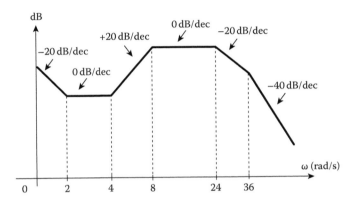

Solution

From the depicted graph, and for $\omega \simeq 0$ ($\omega < 2$), it follows that

$$G(j\omega) = \frac{K}{j\omega} \tag{P7.8.1}$$

Then

$$20\log|G(j\omega)| = 20\log\left|\frac{K}{j\omega}\right| = 20\log|K| - 20\log|j\omega| \tag{P7.8.2}$$

For magnitude of $0\,dB$, we have

$$0 = 20\log|K| - 20\log\omega \tag{P7.8.3}$$

$$\Rightarrow |K| = K = \omega = 8 \tag{P7.8.4}$$

There are zeros at the corner frequencies $\omega_{01} = 2\,rad/s$ and $\omega_{02} = 4\,rad/s$. There are poles at the corner frequencies $\omega_{03} = 8\,rad/s$, $\omega_{04} = 24\,rad/s$, and $\omega_{05} = 36\,rad/s$. Hence, the transfer function is

$$G(j\omega) = \frac{K(1+j(\omega/\omega_{01}))(1+j(\omega/\omega_{02}))}{j\omega(1+j(\omega/\omega_{03}))(1+j(\omega/\omega_{04}))(1+j(\omega/\omega_{05}))} \tag{P7.8.5}$$

By substitution, we have

$$G(s) = \frac{8(1+(s/2))(1+(s/4))}{s(1+(s/8))(1+(s/24))(1+(s/36))}$$ (P7.8.6)

7.9 The following figure illustrates the level-control system of a reservoir.

     a. Draw the Bode plot of the loop transfer function of the system for $N = 1$.

     b. Derive graphically the maximum number of input valves N_{max}, so that the system is stable. Confirm the result with the use of Routh's stability criterion.

     c. Compute the phase margin of the system.

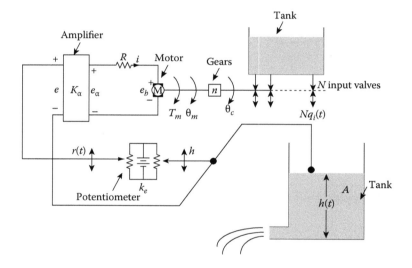

Mathematical Model	Parameters
1. $e(t) = k_e(r(t) - h(t))$	$k_e = 1$ V/m
2. $e_a(t) = k_a e(t)$	$k_a = 50$ V/V
3. $e_a(t) = Ri(t) + e_b(t)$	$R = 10\ \Omega$
4. $e_b(t) = k_b\, \omega_m(t)$	$k_b = 0.075$ V·s/rad
5. $T_m(t) = k_1 i(t)$	$k_1 = 10$ N·m/A
6. $T_m(t) = J\dfrac{d\omega_m(t)}{dt}$	$J = 0.005$ N·m·s²/rad
7. $\omega_m(t) = \dfrac{d\theta_m(t)}{dt}$	$n = 1/100$
8. $\theta_c(t) = n\theta_m(t)$	$k_i = 10$ m³/s·rad
9. $q_i(t) = k_i N \theta_c(t)$	$k_o = 50$ m²/s
10. $q_o(t) = k_o h(t)$	$A = 50$ m²
11. $h(t) = \dfrac{1}{A}\displaystyle\int_0^t (q_i(t) - q_0(t))\, dt$	

Solution

a. We apply Laplace transform to the mathematical model of the system, supposing zero initial conditions. The model is written as

$$(P7.9.1) \overset{LT}{\Rightarrow} E(s) = k_e(R(s) - H(s)) \qquad \text{(P7.9.1')}$$

$$(P7.9.2) \overset{LT}{\Rightarrow} E_a(s) = k_a E(s) \qquad \text{(P7.9.2')}$$

$$(P7.9.3) \overset{LT}{\Rightarrow} E_a(s) = RI(s) + E_b(s) \qquad \text{(P7.9.3')}$$

$$(P7.9.4) \overset{LT}{\Rightarrow} E_b(s) = k_b \Omega_m(s) \qquad \text{(P7.9.4')}$$

$$(P7.9.5) \overset{LT}{\Rightarrow} T_m(s) = k_1 I(s) \qquad \text{(P7.9.5')}$$

$$(P7.9.6) \overset{LT}{\Rightarrow} T_m(s) = Js\Omega_m(s) \qquad \text{(P7.9.6')}$$

$$(P7.9.7) \overset{LT}{\Rightarrow} \Omega_m(s) = s\Theta_m(s) \qquad \text{(P7.9.7')}$$

$$(P7.9.8) \overset{LT}{\Rightarrow} \Theta_c(s) = n\Theta_m(s) \qquad \text{(P7.9.8')}$$

$$(P7.9.9) \overset{LT}{\Rightarrow} Q_i(s) = k_i N \Theta_c(s) \qquad \text{(P7.9.9')}$$

$$(P7.9.10) \overset{LT}{\Rightarrow} Q_o(s) = k_o H(s) \qquad \text{(P7.9.10')}$$

$$(P7.9.11) \overset{LT}{\Rightarrow} H(s) = \frac{Q_i(s) - Q_o(s)}{As} \qquad \text{(P7.9.11')}$$

The block diagram of the system is

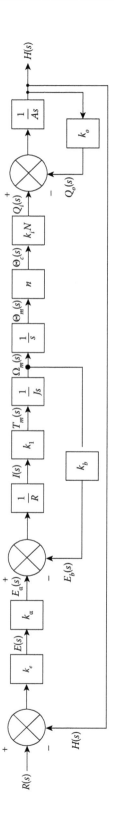

We reduce the block diagram as follows:

$$\frac{\Omega_m(s)}{E_a(s)} = \frac{k_1/RJs}{1+(k_1/RJs)\cdot k_b} = \frac{k_1}{RJs+k_1k_b} \tag{P7.9.12}$$

and

$$\frac{H(s)}{Q_i(s)} = \frac{1/As}{1+(1/As)\cdot k_o} = \frac{1}{As+k_o} \tag{P7.9.13}$$

The open-loop transfer function $G(s)H(s)$ of the system is

$$G(s)H(s) = k_e \cdot k_a \cdot \frac{k_1}{RJs+k_1k_b} \cdot \frac{1}{s} \cdot n \cdot k_i N \cdot \frac{1}{As+k_o} \Rightarrow$$

$$G(s)H(s) = \frac{k_e k_a k_1 k_i Nn}{s(RJs+k_1k_b)(As+k_o)} \tag{P7.9.14}$$

We substitute the parameter values of the system and get

$$(P7.9.14) \Rightarrow G(s)H(s) = \frac{50 \cdot 10 \cdot 10 \cdot (1/100)N}{s(10 \cdot 0.005s + 10 \cdot 0.75)(50s+50)} \Rightarrow$$

$$G(s)H(s) = \frac{20N}{s(s+1)(s+15)} \tag{P7.9.15}$$

For $s = j\omega$ and $N = 1$, we have

$$(P7.9.15) \Rightarrow G(j\omega)H(j\omega) = \frac{20}{j\omega(j\omega+1)(j\omega+15)} \Rightarrow$$

$$G(j\omega)H(j\omega) = \frac{20/15}{j\omega(j(\omega/1)+1)(j(\omega/15)+1)} \tag{P7.9.16}$$

The magnitude of $G(j\omega)H(j\omega)$ in dB is

$$\left|G(j\omega)H(j\omega)\right|_{dB} = 20 \, \log\left|G(j\omega)H(j\omega)\right|$$

$$\overset{(P7.9.16)}{=} 20 \, \log\left(\frac{20}{15}\right) - 20 \, \log|j\omega| - 20 \, \log\left|j\frac{\omega}{1}+1\right| - 20 \, \log\left|j\frac{\omega}{15}+1\right|$$

$$= 20 \, \log\left(\frac{20}{15}\right) - 20 \, \log\omega - 20 \, \log\sqrt{1+\omega^2} - 20 \, \log\sqrt{1+\left(\frac{\omega}{15}\right)^2} \tag{P7.9.17}$$

The phase of $G(j\omega)H(j\omega)$ in degrees is

$$\sphericalangle(G(j\omega)H(j\omega)) = -\sphericalangle(j\omega) - \sphericalangle\left(j\frac{\omega}{1}+1\right) - \sphericalangle\left(j\frac{\omega}{15}+1\right) \Rightarrow$$

$$\sphericalangle(G(j\omega)H(j\omega)) = -\tan^{-1}\frac{\omega}{0} - \tan^{-1}\frac{\omega}{1} - \tan^{-1}\frac{\omega}{15} \Rightarrow$$

$$\sphericalangle(G(j\omega)H(j\omega)) = -90° - \tan^{-1}\omega - \tan^{-1}\frac{\omega}{15} \qquad\qquad (P7.9.18)$$

The corner frequencies are $\omega_{01} = 1\,\text{rad/s}$ at the pole $s = -1$, and $\omega_{02} = 15\,\text{rad/s}$ at the pole $s = -15$.

We plot the magnitude curve approximately with the use of asymptotes as follows:

1. The term $-20\log\omega$, that is, $(j\omega)^{-1}$, is a straight line with a slope of $-20\,\text{dB/dec}$.

2. The term $-20\log\sqrt{1+\omega^2}$, that is, $(1 + j\omega)^{-1}$, is activated at the corner frequency $\omega_{01} = 1\,\text{rad/s}$ and contributes with a slope of $-20\,\text{dB/dec}$. Thus, the total slope is $-20\,\text{dB/dec} + (-20\,\text{dB/dec}) = -40\,\text{dB/dec}$.

3. The term $-20\log\sqrt{1+(\omega/15)^2}$, that is, $(1 + j(\omega/15))^{-1}$, is activated at the corner frequency of $\omega_{02} = 15\,\text{rad/s}$ and contributes by a slope of $-20\,\text{dB/dec}$. The slope of the magnitude curve for $\omega > 15\,\text{rad/s}$ is $-40\,\text{dB/dec} + (-20\,\text{dB/dec}) = -60\,\text{dB/dec}$.

4. For $\omega = 0.1\,\text{rad/s}$, the approximate curve begins at $22.5\,\text{dB}$ ($= 20\log(20/15) - 20\log 0.1 = 2.5\,\text{dB} - (-20\,\text{dB})$).

We plot the exact phase curve by constructing the following table of values, based on the relationship (P7.9.18):

ω	$(j\omega)^{-1}$	$\left(J\frac{\omega}{1}+1\right)^{-1}$	$\left(J\frac{\omega}{15}+1\right)^{-1}$	$G(j\omega)H(j\omega)$
0.1	−90°	−5.7°	−0.4°	−96.1°
0.5	−90°	−26.6°	−1.9°	−118.5°
1	−90°	−45°	−3.8°	−138.8°
5	−90°	−78.7°	−18.4°	−187.1°
10	−90°	−84.3°	−33.7°	−208°
15	−90°	−86.2°	−45°	−221.2°
50	−90°	−88.9°	−73.3°	−252.2°
100	−90°	−89.4°	−81.5°	−260.9°

The magnitude (approximate and real) and phase plots are depicted below

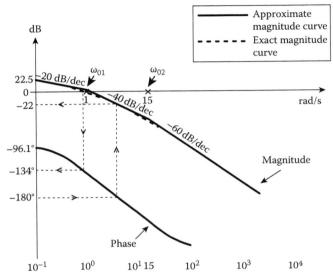

b. We observe from the Bode diagram that the gain that corresponds to phase $\varphi = -180°$ is $-22\,\text{dB}$. Hence, $20 \log |N| = 22 \Rightarrow N = 10^{22/20} \simeq 12$.

The maximum number of input valves that can be inserted, so that the system is stable, is

$$N_{max} \simeq 12 \qquad (P7.9.19)$$

We confirm our result by applying Routh's stability criterion.

The characteristic equation of the system is:

$$1 + G(s)H(s) = 0 \Rightarrow 1 + \frac{20\,N}{s(s+1)(s+15)} = 0 \Rightarrow$$

$$s(s+1)(s+15) + 20N = 0 \Rightarrow 16s^2 + 15s + 20N = 0 \qquad (P7.9.20)$$

Routh's tabulation is

s^3	1	15
s^2	16	$20N$
s^1	$\dfrac{240 - 20N}{16}$	
s^0	$20\,N$	

In order to find N_{max}, we set the term of row s^1 equal to zero:

$$\frac{240 - 20N_{max}}{16} = 0 \Rightarrow N_{max} = \frac{240}{20} = 12$$

As expected, the number of valves is the same.

c. From the magnitude diagram we observe that for $|G(j\omega)H(j\omega)| = 0\,\mathrm{dB}$ the angle is equal to $-134°$. The phase margin is equal to

$$\sphericalangle\varphi_\alpha = 180° - \left|-134°\right| = 180° - 134° = 46° > 0 \tag{P7.9.21}$$

The system is stable for $0 < N < 12$.

7.10 Assuming that $K = 1$, plot the Bode diagrams of the open-loop transfer functions for the systems shown in Figure (a) and (b). Are the closed-loop systems stable?

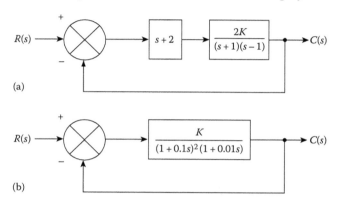

(a)

(b)

Solution

a. The open-loop transfer function of the system shown in Figure (a) is

$$G(s)H(s) = \frac{2K(s+2)}{(s+1)(s-1)} \tag{P7.10.1}$$

For $s = j\omega$ and $K = 1$, from the relationship (P7.10.1), we have

$$G(j\omega)H(j\omega) = \frac{2(j\omega+2)}{(j\omega+1)(j\omega-1)} = \frac{4(j(\omega/2)+1)}{(j(\omega/1)+1)(j(\omega/1)-1)} \tag{P7.10.2}$$

The magnitude of $G(j\omega)H(j\omega)$ in dB is

$$|G(j\omega)H(j\omega)|_{\mathrm{dB}} \overset{(\mathrm{P7.10.2})}{=} 20\log\left|\frac{4(j(\omega/2)+1)}{(j(\omega/1)+1)(j(\omega/1)-1)}\right|$$

$$= 20\log\left|4\left(j\frac{\omega}{2}+1\right)\right| - 20\log\left|\left(j\frac{\omega}{1}+1\right)\left(j\frac{\omega}{1}-1\right)\right|$$

$$= 20\log 4 + 20\log\left|\left(j\frac{\omega}{2}+1\right)\right| - 20\log\left|\left(j\frac{\omega}{1}+1\right)\right| - 20\log\left|\left(j\frac{\omega}{1}-1\right)\right|$$

$$= 20\log 4 + 20\log\sqrt{1+\left(\frac{\omega}{2}\right)^2} - 40\log\sqrt{1+\left(\frac{\omega}{1}\right)^2} \tag{P7.10.3}$$

The phase of $G(j\omega)H(j\omega)$ in degrees is

$$\sphericalangle G(j\omega)H(j\omega) = \sphericalangle\left(\frac{4(j(\omega/2)+1)}{(j(\omega/1)+1)(j(\omega/1)-1)}\right)$$

$$= \sphericalangle\left(4\left(j\frac{\omega}{2}+1\right)\right) - \sphericalangle\left(\left(j\frac{\omega}{1}+1\right)\left(j\frac{\omega}{1}-1\right)\right)$$

$$= \sphericalangle(4) + \sphericalangle\left(j\frac{\omega}{2}+1\right) - \sphericalangle\left(j\frac{\omega}{1}+1\right) - \sphericalangle\left(j\frac{\omega}{1}-1\right)$$

$$= \tan^{-1}\frac{\omega}{2} - \tan^{-1}\frac{\omega}{1} - \left(180° - \tan^{-1}\frac{\omega}{1}\right) \Rightarrow$$

$$\sphericalangle G(j\omega)H(j\omega) = \tan^{-1}\frac{\omega}{2} - 180° \qquad\qquad (P7.10.4)$$

The corner frequency is $\omega_{01} = 1\,\text{rad/s}$ at the poles $s = 1$ and $s = -1$, and $\omega_{02} = 2\,\text{rad/s}$ at the zero $s = -2$.

In order to plot the approximate magnitude curve we proceed as follows:

1. For $\omega = 0.1\,\text{rad/s}$, the magnitude curve begins at $20\log 4 = 12\,\text{dB}$ and has a slope of $0\,\text{dB/dec}$, due to the constant term $20\log 4$.

2. The terms $-20\log|(j(\omega/1) + 1)|$ and $-20\log|(j(\omega/1) - 1)|$ are activated at the corner frequency $\omega_{01} = 1\,\text{rad/s}$, and contribute with a slope of $-40\,\text{dB/dec}$. Consequently, for $\omega \geq 1\,\text{rad/s}$, the magnitude curve has a slope of $-40\,\text{dB/dec}$ ($=0\,\text{dB/dec} + (-40\,\text{dB/dec})$).

3. The term $20\log\sqrt{1+(\omega/2)^2}$ is activated at the corner frequency $\omega_{02} = 2\,\text{rad/s}$ and contributes to the slope by $+20\,\text{dB/dec}$. Therefore, for $\omega \geq 2\,\text{rad/s}$, the total magnitude curve has a slope of $-40\,\text{dB/dec} + 20\,\text{dB/dec} = -20\,\text{dB/dec}$.

In order to plot the real phase curve, we construct the following table of values, based on the relationship (P7.10.4):

ω	$-180°$	$\tan^{-1}\dfrac{\omega}{2}$	$G(j\omega)H(j\omega)$
0.1	$-180°$	2.9°	$-177.1°$
0.5	$-180°$	14°	$-166°$
1	$-180°$	26.6°	$-153.4°$
2	$-180°$	45°	$-135°$
5	$-180°$	68.2°	$-111.8°$
10	$-180°$	78.7°	$-101.3°$

The Bode diagram of the open-loop transfer function is

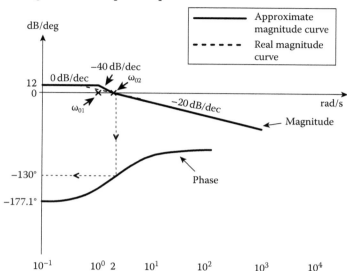

For $|G(j\omega)H(j\omega)|_{dB} = 0$, we have $\omega = 2.2\,\mathrm{rad/s}$ and the phase is approximately equal to $-130°$. Consequently, the phase margin is $\varphi_\alpha = 180° - |-130°| = 180° - 130° = 50° > 0$.

The curve of $G(j\omega)H(j\omega)$ approaches $-180°$, as ω is reduced and $\sphericalangle G(j\omega_c)H(j\omega_c) = -180°$ for $\omega_c = 0\,\mathrm{rad/s}$.

In this case, the gain margin is $-12\,\mathrm{dB} < 0$.

A negative gain margin does not necessarily mean that the system is unstable, and since $\varphi_\alpha > 0$ we conclude that the closed-loop system is stable.

b. The open-loop transfer function of the system shown in Figure (b) is

$$G(s)H(s) = \frac{K}{(1+0.1s)^2(1+0.01s)} \tag{P7.10.5}$$

For $s = j\omega$ and $K = 1$, from the relationship (P7.10.5), we get

$$G(j\omega)H(j\omega) = \frac{1}{(1+j(\omega/10))^2(1+j(\omega/100))} \tag{P7.10.6}$$

The magnitude of $G(j\omega)H(j\omega)$ in dB is

$$|G(j\omega)H(j\omega)|_{dB} \overset{(P7.10.6)}{=} 20\log 1 - 20\log\left|\left(1+j\frac{\omega}{10}\right)^2\left(1+j\frac{\omega}{100}\right)\right|$$

$$= -40\log\sqrt{1+\left(\frac{\omega}{10}\right)^2} - 20\log\sqrt{1+\left(\frac{\omega}{100}\right)^2} \tag{P7.10.7}$$

The phase of $G(j\omega)H(j\omega)$ in degrees is

$$\sphericalangle G(j\omega)H(j\omega) \overset{(P7.10.6)}{=} -2\tan^{-1}\frac{\omega}{10} - \tan^{-1}\frac{\omega}{100} \tag{P7.10.8}$$

The corner frequencies are $\omega_{01} = 10\,\text{rad/s}$ for the double pole $s = -10$ and $\omega_{02} = 100\,\text{rad/s}$ for the pole $s = -100$.

The approximate magnitude curve is plotted as usual:

1. For $\omega = 1\,\text{rad/s}$ the magnitude curve begins from $0\,\text{dB}$ and has a slope of $0\,\text{dB/dec}$.

2. The term $-40\log\sqrt{1+(\omega/10)^2}$, that is, $(1 + j(\omega/10))^{-2}$, is activated at the corner frequency $\omega_{01} = 10\,\text{rad/s}$ and contributes with a slope of $-40\,\text{dB/dec}$. Thus, for $\omega \geq 10\,\text{rad/s}$, the slope of the magnitude curve is $0\,\text{dB/dec} + (-40\,\text{dB/dec}) = -40\,\text{dB/dec}$.

3. The term $-20\log\sqrt{1+(\omega/100)^2}$, that is, $(1 + j(\omega/100))^{-1}$, is activated at the frequency $\omega_{02} = 100\,\text{rad/s}$ and contributes with a slope of $-20\,\text{dB/dec}$. Hence, for $\omega \geq 100\,\text{rad/s}$, the total slope of the magnitude curve is $-40\,\text{dB/dec} + (-20\,\text{dB/dec}) = -60\,\text{dB/dec}$.

In order to plot the phase curve, we construct the based on the relationship (P7.10.8):

ω	$-2\tan^{-1}\dfrac{\omega}{10}$	$-2\tan^{-1}\dfrac{\omega}{100}$	$G(j\omega)H(j\omega)$
1	$-11.4°$	$-0.6°$	$-12°$
5	$-53.1°$	$-2.9°$	$-56°$
10	$-90°$	$-5.7°$	$-95.7°$
20	$-127°$	$-11.3°$	$-138.3°$
50	$-157.4°$	$-26.6°$	$-184°$
100	$-168.6°$	$-45°$	$-213.6°$
500	$-177.7°$	$-78.7°$	$-256.4°$

The following figure depicts the Bode diagram of the loop transfer function of the system shown in Figure (b):

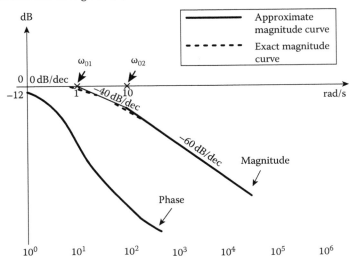

7.11 Describe analytically the design process of the Bode plot for the following transfer function:

$$G(s) = \frac{K(2s+1)}{s^2(0.5s+1)}, \quad \text{where } K = 30$$

Solution

1. Magnitude curve

$$G(s) = \frac{K(2s+1)}{s^2(0.5s+1)} \xrightarrow{s=j\omega} G(j\omega) = \frac{K(2j\omega+1)}{(j\omega)^2(0.5j\omega+1)} \Leftrightarrow$$

$$G(j\omega) = \frac{K(j(\omega/0.5)+1)}{(j\omega)^2(j(\omega/2)+1)}$$

(P7.11.1)

The terms of the transfer function, after the frequency has been increased, are as follows:

A gain constant $K = 30$, a double pole at the axis origin, a zero at the frequency 0.5 rad/s, and a pole at the frequency 2 rad/s.

We initially plot the distinct diagrams of magnitude for every term that corresponds to the poles, zeros, and to the gain constant.

The constant term is $20 \log_{10} 30 = 29.5$ dB.

The magnitude curve that corresponds to the pole at the axes origin begins from zero frequency and approaches infinity. It has a slope of -40 dB/dec, as it is squared and it intersects the straight line of 0 dB, at the frequency 1 rad/s.

The magnitude plot that corresponds to the zero at the frequency 0.5 rad/s has a slope of 20 dB/dec. The slope of the diagram for lower frequencies is 0 dB.

The magnitude diagram that corresponds to the pole at the frequency 2 rad/s has a slope equal to -20 dB/dec. The slope of the diagram for lower frequencies, from the frequency curvature, is 0 dB.

The final approximate magnitude diagram is

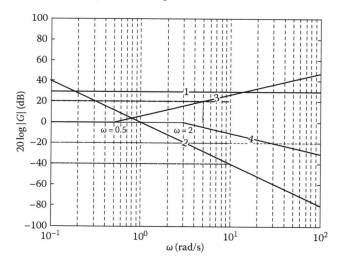

The final magnitude diagram can be designed by drawing individually each part of the diagram, starting with the term that corresponds to the lowest frequency. To achieve that, we add algebraically the previous diagrams:

- For $0.1 \leq \omega < 0.5$, the asymptotic curve is described by the expression $29.5 - 40 \log \omega$.
- For $\omega = 0.1 \, \text{rad/s}$, the magnitude is $29.5 - 40 \log 0.1 = 69.5 \, \text{dB}$, and the curve slope is $-40 \, \text{dB/dec}$.
- For $0.5 \leq \omega < 2$, the asymptotic curve is described by the expression $29.5 - 40 \log \omega + 20 \log(\omega/0.5)$.
- For $\omega = 0.5 \, \text{rad/s}$, the magnitude is $29.5 - 40 \log 0.5 + 20 \log(0.5/0.5) = 41.5 \, \text{dB}$, and the curve slope is $-20 \, \text{dB/dec}$.
- For $2 \leq \omega$, the asymptotic curve is described by the expression $29.5 - 40 \log \omega + 20 \log(\omega/0.5) - 20 \log(\omega/2)$.
- For $\omega = 2 \, \text{rad/s}$, the magnitude is $29.5 - 40 \log 2 + 20 \log(2/0.5) - 20 \log(2/2) = 29.5 - 12 + 12 + 0 = 29.5 \, \text{dB}$, and the curve slope is $-40 \, \text{dB/s}$.

For $\omega = 20 \, \text{rad/s}$, as the frequency is multiplied by 10, and the slope is $-40 \, \text{dB/dec}$, the magnitude is reduced by $40 \, \text{dB}$; therefore, it is $29.5 - 40 = -10.5 \, \text{dB}$.

In order to draw the exact diagram, we calculate the real values of magnitude from the expression $20 \log |G(j\omega)|$ at the aforementioned frequencies.

ω (rad/s)	0.1	0.5	2	20
$20 \log |G(j\omega)|$ dB	69.7019	44.3306	26.7954	−10.4981

2. Phase curve

The phase curve of the system is also drawn by adding sequentially the values that correspond to each one of the distinct terms.

The segments of the approximate curves are as follows:

The phase angle that corresponds to the constant term is equal to $0°$. The phase angle that corresponds to the double pole at the axes origin is $(-90°) \cdot 2 = -180°$.

The linear approach of the phase for the zero at the frequency $0.5 \, \text{rad/s}$ is $0°$ for $\omega < 0.05 \, \text{rad/s}$, $45°$ for $\omega = 0.5 \, \text{rad/s}$, and $90°$ for $\omega > 5 \, \text{rad/s}$. The linear approach is a straight line that connects these three points.

The phase diagram of the distinct terms is given below:

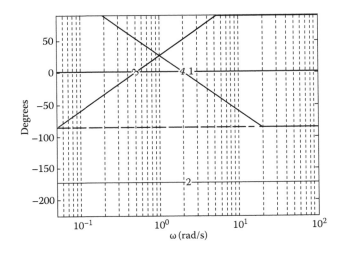

In order to find the exact values of phase, we use the following formula:

$$\angle G(j\omega) = \angle \frac{K(j(\omega/0.5)+1)}{j\omega^2(j(\omega/2)+1)} = \angle K + \angle\left(j\frac{\omega}{0.5}+1\right) - \angle j\omega^2 - \angle\left(j\frac{\omega}{2}+1\right) \Leftrightarrow$$

$$\angle G(j\omega) = 0° + \tan^{-1}\frac{\omega}{0.5} - 2*90° - \tan^{-1}\frac{\omega}{2}$$

The Bode diagram of the given system is the following

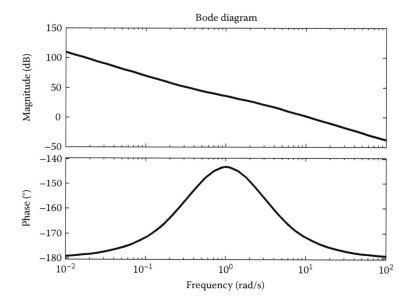

7.12 The magnitude plot of the frequency response depicted in the following figure refers to the open-loop transfer system of a unity-feedback control system.

a. For $A = 10$, find the loop transfer function $G(s)$ of the system.

b. Write down the magnitude and phase equations and then plot the curves and the magnitude and phase asymptotes of $G(j\omega)$ in the Bode diagram. Calculate the gain and phase margins. For which value of K is the phase margin 45°?

c. Find the closed-loop transfer function $F(s)$ of the system. Compute the values of J, ω_n, and ω_d so that the phase margin is 45°.

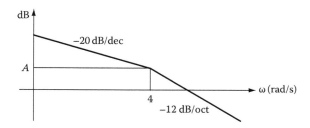

Solution

a. The loop transfer function that results from the diagram is

$$T(s) = \frac{K}{s(1+\tau s)} \qquad \text{(P7.12.1)}$$

We know that $A = 10$, and we derive from the figure that $\tau = 1/4 = 0.25\,\text{s}$.
The transfer function $T(s)$ is found as follows:

$$20 \log\left(\frac{K}{\omega}\right) = 10\,\text{dB} \Rightarrow 20 \log\left(\frac{K}{4}\right) = 10\,\text{dB} \Rightarrow$$

$$20 \log K - 20 \log 4 = 10\,\text{dB} \Rightarrow$$

$$\Rightarrow 20 \log K = 10 + 12 = 22\,\text{dB} \Rightarrow K = 10^{22/20} = 12.59 \Rightarrow$$

$$T(s) = \frac{K}{s(1+\tau s)} = \frac{12.59}{s(1+0.25s)} \qquad \text{(P7.12.2)}$$

b. The magnitude and phase equations are

$$|T(j\omega)|_{\text{dB}} = 20 \log\left[\frac{12.59}{\omega\sqrt{1+0.25^2\omega^2}}\right] \qquad \text{(P7.12.3)}$$

$$\varphi = -90° - \tan^{-1} 0.25\omega \qquad \text{(P7.12.4)}$$

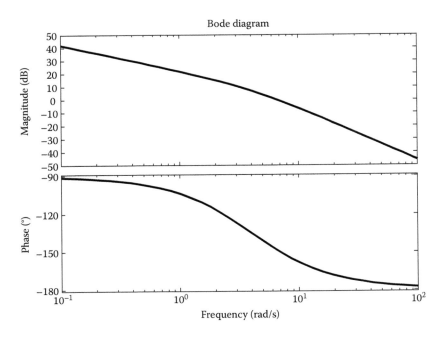

Bode diagram

From the magnitude and phase diagrams, we find that:

For $\omega = 6.56\,\text{rad/s}$, the phase margin is $31.4°$.

For $\omega = 4\,\text{rad/s}$, the phase margin is $45°$ and $T_{dB} = 7\,\text{dB}$. Therefore,

$$K_a\,\text{dB} = 20\log K\,\text{dB} - 7\,\text{dB} = 20\log 12.59 - 7\,\text{dB} = 22 - 7 = 15\,\text{dB} \Rightarrow$$

$$K_a = 10^{15/20} = 5.62 \tag{P7.12.5}$$

c. The closed-loop transfer function is

$$F(s) = \frac{G(s)}{1+G(s)} = \frac{K/(s(1+\tau s))}{1+(K/(s(1+\tau s)))} = \frac{K}{\tau s^2 + s + K}$$

$$= \frac{1}{(\tau/K)s^2 + (1/K)s + 1} = \frac{1}{(1/\omega_n^2)s^2 + (2J/\omega_n)s + 1} = \frac{1}{(0.25/5.62)s^2 + (1/5.62)s + 1} \tag{P7.12.6}$$

We compute J, ω_n, and ω_d as follows:

$$\frac{\tau}{K} = \frac{1}{\omega_n^2} = \frac{0.25}{5.62} \Rightarrow \omega_n = 4.74\,\text{rad/s} \tag{P7.12.7}$$

$$\frac{1}{K} = \frac{2J}{\omega_n} = \frac{1}{5.62} \Rightarrow J = \frac{\omega_n}{2 \cdot 5.62} = \frac{4.74}{2 \cdot 5.62} = 0.42 \tag{P7.12.8}$$

$$\omega_d = \omega_n\sqrt{1 - 2J^2} = 4.74\sqrt{1 - 2 \cdot 0.42^2} = 3.81\,\text{rad/s} \tag{P7.12.9}$$

8

State-Space Representation of Control Systems

8.1 Introduction

State-space representation is used for analyzing various types (e.g., linear and nonlinear, time-invariant or time-variant) of control systems. The system is modeled with **a set of first-order differential equations** that describe the system state.

The **state variables** describe the future response of a system if the present state of the system, the inputs, and the equations that describe its operation are known. The state variables cannot be always observed or measured, although they influence the behavior of the system. In other words, the state variables determine how the system evolves given the present state.

The **state (differential) equation** provides the relationship between the inputs, the system state, and the rate of change of the system state.

Consider the linear **time-invariant** multiple-input, multiple-output (MIMO) system shown in Figure 8.1.

The dynamic equations that describe the system, that is, the state equations, are

$$\dot{x}(t) = Ax(t) + Bu(t)$$

$$y(t) = Cx(t) + Du(t) \tag{8.1}$$

$$x(t_o) = x(0)$$

The matrices A, B, C, D are called **state-space matrices**. Matrix A is a square matrix of $n \times n$ dimensions. It is called **state matrix**. The dimensions of matrix B are $n \times r$ and B is called **input matrix**. Matrix C has dimensions $m \times n$ and it is called **output matrix**. Matrix D has dimensions $m \times r$ and it is called **feedthrough** or **feedforward matrix**.

In the case of single-input, single-output (SISO) systems, where $r = m = 1$, the system is described by the following state equations:

$$\dot{x}(t) = Ax(t) + bu(t)$$

$$y(t) = c^T x(t) + du(t) \tag{8.2}$$

$$x(0) = x_o$$

where
   c is a column vector with n elements
   b is a column vector with n elements
   d is a scalar quantity
   $x(0) = x_o$ is the column vector of the initial conditions of state variables

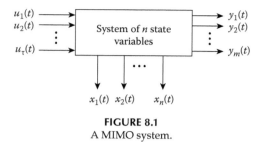

FIGURE 8.1
A MIMO system.

8.2 Eigenvalues and Eigenvectors

The elements of the state matrix depend on the components of the system.

Consider a nth order system. There are column vectors $x = x_i$ ($i = 1, 2, \ldots, n$) and real or complex valued parameters λ that satisfy the equation

$$Ax = \lambda x \Rightarrow (\lambda I - A)x = 0 \qquad (8.3)$$

The matrix $(\lambda I_n - A)$ is called characteristic matrix of the system. The values of the parameter λ that satisfy the equation $(\lambda I_n - A)x = 0$ are called eigenvalues or characteristic values and they form a column vector.

The characteristic polynomial of the system can be found, if we set the determinant of the characteristic matrix of the system equal to zero.

$$P(\lambda) = \det(\lambda I - A) = \lambda^n + a_{n-1}\lambda^{n-1} + \cdots a_1\lambda + a_0 = 0 \qquad (8.4)$$

The roots of $P(\lambda)$, that is, the eigenvalues of $P(\lambda)$, are the **poles** of the closed-loop system.

If the roots of the characteristic equation of the system $\det(\lambda I - A) = 0$ are distinct, that is, $-\lambda_1 \neq -\lambda_2 \neq \cdots \neq -\lambda_n$, then the state equations of the system can be **decoupled** from each other. This means that a system can be described by n simple and independent from each other first-order differential equations. This technique simplifies the solving procedure of state equations.

It is achieved by the linear transformation:

$$x = Px^* \qquad (8.5)$$

The system matrix then becomes diagonal, that is,

$$A = P^{-1}AP = \Lambda = \begin{bmatrix} a_{11}^* & 0 & . & . & 0 \\ 0 & a_{22}^* & . & . & 0 \\ . & . & . & . & 0 \\ 0 & 0 & . & a_{(n-1)(n-1)}^* & 0 \\ 0 & 0 & . & 0 & a_{nn}^* \end{bmatrix} = \begin{bmatrix} -\lambda_1 & 0 & . & . & 0 \\ 0 & -\lambda_2 & . & . & 0 \\ . & . & . & . & 0 \\ 0 & 0 & . & -\lambda_{n-1} & 0 \\ 0 & 0 & . & 0 & -\lambda_n \end{bmatrix}$$

$$(8.6)$$

where P is a matrix of dimensions $n \times n$, whose columns are the n linear independent eigenvectors of the matrix A.

$$P = X = \begin{bmatrix} X_1 & X_2 & . & . & X_n \end{bmatrix} = \begin{bmatrix} x_{11} & x_{12} & & x_{1n} \\ x_{21} & x_{22} & & x_{21} \\ & & & \\ x_{n1} & x_{n2} & & x_{nn} \end{bmatrix} \tag{8.7}$$

8.3 State-Space Representation of Dynamic Systems

1. Suppose that a system is described by the following linear differential equation:

$$y^{(n)} + a_1 y^{(n-1)} + \cdots + a_{n-1}\dot{y} + a_n y = u \tag{8.8}$$

We consider the following state variables:

$$\left. \begin{array}{c} x_1 = y \\ x_2 = \dot{y} \\ \vdots \\ x_n = y^{(n-1)} \end{array} \right\} \tag{8.9}$$

From (8.6), Equation 8.5 is written as

$$\dot{x}_n = -a_n x_1 - \cdots - a_1 x_n + u \tag{8.10}$$

The state equations and the output of the system are in the **controllable canonical form**:

$$\begin{bmatrix} \dot{x}_1 \\ \dot{x}_2 \\ \vdots \\ \dot{x}_n \end{bmatrix} = \begin{bmatrix} 0 & 1 & 0 & \cdots & 0 \\ 0 & 0 & 1 & \cdots & 0 \\ \vdots & \vdots & \vdots & \vdots & \vdots \\ 0 & 0 & 0 & \cdots & 1 \\ -a_n & -a_{n-1} & -a_{n-2} & \cdots & -a_1 \end{bmatrix} \cdot \begin{bmatrix} x_1 \\ x_2 \\ \vdots \\ x_n \end{bmatrix} + \begin{bmatrix} 0 \\ 0 \\ \vdots \\ 1 \end{bmatrix} u$$

$$y = \begin{bmatrix} 1 & 0 & \cdots & 0 \end{bmatrix} \cdot \begin{bmatrix} x_1 \\ x_2 \\ \vdots \\ x_n \end{bmatrix} \tag{8.11}$$

The system matrix is called **companion matrix**. All elements of the upper diagonal are equal to one, the elements of the last row are the coefficients of the characteristic polynomial in ascending order with a negative sign, while the rest of the elements are zeros.

In the case of distinct eigenvalues $(-\lambda_1, -\lambda_2, \ldots, -\lambda_n)$, the matrix that transforms the controllability canonical form of the system to **diagonal canonical form** is called Vandermonde matrix and is given by

$$P = \begin{bmatrix} 1 & 1 & 1 & . & 1 \\ -\lambda_1 & -\lambda_2 & -\lambda_3 & . & -\lambda_n \\ (-\lambda_1)^2 & (-\lambda_2)^2 & (-\lambda_3)^2 & . & (-\lambda_n)^2 \\ . & . & . & . & . \\ (-\lambda_1)^{n-1} & (-\lambda_2)^{n-1} & (-\lambda_3)^{n-1} & & (-\lambda_n)^{n-1} \end{bmatrix} \tag{8.12}$$

The following figure depicts the implementation block diagram of the relationships given in (8.11).

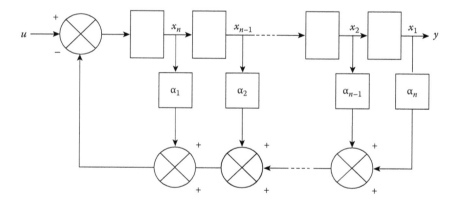

2. Consider now a differential equation of the form

$$y^{(n)} + a_1 y^{(n-1)} + \cdots + a_{n-1}\dot{y} + a_n y = b_0 u^{(n)} + b_1 u^{(n-1)} + \cdots + b_{n-1}\dot{u} + b_n u \tag{8.13}$$

The state variables are the following:

$$\left. \begin{aligned} x_1 &= y - \beta_0 u \\ x_2 &= \dot{y} - \beta_0 \dot{u} - \beta_1 u = \dot{x}_1 - \beta_1 u \\ x_3 &= \ddot{y} - \beta_0 \ddot{u} - \beta_1 \dot{u} - \beta_2 u = \dot{x}_2 - \beta_2 u \\ &\vdots \\ x_n &= y^{(n-1)} - \beta_0{}^{(n-1)}u - \beta_1{}^{(n-2)}u - \cdots - \beta_{n-2}\dot{u} - \beta_{n-1}u = \dot{x}_{n-1} - \beta_{n-1}u \end{aligned} \right\} \tag{8.14}$$

where $\beta_0, \beta_1, \ldots, \beta_n$ are determined by the relationships

$$\left.\begin{aligned}
\beta_o &= b_0 \\
\beta_1 &= b_1 - a_1\beta_0 \\
\beta_2 &= b_2 - a_1\beta_1 - a_2\beta_0 \\
&\vdots \\
\beta_n &= b_n - a_1\beta_{n-1} - \cdots - a_{n-1}\beta_1 - a_n\beta_0
\end{aligned}\right\} \qquad (8.15)$$

Moreover,

$$\left.\begin{aligned}
\dot{x}_1 &= x_2 + \beta_1 u \\
\dot{x}_2 &= x_3 + \beta_2 u \\
&\vdots \\
\dot{x}_{n-1} &= x_n + \beta_{n-1} u \\
\dot{x}_n &= -a_n x_1 - a_{n-1}x_2 - \cdots - a_1 x_n + \beta_n u
\end{aligned}\right\} \qquad (8.16)$$

The state equations are

$$\begin{bmatrix} \dot{x}_1 \\ \dot{x}_2 \\ \vdots \\ \dot{x}_n \end{bmatrix} = \begin{bmatrix} 0 & 1 & 0 & \cdots & 0 \\ 0 & 0 & 1 & \cdots & 0 \\ \vdots & \vdots & \vdots & \vdots & \vdots \\ 0 & 0 & 0 & \cdots & 1 \\ -a_n & -a_{n-1} & -a_{n-2} & \cdots & -a_1 \end{bmatrix} \cdot \begin{bmatrix} x_1 \\ x_2 \\ \vdots \\ x_n \end{bmatrix} + \begin{bmatrix} \beta_1 \\ \beta_2 \\ \vdots \\ \beta_n \end{bmatrix} u$$

$$y = \begin{bmatrix} 1 & 0 & \cdots & 0 \end{bmatrix} \cdot \begin{bmatrix} x_1 \\ x_2 \\ \vdots \\ x_n \end{bmatrix} + \beta_0 u \qquad (8.17)$$

Note that the transfer function of the system is

$$G(s) = \frac{b_0 s^n + b_1 s^{n-1} + \cdots + b_{n-1}s + b_n}{s^n + a_1 s^{n-1} + \cdots + a_{n-1}s + a_n} \qquad (8.18)$$

The state equations given by the relationships (8.16) and (8.17) are in **phase-variable canonical form**, while the state equations are called **phase variables**.

The implementation block diagram of relationships (8.16) and (8.17) is depicted below.

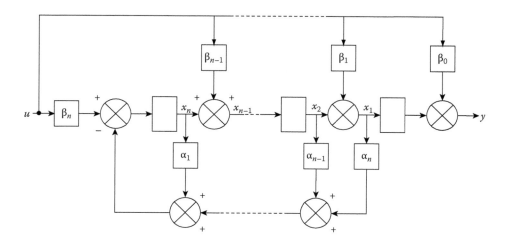

3. Consider a system with the following transfer function:

$$G(s) = \frac{b_n s^n + b_{n-1} s^{n-1} + \cdots + b_1 s + b_0}{s^n + a_{n-1} s^{n-1} + \cdots + a_1 s + a_0} = \frac{Y(s)}{U(s)} \qquad (8.19)$$

The state variables are

$$\dot{x}_1(t) = -a_0 x_n(t) + (b_0 - a_0 b_n) u(t)$$

$$\dot{x}_2(t) = x_1(t) - a_1 x_n(t) + (b_1 - a_1 b_n) u(t)$$

$$\cdots\cdots\cdots\cdots\cdots\cdots\cdots\cdots\cdots\cdots\cdots\cdots \qquad (8.20)$$

$$\dot{x}_{n-1}(t) = x_{n-2}(t) - a_{n-2} x_n(t) + (b_{n-2} - a_{n-2} b_n) u(t)$$

$$\dot{x}_n(t) = x_{n-1}(t) - a_{n-1} x_n(t) + (b_{n-1} - a_{n-1} b_n) u(t)$$

The output is

$$y(t) = b_n u(t) + x_n(t) \qquad (8.21)$$

Based on the relationships (8.20) and (8.21) we write the state equations in observable canonical form:

$$
\begin{bmatrix} \dot{x}_1(t) \\ \dot{x}_2(t) \\ . \\ \dot{x}_{n-1}(t) \\ \dot{x}_n(t) \end{bmatrix} = \begin{bmatrix} 0 & 0 & . & 0 & -a_0 \\ 1 & 0 & . & 0 & -a_1 \\ 0 & 1 & . & 0 & -a_2 \\ . & . & . & . & . \\ 0 & 0 & . & 1 & -a_{n-1} \end{bmatrix} \begin{bmatrix} x_1(t) \\ x_2(t) \\ . \\ x_{n-1}(t) \\ x_n(t) \end{bmatrix} + \begin{bmatrix} b_0 - a_0 b_n \\ b_1 - a_1 b_n \\ . \\ b_{n-2} - a_{n-2} b_n \\ b_{n-1} - a_{n-1} b_n \end{bmatrix} u(t)
$$

and (8.22)

$$
y(t) = \begin{bmatrix} 0 & 0 & . & 1 \end{bmatrix} \begin{bmatrix} x_1(t) \\ x_2(t) \\ . \\ x_{n-1}(t) \\ x_n(t) \end{bmatrix} + b_n u(t)
$$

The system matrix in the observable canonical form is the transpose of the system matrix in the controllable canonical form.

4. Suppose that the transfer function of the system is

$$
G(s) = \frac{b_m s^n + b_{m-1} s^{n-1} + \cdots + b_1 s + b_0}{s^n + a_{n-1} s^{n-1} + \cdots + a_1 s + a_0} = \frac{Y(s)}{U(s)}, \quad m \le n \tag{8.23}
$$

In the case of distinct poles, $G(s)$ is written as a sum of partial fractions, that is,

$$
G(s) = \frac{Y(s)}{U(s)} = \frac{k_1}{s + p_1} + \frac{k_2}{s + p_2} + \cdots + \frac{k_n}{s + p_n} \tag{8.24}
$$

The coefficients $k_i (i = 1, 2, \ldots, n)$ are calculated with the use of Heaviside's formula:

$$
k_i = \lim_{s \to -p_i} \frac{Y(s)}{U(s)} (s + p_i) \tag{8.25}
$$

The state variables are

$$
\dot{x}_1(t) = -p_1 x_1(t) + k_1 u(t)
$$

$$
\dot{x}_1(t) = -p_2 x_2(t) + k_2 u(t)
$$

$$
\cdots\cdots\cdots\cdots\cdots\cdots\cdots
$$

$$
\dot{x}_n(t) = -p_n x_n(t) + k_n u(t) \tag{8.26}
$$

As the state variables are decoupled, the state matrix is diagonal:

$$
\begin{bmatrix} \dot{x}_1(t) \\ \dot{x}_2(t) \\ \cdot \\ \dot{x}_{n-1}(t) \\ \dot{x}_n(t) \end{bmatrix} = \begin{bmatrix} -p_1 & 0 & 0 & \cdot & 0 \\ 0 & -p_2 & 0 & \cdot & 0 \\ 0 & 0 & \cdot & \cdot & 0 \\ 0 & \cdot & 0 & -p_{n-1} & \cdot \\ 0 & 0 & \cdot & 0 & -p_n \end{bmatrix} \begin{bmatrix} x_1(t) \\ x_2(t) \\ \cdot \\ x_{n-1}(t) \\ x_n(t) \end{bmatrix} + \begin{bmatrix} k_1 \\ k_2 \\ \cdot \\ k_{n-1} \\ k_n \end{bmatrix} u(t)
\tag{8.27}
$$

The output is given by

$$
y(t) = \begin{bmatrix} 1 & 1 & \cdot & 1 \end{bmatrix} \begin{bmatrix} x_1(t) \\ x_2(t) \\ \cdot \\ x_{n-1}(t) \\ x_n(t) \end{bmatrix}
\tag{8.28}
$$

5. If the system matrix has multiple eigenvalues, then it cannot be diagonalized. There is, however, a diagonal form called **Jordan canonical form**. A square matrix of dimensions $n \times n$ is in Jordan canonical form if

$$
J = \begin{bmatrix} J_{b_1} & 0 & 0 & \cdot & 0 \\ 0 & J_{b_2} & 0 & \cdot & 0 \\ 0 & 0 & \cdot & \cdot & 0 \\ 0 & \cdot & 0 & J_{b_{n-1}} & \cdot \\ 0 & 0 & \cdot & 0 & J_{b_n} \end{bmatrix}
\tag{8.29}
$$

The partial Jordan submatrices, called the Jordan blocks, are given by:

$$
J_{b_i} = \begin{bmatrix} -\lambda_j & 1 & 0 & \cdot & 0 \\ 0 & -\lambda_j & 1 & \cdot & 0 \\ 0 & 0 & \cdot & \cdot & 0 \\ 0 & \cdot & 0 & -\lambda_j & 1 \\ 0 & 0 & \cdot & 0 & -\lambda_j \end{bmatrix}
\tag{8.30}
$$

Remarks:

- The elements of the main diagonal of the Jordan matrix are the eigenvalues of the system.
- The order of the Jordan submatrices is equal to the multiplicity of the corresponding eigenvalue, and their number is equal to the number of the linear independent eigenvectors.

8.3.1 Relationship between State Equations and Transfer Function

Suppose that a system is described by the state equations given in (8.1). The transfer function matrix of the system is computed by the relationship

$$G(s) = C(sI - A)^{-1}B + D = \frac{C \cdot \operatorname{adj}(sI - A) \cdot B + D \cdot \det(sI - A)}{\det(sI - A)} \tag{8.31}$$

Based on (8.31) and for any output $Y_j(s)$, it holds that

$$Y_j(s) = \sum_{i=1}^{p} G_{ji}(s)U_i(s), \quad j = 1, \ldots, q \tag{8.32}$$

8.4 Solving State Equations

The dynamic (state) equations of MIMO systems are of the form

$$\dot{x}(t) = Ax(t) + Bu(t)$$
$$y(t) = Cx(t) + Du(t) \tag{8.33}$$
$$x(t_o) = x(0)$$

This is a system of n first-order differential equations, the solution of which provides the system state $x(t)$. The state $x(t)$ is analyzed as

$$x(t) = x_{zi}(t) + x_{zs}(t) \tag{8.34}$$

where
   $x_{zi}(t)$ is the **zero-input** or **initial condition response**, that is, the system response if it is excited only by the initial conditions
   $x_{zs}(t)$ is the **zero-state response**, which is the system response due to the system inputs

8.4.1 Solving the Homogeneous Equation $\dot{x}(t) = Ax(t)$

The solution of the homogeneous equation

$$\left. \begin{array}{l} \dot{x}(t) = Ax(t) \\ x(0) = x_o \end{array} \right\} \tag{8.35}$$

is

$$x(t) = e^{At}x(0) \tag{8.36}$$

The exponential matrix e^{At} is denoted by $\Phi(t)$ and it is called **state-transition matrix**. It represents the response of the system to its initial conditions, that is, the initial condition response. It is given by

$$\Phi(t) = e^{At} = L^{-1}\{(sI - A)^{-1}\} \tag{8.37}$$

There are several methods for computing the state-transition matrix. Some of them are the following:

1. The state-transition matrix results from the inverse Laplace transform of $(sI - A)^{-1}$.
2. The state-transition matrix can be found also by expressing the matrix e^{At} into power series:

$$\Phi(t) = e^{At} = \sum_{k=0}^{\infty} \frac{A^k t^k}{k!} = I + At + \frac{A^2 t^2}{2!} + \cdots + \frac{A^{k-1} t^{k-1}}{(k-1)!} + \cdots \tag{8.38}$$

3. If matrix A has distinct eigenvalues, then the eigenvectors $u_1, u_2, \ldots, u_n$ are linear independent from each other and

$$A u_i = \lambda_i u_i \tag{8.39}$$

The matrix of the eigenvectors transforms the matrix A into a diagonal matrix:

$$M = \begin{bmatrix} u_1 & \vdots & u_2 & \vdots & \cdots & \vdots & u_n \end{bmatrix} \tag{8.40}$$

The state-transition matrix is now given by

$$\Phi(t) = e^{At} = M \begin{bmatrix} e^{-\lambda_1 t} & & & 0 \\ & e^{-\lambda_2 t} & & \\ & & \ddots & \\ 0 & & & e^{-\lambda_n t} \end{bmatrix} M^{-1} \tag{8.41}$$

4. If the characteristic polynomial has distinct roots, then the state-transition matrix, from **Sylvester's theorem**, is given by

$$\Phi(t) = e^{At} = \sum_{j}^{n} e^{-\lambda_j t} P_j(\lambda) \tag{8.42}$$

where

$$P_j(\lambda) = \frac{\sum_{i=1}^{n} (A + \lambda_i I)}{\sum_{i=1}^{n} (-\lambda_j + \lambda_i)} \quad i \ne j \tag{8.43}$$

5. If the characteristic polynomial has distinct roots, from the **Cayley–Hamilton theorem**, the state-transition matrix is given by

$$\Phi(t) = e^{At} = \sum_{i}^{n-1} a_i(t) A^i \tag{8.44}$$

For any distinct eigenvalue $(-\lambda_j)$ it holds that

$$e^{-\lambda_j(t)} = \sum_{i=0}^{n-1} a_i(t)(-\lambda_j)^i, \quad j = 1, 2, \ldots, n \tag{8.45}$$

8.4.2 General Solution of State Equations

The general solution for the state vector is

$$x(t) = \Phi(t)x(0) + \int_0^t \Phi(t-\lambda)bu(\lambda)\,d\lambda \tag{8.46}$$

or

$$x(t) = L^{-1}\{X(s)\} = L^{-1}\{(sI - A)^{-1}x_0 + (sI - A)^{-1}bU(s)\} \tag{8.47}$$

The output vector is determined by the relationship

$$y(t) = Cx(t) + Du(t) \tag{8.48}$$

8.5 Block Diagrams of State Equations

Consider a MIMO system described by the state equations of (8.1). By applying Laplace transform to the system we get

$$sX(s) - x(0) = AX(s) + BU(s) \tag{8.49}$$

$$Y(s) = CX(s) + DU(s) \tag{8.50}$$

Figures 8.2 and 8.3 depict the block diagrams of the state equations in the time domain and in the s-domain, respectively.

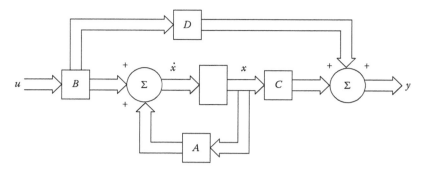

FIGURE 8.2
Block diagram of the state equations in the time domain.

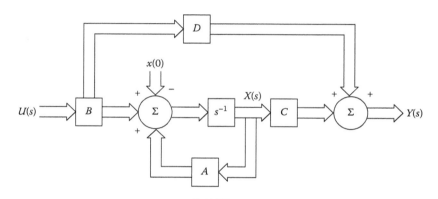

FIGURE 8.3
Block diagram of the state equations in the s-domain.

8.6 Signal Flow Block Diagrams of State Equations

The following figure depicts the signal flow diagram of the system shown in Figure 8.3.

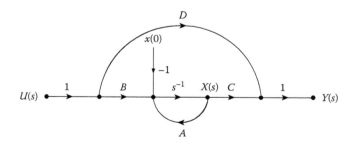

Consider now a SISO system described by a differential equation of the form

$$y^{(n)} + a_{n-1}y^{(n-1)} + \cdots + a_1 y^{(1)} + a_0 y = u(t) \tag{8.51}$$

The signal flow diagram of the state equations for nonzero initial conditions $y^{(k)}(0)$ is depicted below.

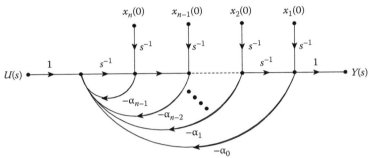

8.7 Controllability and Observability

A system is **controllable** on the time interval $[t_o, t_f]$, if an external input $u(t)$ can move the internal state of the system from any initial state $x(t_o)$ to any other final state $x(t_f)$ in a finite time interval $t_f - t_o$.

This means that with the appropriate input signal we can achieve any desirable system state. The **controllability matrix** S is given by

$$S = \begin{bmatrix} B & \vdots & AB & \vdots & \cdots & \vdots & A^{n-1}B \end{bmatrix} \tag{8.52}$$

A system is controllable if rank $[S] = n$.

A system is called **observable** in the time interval $[t_o, t_f]$ if the initial state $x(t_o)$ can be derived from the output $y(t)$ observed on the finite time interval $t_f - t_o$.

The output vector $y(t)$ is observable if rank$[Q] = m$, where

$$Q = \begin{bmatrix} D & \vdots & CB & \vdots & CAB & \vdots & \cdots & \vdots & CA^{n-1}B \end{bmatrix} \tag{8.53}$$

The state vector $x(t)$ is **observable** if rank$[R^T] = n$, where

$$R^T = \begin{bmatrix} C^T & \vdots & A^T C^T & \vdots & (A^T)^2 C^T & \vdots & \cdots & \vdots & (A^T)^{n-1} C^T \end{bmatrix} \tag{8.54}$$

The matrix R^T is called **observability matrix**.

Note that a system that can be written in phase-variable canonical form is always observable. It is also important to note that for complete controllability and observability there must be no pole-zero cancellations in the system transfer function.

8.8 Modern Control System Design Methods

8.8.1 Placement of Eigenvalues with State Feedback

If a proper set of coefficients k_i is chosen for the feedback loops of the state variables, then a new set of eigenvalues can be established for every controllable system.

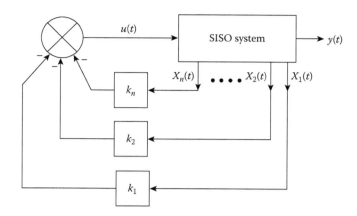

The following equations (known as feedback law) hold for the system depicted above:

$$u = -Kx(t) \tag{8.55}$$

$$K = [k_1 k_2 \cdots k_n] \tag{8.56}$$

A proper selection of matrix K can turn an unstable and relatively slow open-loop system to a stable and fast closed-loop system.

The linear time-invariant SISO system becomes

$$\dot{x} = (A - BK)x = A_f x \tag{8.57}$$

The characteristic equation of the closed-loop system is

$$\left| sI - A_f \right| = \left| sI - A + BK \right| = 0 \tag{8.58}$$

Consider a controllable system with transfer function

$$G(s) = \frac{b_{n-1}s^{n-1} + \cdots + b_1 s + b_0}{s^n + a_{n-1}s^{n-1} + \cdots + a_1 s + a_0} \tag{8.59}$$

The state equations of this system are

$$\dot{x}(t) = \begin{bmatrix} 0 & 1 & 0 & \cdots & 0 \\ 0 & 0 & 1 & \cdots & 0 \\ \vdots & \vdots & \vdots & \cdots & \vdots \\ 0 & 0 & 0 & \cdots & 1 \\ -a_0 & -a_1 & -a_2 & \cdots & -a_{n-1} \end{bmatrix} x(t) + \begin{bmatrix} 0 \\ 0 \\ \vdots \\ 1 \end{bmatrix} u(t)$$

$$y(t) = [b_0 \quad b_1 \quad b_2 \quad \cdots \quad b_{n-1}]x(t) \tag{8.60}$$

The desired characteristic equation of the closed-loop system is

$$\Phi(s) = s^n + a_{n-1}s^{n-1} + \cdots + a_1 s + a_0 = 0 \tag{8.61}$$

and recall that

$$u(t) = -Kx(t) \tag{8.62}$$

J. E. Ackermann developed a way for computing the vector K. **Ackermann's formula** is

$$K = [0 \quad 0 \quad \ldots \quad 0 \quad 1][B \quad AB \quad \cdots \quad A^{n-2}B \quad A^{n-1}B]^{-1}\Phi(A) \tag{8.63}$$

where

$$\Phi(A) = A^n + a_{n-1}A^{n-1} + \cdots + a_1 A + a_0 I \tag{8.64}$$

With the use of Ackermann's formula we can place the poles of the system at the desirable positions. All the roots of the characteristic equation of a system can be placed at any location on the complex plane, insofar as the system is both controllable and observable.

8.8.2 Pole Placement Procedure

1. We examine the controllability of the system state vector.
2. From the characteristic polynomial for the matrix A we have

$$|sI - A| = s^n + a_1 s^{n-1} + \cdots + a_{n-1}s + a_n \tag{8.65}$$

   We determine the values $a_1, a_2, \ldots, a_{n-1}, a_n$.
3. We find the matrix T, which transforms the state equations in the controllable canonical form. If the system is already in controllable canonical form, then $T = I$. It holds that

$$T = SW \tag{8.66}$$

where

$$S = [B \quad \vdots \quad AB \quad \vdots \quad \cdots \quad \vdots \quad A^{n-1}B]$$

$$W = \begin{bmatrix} a_{n-1} & a_{n-2} & \cdots & a_1 & 1 \\ a_{n-2} & a_{n-3} & \cdots & 1 & 0 \\ \vdots & \vdots & \vdots & \vdots & \vdots \\ a_1 & 1 & \cdots & 0 & 0 \\ 1 & 0 & \cdots & 0 & 0 \end{bmatrix} \left. \right\} \tag{8.67}$$

4. Denoting by $\mu_1, \ldots, \mu_n$ the desired eigenvalues, we have

$$(s - \mu_1)(s - \mu_2) \cdots (s - \mu_n) = s^n + b_1 s^{n-1} + \cdots + b_{n-1}s + b_n \qquad (8.68)$$

From (8.68) we compute the coefficients $b_1, b_2, \ldots, b_{n-1}, b_n$.

5. The state-feedback matrix K is calculated by the formula

$$K = [b_n - a_n \quad \vdots \quad b_{n-1} - a_{n-1} \quad \vdots \quad \cdots \quad \vdots \quad b_2 - a_2 \quad \vdots \quad b_1 - a_1]T^{-1} \qquad (8.69)$$

8.8.3 Decoupling State-Feedback Inputs–Outputs

If a system has an equal number of inputs and outputs ($r = m$), then there is a state-feedback matrix (Falb–Wolovich theorem) such that **one input of the closed-loop system affects only one of its outputs**, that is, $y_i = f(u_i)$. The state-feedback matrix is given by

$$K = -(B^*)^{-1} \cdot A^* \qquad (8.70)$$

where

$$B^* = \begin{bmatrix} C_1 A^{d_1} B \\ \vdots \\ C_2 A^{d_2} B \\ \cdots \\ \vdots \\ C_r A^{d_r} B \end{bmatrix}, \quad \left| B^* \right| \neq 0 \qquad (8.71)$$

and

$$A^* = \begin{bmatrix} C_1 A^{d_1+1} \\ \vdots \\ C_2 A^{d_2+1} \\ \cdots \\ \vdots \\ C_r A^{d_r+1} \end{bmatrix} \qquad (8.72)$$

C_i is the ith row of the matrix C

d_i are integer numbers that can be computed as follows:

$$d_i = \begin{bmatrix} \min j & : & C_i A^j B \neq 0, j = 0,1,2,\ldots,n-1 \\ n-1 & \text{if} & C_i A^j B = 0 \quad \text{for all } j \end{bmatrix} \qquad (8.73)$$

The purpose of this procedure is to make each output subject to only one input. In other words, the goal is to convert one MIMO system into many SISO systems. In this way the system analysis is significantly simplified.

8.9 State Observers

Sometimes it is not possible to get measurements of all the system variables. In this case we have to estimate somehow the state variables. One technique that can be applied to a controllable and observable system is to sequentially differentiate some of the state variables in order to derive the other variables. However, using differentiators is not a good practice as they increase the potential measurement noise.

Alternatively, we can design a system that estimates the state variables based on some available measurements of the input or output variables.

A **state observer** estimates (or observes) the state variables by comparing the difference between the measured and the estimated states, and by using feedback, the error converges asymptotically to zero. A state observer (depending on the type) can estimate all system states or only a minimum required number of state variables.

The block diagram of a state observer is depicted below.

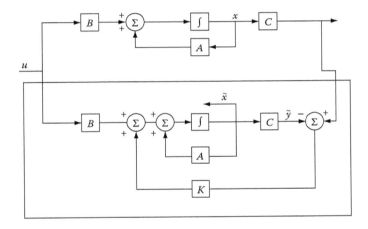

Consider a system described by the state equations

$$\dot{x} = Ax + Bu$$
$$y = Cx$$

(8.74)

Moreover, $\tilde{x}$ is the estimate of the state vector x derived from the model

$$\dot{\tilde{x}} = A\tilde{x} + Bu + K(y - C\tilde{x}) = A\tilde{x} + Bu + K(y - \tilde{y})$$

(8.75)

where
$\tilde{y}$ is the estimated output
$(y - \tilde{y})$ is the correction term that follows the state $\tilde{x}$

From (8.75) we get:

$$\dot{\tilde{x}} = (A - KC)\tilde{x} + Bu + Ky$$

(8.76)

The difference between the real and the estimated values of the state vector is called estimation error and is given by

$$e(t) = x(t) - \tilde{x}(t) \tag{8.77}$$

For the estimation error we have

$$\dot{e}(t) = \dot{x}(t) - \dot{\tilde{x}}(t) = (Ax + Bu) - [(A - KC)\tilde{x} + Bu + KCx] \Rightarrow$$
$$\dot{e}(t) = A(x - \tilde{x}) - KC(x - \tilde{x}) = [A - KC]e(t) \tag{8.78}$$

By solving (8.78) we get

$$e(t) = e(0)e^{(A-KC)t} \tag{8.79}$$

If the eigenvalues of the matrix $A - KC$ are on the left-half plane, then as $t \to \infty$ we get $e(t) \to 0$ for any $e(0) = x(0) - \tilde{x}(0)$.

The design of a state observer can be done according to the following methods:

1. By directly substituting K at the desired characteristic polynomial.
   If the system is of order less than three, K is computed according to the following relationships:

$$K = [k_1 \quad k_2 \quad \ldots \quad k_n]^T \tag{8.80}$$

$$|sI - (A - KC)| = \prod_{i=1}^{n}(s - \lambda_i) \tag{8.81}$$

2. From Ackermann's formula
   The matrix that resolved the pole placement problem is given by relationship (8.63). Therefore,

$$K = \Phi(A) \begin{bmatrix} C \\ CA \\ \cdot \\ \cdot \\ \cdot \\ CA^{n-2} \\ CA^{n-1} \end{bmatrix}^{-1} \begin{bmatrix} 0 \\ 0 \\ \cdot \\ \cdot \\ \cdot \\ 0 \\ 1 \end{bmatrix} \tag{8.82}$$

where $\varphi(s) = \prod_{i=1}^{n}(s - \lambda_i)$ is the desired characteristic polynomial.

3. With the similarity transformation
   A system can be expressed in observable canonical form by the similarity transformation:

$$Q = (WN^*)^{-1} \tag{8.83}$$

where
    N is the observability matrix
    * denotes a complex conjugate transposed matrix

Matrix W is computed by Equation 8.84 and $a_1, a_2, \ldots, a_{n-1}$ are the coefficients of the characteristic polynomial of the initial system (relationship (8.85)):

$$W = \begin{bmatrix} a_1 & a_2 & . & . & a_{n-1} & 1 \\ a_2 & a_3 & . & . & 1 & 0 \\ . & . & . & . & . & . \\ . & . & . & . & . & . \\ a_{n-1} & 1 & . & . & 0 & 0 \\ 1 & 0 & . & . & 0 & 0 \end{bmatrix} \tag{8.84}$$

$$|sI - A| = s^n + a_{n-1}s^{n-1} + \cdots + a_1 s + a_0 \tag{8.85}$$

Matrix K is now given by

$$K = Q \begin{bmatrix} \gamma_0 - a_0 \\ \gamma_1 - a_1 \\ . \\ . \\ \gamma_{n-1} - a_{n-1} \end{bmatrix} \tag{8.86}$$

In practice, the state cannot be measured, thus we have to design a state observer, where the estimated state $\tilde{x}(t)$ is given as feedback.

Consequently, we first define the state-feedback matrix K, which is associated with the pole placement, and then we derive the observer matrix K_e. Such a system is depicted in the following block diagram.

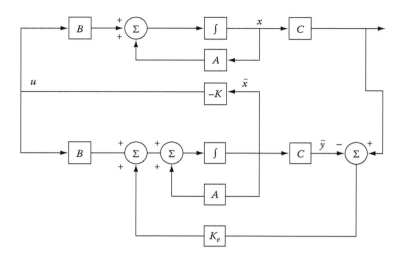

Suppose that the feedback of the observed state is ruled by

$$u(t) = -K\tilde{x}(t) \tag{8.87}$$

The state equation becomes

$$\dot{x} = Ax - BK\tilde{x} = (A - BK)x + BK(x - \tilde{x}) \tag{8.88}$$

$$\overset{(8.77)}{\Rightarrow} \dot{x} = Ax - BK\tilde{x} = (A - BK)x + BKe \tag{8.89}$$

The error of the observer is given by (8.78), but instead of K we have K_e:

$$\dot{e}(t) = [A - K_e C]e(t) \tag{8.90}$$

From (8.89) and (8.90) we derive the model of the control system with observed state feedback:

$$\begin{bmatrix} \dot{x} \\ \dot{e} \end{bmatrix} = \begin{bmatrix} A - BK & BK \\ 0 & A - K_e C \end{bmatrix} \begin{bmatrix} x \\ e \end{bmatrix} \tag{8.91}$$

The characteristic equation is

$$\begin{bmatrix} sI - A + BK & -BK \\ 0 & sI - A + K_e C \end{bmatrix} = 0 \Rightarrow \tag{8.92}$$

$$|sI - A + BK||sI - A + K_e C| = 0$$

From (8.92) we conclude that the poles of the control system with observed state feedback are the poles obtained by the pole placement procedure and the poles obtained by the design of the observer. Thus, these two procedures are independent and can be performed separately.

Formulas

TABLE F8.1

State Equations of Linear Systems

S/N	System Category	Representation in State Space	Dimensions of Matrices
1	Multiple-input, multiple-output systems	$\dot{x}(t) = Ax(t) + Bu(t)$	$A: (n \times n)$
		$y(t) = Cx(t) + Du(t)$	$B: (n \times r)$
		$x(t_0) = x_0 = x(0)$	$C: (m \times n)$
			$D: (m \times r)$
			$r, m > 1$
2	Multiple-input, single-output systems	$\dot{x}(t) = Ax(t) + Bu(t)$	$A: (n \times n)$
		$y(t) = c^T x(t) + d^T u(t)$	$B: (n \times r)$
		$x(0) = x_0$	$c: (n \times 1)$
			$d: (m \times 1)$
			$r > 1$
			$m = 1$
3	Single-input, multiple-output systems	$\dot{x}(t) = Ax(t) + bu(t)$	$A: (n \times n)$
		$y(t) = Cx(t) + du(t)$	$b: (n \times 1)$
		$x(0) = x_0$	$C: (m \times n)$
			$d: (m \times 1)$
			$r = 1$
			$m > 1$
4	Single-input, single-output systems	$\dot{x}(t) = Ax(t) + bu(t)$	$A: (n \times n)$
		$y(t) = c^T x(t) + du(t)$	$b: (n \times 1)$
		$x(0) = x_0$	$c: (n \times 1)$
			$d: (1 \times 1)$

TABLE F8.2

State Space Representation in Canonical Form

1. Differential equation of the form

$$y^{(n)} + a_1 y^{(n-1)} + \cdots + a_{n-1} y^{(1)} + a_n y = u$$

Phase-variable canonical form

$$\begin{bmatrix} \dot{x}_1 \\ \dot{x}_2 \\ \vdots \\ \dot{x}_{n-1} \\ \dot{x}_n \end{bmatrix} = \begin{bmatrix} 0 & 1 & 0 & \cdots & 0 \\ 0 & 0 & 1 & \cdots & 0 \\ \vdots & \vdots & \vdots & \vdots & \vdots \\ 0 & 0 & 0 & \cdots & 1 \\ -a_n & -a_{n-1} & -a_{n-2} & \cdots & -a_1 \end{bmatrix} \begin{bmatrix} x_1 \\ x_2 \\ \vdots \\ x_{n-1} \\ x_n \end{bmatrix} + \begin{bmatrix} 0 \\ 0 \\ \vdots \\ 0 \\ 1 \end{bmatrix} u$$

$$y = [1 \quad 0 \quad \cdots \quad 0] \begin{bmatrix} x_1 \\ x_2 \\ \vdots \\ x_n \end{bmatrix}$$

2(a). Differential equation of the form

$$y^{(n)} + a_1 y^{(n-1)} + \cdots + a_{n-1} y^{(1)} + a_n y = b_0 u^{(n)} + b_1 u^{(n-1)} + \cdots + b_{n-1} u^{(1)} + b_n u$$

Phase-variable canonical form

$$\begin{bmatrix} \dot{x}_1 \\ \dot{x}_2 \\ \vdots \\ \dot{x}_{n-1} \\ \dot{x}_n \end{bmatrix} = \begin{bmatrix} 0 & 1 & 0 & \cdots & 0 \\ 0 & 0 & 1 & \cdots & 0 \\ \vdots & \vdots & \vdots & \vdots & \vdots \\ 0 & 0 & 0 & \cdots & 1 \\ -a_n & -a_{n-1} & -a_{n-2} & \cdots & -a_1 \end{bmatrix} \begin{bmatrix} x_1 \\ x_2 \\ \vdots \\ x_{n-1} \\ x_n \end{bmatrix} + \begin{bmatrix} \beta_1 \\ \beta_2 \\ \vdots \\ \beta_{n-1} \\ \beta_n \end{bmatrix} u$$

$$y = [1 \quad 0 \quad \cdots \quad 0] \begin{bmatrix} x_1 \\ x_2 \\ \vdots \\ x_n \end{bmatrix} + \beta_0 u$$

where

$$\beta_0 = b_0$$

$$\beta_1 = b_1 - \alpha_1 \beta_0$$

$$\vdots$$

$$\beta_n = b_n - \alpha_1 \beta_{n-1} - \cdots - \alpha_{n-1} \beta_1 - \alpha_n \beta_0$$

2(b). Differential equation of the form

$$y^{(n)} + a_1 y^{(n-1)} + \cdots + a_{n-1} y^{(1)} + a_n y = b_0 u^{(n)} + b_1 u^{(n-1)} + \cdots + b_{n-1} u^{(1)} + b_n u$$

Controllable canonical form

$$\begin{bmatrix} \dot{x}_1 \\ \dot{x}_2 \\ \vdots \\ \dot{x}_{n-1} \\ \dot{x}_n \end{bmatrix} = \begin{bmatrix} 0 & 1 & 0 & \cdots & 0 \\ 0 & 0 & 1 & \cdots & 0 \\ \vdots & \vdots & \vdots & \vdots & \vdots \\ 0 & 0 & 0 & \cdots & 1 \\ -a_n & -a_{n-1} & -a_{n-2} & \cdots & -a_1 \end{bmatrix} \begin{bmatrix} x_1 \\ x_2 \\ \vdots \\ x_{n-1} \\ x_n \end{bmatrix} + \begin{bmatrix} 0 \\ 0 \\ \vdots \\ 0 \\ 1 \end{bmatrix} u$$

$$y = [b_n - a_n b_0 \quad \vdots \quad b_{n-1} - a_{n-1} b_0 \quad \vdots \quad \cdots \quad \vdots \quad b_1 - a_1 b_0] \begin{bmatrix} x_1 \\ x_2 \\ \vdots \\ x_n \end{bmatrix} + b_0 u$$

TABLE F8.2 (continued)

State Space Representation in Canonical Form

2(c). Differential equation of the form

$$y^{(n)} + a_1 y^{(n-1)} + \cdots + a_{n-1} y^{(1)} + a_n y = b_0 u^{(n)} + b_1 u^{(n-1)} + \cdots + b_{n-1} u^{(1)} + b_n u$$

Observable canonical form

$$\begin{bmatrix} \dot{x}_1 \\ \dot{x}_2 \\ \vdots \\ \dot{x}_n \end{bmatrix} = \begin{bmatrix} 0 & 0 & 0 & \cdots & -a_n \\ 1 & 0 & 0 & \cdots & -a_{n-1} \\ \vdots & \vdots & & \vdots & \vdots \\ 0 & 0 & \cdots & 1 & \cdots & -a_1 \end{bmatrix} \begin{bmatrix} x_1 \\ x_2 \\ \vdots \\ x_n \end{bmatrix} + \begin{bmatrix} b_n - a_n b_0 \\ b_{n-1} - a_{n-1} b_0 \\ \vdots \\ b_1 - a_1 b_0 \end{bmatrix} u$$

$$y = \begin{bmatrix} 0 & 0 & \cdots & 0 & 1 \end{bmatrix} \begin{bmatrix} x_1 \\ x_2 \\ \vdots \\ x_n \end{bmatrix} + b_0 u$$

3(a). Transfer function of the form

$$\frac{Y(s)}{U(s)} = \frac{b_0 s^n + b_1 s^{n-1} + \cdots + b_{n-1} s + b_n}{(s - p_1)(s - p_2) \cdots (s - p_n)}$$

$$= b_0 + \frac{c_1}{s - p_1} + \frac{c_2}{s - p_2} + \cdots + \frac{c_n}{s - p_n}$$

Diagonal canonical form

$$\begin{bmatrix} \dot{x}_1 \\ \dot{x}_2 \\ \vdots \\ \dot{x}_n \end{bmatrix} = \begin{bmatrix} p_1 & & & 0 \\ & p_2 & & \\ & & \ddots & \\ 0 & & & p_n \end{bmatrix} \begin{bmatrix} x_1 \\ x_2 \\ \vdots \\ x_n \end{bmatrix} + \begin{bmatrix} 1 \\ 1 \\ \vdots \\ 1 \end{bmatrix} u$$

$$y = \begin{bmatrix} c_1 & c_2 & \cdots & c_n \end{bmatrix} \begin{bmatrix} x_1 \\ x_2 \\ \vdots \\ x_n \end{bmatrix} + b_0 u$$

3(b). Transfer function of the form

$$\frac{Y(s)}{U(s)} = \frac{b_0 s^n + b_1 s^{n-1} + \cdots + b_{n-1} s + b_n}{(s - p_1)^3 (s - p_4)(s - p_5) \cdots (s - p_n)}$$

Jordan canonical form

$$\begin{bmatrix} \dot{x}_1 \\ \dot{x}_2 \\ \dot{x}_3 \\ \dot{x}_4 \\ \vdots \\ \dot{x}_n \end{bmatrix} = \begin{bmatrix} p_1 & 1 & 0 & 0 & \cdots & 0 \\ 0 & p_1 & 1 & \vdots & & \vdots \\ 0 & 0 & p_1 & 0 & & 0 \\ 0 & \cdots & 0 & p_4 & \cdots & 0 \\ \vdots & & \vdots & \vdots & \ddots & \\ 0 & \cdots & 0 & 0 & & p_n \end{bmatrix} \begin{bmatrix} x_1 \\ x_2 \\ x_3 \\ x_4 \\ \vdots \\ x_n \end{bmatrix} + \begin{bmatrix} 0 \\ 0 \\ 1 \\ 1 \\ \vdots \\ 1 \end{bmatrix} u$$

$$y = \begin{bmatrix} c_1 & c_2 & \cdots & c_n \end{bmatrix} \begin{bmatrix} x_1 \\ x_2 \\ \vdots \\ x_n \end{bmatrix} + b_0 u$$

TABLE F8.3

Transformations of the State Vector: Special Forms of State Equations

S/N	Formulas	Remarks
1	$x=Tz \qquad (|T|\neq 0)$ $\dot{x}=Ax+Bu \qquad\qquad \dot{z}=A^*z+B^*u$ $y=Cx+Du \quad \overset{*}{\Leftrightarrow} \quad y=C^*z+D^*u$ $x(0)=x_0 \qquad\qquad z(0)=z_0$ $A^*=T^{-1}AT \;\vdots\; C^*=CT \;\vdots\; z(0)=T^{-1}x(0)$ $B^*=T^{-1}B \;\vdots\; D^*=D \;\vdots$	Transformed state-space model (A^*, B^*, C^*, D^*) or by the linear transformation $x=Tz$
2	$A^*=\begin{bmatrix} 0 & 1 & 0 & \cdots & 0 \\ 0 & 0 & 1 & \cdots & 0 \\ \vdots & \vdots & \vdots & \vdots & \vdots \\ 0 & 0 & 0 & \cdots & 1 \\ -a_0^* & -a_1^* & -a_2^* & \cdots & -a_{n-1}^* \end{bmatrix}$ $b^*=\begin{bmatrix}0\\0\\\vdots\\1\end{bmatrix}$ $T=p^{-1}$ $P=\begin{bmatrix}q\\ \cdots\\ qA\\ \cdots\\ \vdots\\ qA^{n-1}\end{bmatrix}$	Single-input system (A^*, B^*, C^*, D^*) or In phase-variable canonical form q: the last row of the controllability matrix S $S=\begin{bmatrix} b & \vdots & Ab & \vdots & \cdots & \vdots & A^{(n-1)}b \end{bmatrix}$ $|S|\neq 0$
3	$\Lambda=T^{-1}A^*T$ $T=\begin{bmatrix} 1 & 1 & \cdots & 1 \\ \lambda_1 & \lambda_2 & \cdots & \lambda_n \\ \vdots & \vdots & \vdots & \vdots \\ \lambda_1^{n-1} & \lambda_2^{n-1} & \cdots & \lambda_n^{n-1} \end{bmatrix}$	From the phase-variable canonical form to the diagonal canonical form Λ: the diagonalized matrix T: the Vandermonde matrix $(|T|\neq 0)$

TABLE F8.4

Transfer Function Matrix—Solution of State Equation—Controllability—Observability

S/N	Comments	Formulas
1	Transfer function $G(s)$	$G(s) = C(sI - A)^{-1} B + D$
2	Solution of the homogeneous state equation $\dot{x}(t) = Ax(t)$ $x(0) = x_0$	$x(t) = e^{At}x(0)$
3	State-transition matrix $\Phi(t)$	$\Phi(t) = e^{At} = L^{-1}\{(sI - A)^{-1}\}$

or

$$\Phi(t) = e^{At} = I + At + \frac{A^2 t^2}{2!} + \cdots + \frac{A^{k-1} t^{k-1}}{(k-1)!} + \cdots$$

Special cases:

a. Matrix A has distinct eigenvalues and is diagonal

$$\Phi(t) = e^{At} = \begin{bmatrix} e^{\lambda_1 t} & & & 0 \\ & e^{-\lambda_2 t} & & \\ & & \ddots & \\ 0 & & & e^{-\lambda_n t} \end{bmatrix}$$

b. Matrix A has distinct eigenvalues and linearly independent eigenvectors

$$\Phi(t) = e^{At} = M \begin{bmatrix} e^{-\lambda_1 t} & & & 0 \\ & e^{-\lambda_2 t} & & \\ & & \ddots & \\ 0 & & & e^{-\lambda_n t} \end{bmatrix} M^{-1}$$

$$M = [u_1 \;\vdots\; u_2 \;\vdots\; \cdots \;\vdots\; u_n]$$

$$Au_i = \lambda_i u_i$$

| 4 | General solution of the state equation | $x(t) = \Phi(t)x(0) + \displaystyle\int_0^t \Phi(t - \lambda)bu(\lambda)\,d\lambda$ |

or

$$x(t) = L^{-1}\{(sI - A)^{-1}(x_0 + bU(s))\}$$

5	Controllability of the state vector	$S = [B \;\vdots\; AB \;\vdots\; \cdots \;\vdots\; A^{n-1}B]$ $\mathrm{rank}[S] = n$
6	Observability of the state vector	$R^T = [C^T \;\vdots\; A^T C^T \;\vdots\; \cdots \;\vdots\; (A^T)^{n-1}C^T]$ $\mathrm{rank}[R^T] = n$
7	Observability of the output vector	$Q = [D \;\vdots\; CB \;\vdots\; CAB \;\vdots\; \cdots \;\vdots\; CA^{n-1}B]$ $\mathrm{rank}[Q] = n$
8	Controllability of the output vector	$M = [CB \;\vdots\; CAB \;\vdots\; \ldots \;\vdots\; CA^{n-1}B \;\vdots\; D]$ $\mathrm{rank}[M] = n$

TABLE F8.5

Modern Design Methods of Control Systems

S/N	Method	Types	Remarks
1	Placement of the eigenvalues with state feedback	$u = -Kx$ $K = \begin{bmatrix} k_1 k_2 \cdots k_n \end{bmatrix}$ $\dot{x} = (A - BK)x$ $|sI - A + BK| = 0$	State feedback control law State-feedback gain matrix New state equation Characteristic equation
	First way for finding the matrix K	$K = \begin{bmatrix} 0 & 0 & \cdots & 0 & 1 \end{bmatrix} [B \quad AB \quad \cdots \quad A^{n-2}B \quad A^{n-1}B]^{-1} \Phi(\prime)$ $\Phi(A) = A^n + a_{n-1}A^{n-1} + \cdots + a_1 A + a_0 I$ $G(s) = \dfrac{b_{n-1}s^{n-1} + \cdots + b_1 s + b_0}{s^n + a_{n-1}s^{n-1} + \cdots + a_1 s + a_0}$	Ackermann's formula
	Second way for finding the matrix K	$K = \begin{bmatrix} b_n - a_n & : & b_{n-1} - a_{n-1} & : & \cdots & : & b_2 - a_2 & : & b_1 - a_1 \end{bmatrix} T^{-1}$ $|sI - A| = s^n + a_1 s^{n-1} + \cdots + a_{n-1}s + a_n$ $(s - \mu_1)(s - \mu_2) \cdots (s - \mu_n) = s^n + b_1 s^{n-1} + \cdots + b_{n-1}s + b_n$ $T = SW$ $S = [B \quad : \quad AB \quad : \quad \cdots \quad : \quad A^{n-1}B]$ $W = \begin{bmatrix} a_{n-1} & a_{n-2} & \cdots & a_1 & 1 \\ a_{n-2} & a_{n-3} & \cdots & 1 & 0 \\ \vdots & \vdots & \vdots & \vdots & \vdots \\ a_1 & 1 & \cdots & 0 & 0 \\ 1 & 0 & \cdots & 0 & 0 \end{bmatrix}$	μ_i: desirable eigenvalues T: transformation matrix of the state equations in the controllable canonical form

2   Input-output decoupling with state feedback

$$K = -(B^*)^{-1} \cdot A^*$$

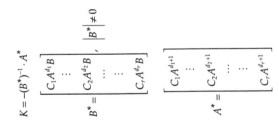

$$B^* = \begin{bmatrix} C_1 A^{d_1} B \\ \vdots \\ C_2 A^{d_2} B \\ \vdots \\ C_r A^{d_r} B \end{bmatrix}, \quad \left| B^* \right| \neq 0$$

$$A^* = \begin{bmatrix} C_1 A^{d_1+1} \\ \vdots \\ C_2 A^{d_2+1} \\ \vdots \\ C_r A^{d_r+1} \end{bmatrix}$$

C_i is the ith row of the matrix C

$$d_i = \begin{bmatrix} \min j : C_i A^j B \neq 0, \, j = 0,1,2,\dots,n-1 \\ n-1 \quad \text{if } C_i A^j B = 0, \text{ for all } j \end{bmatrix}$$

Problems

8.1 Find the state-space representation of the electric circuit of the system shown in the figure below. Suppose zero initial conditions.

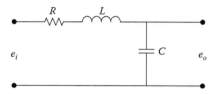

Solution

We apply Kirchhoff's voltage law at the circuit and get

$$L\frac{di(t)}{dt} + Ri(t) + \frac{1}{C}\int_0^t i(t)dt = e_i(t) \tag{P8.1.1}$$

$$\frac{1}{C}\int_0^t i(t)dt = e_o(t) \tag{P8.1.2}$$

We apply Laplace transform to the previous set of equations. The initial conditions are zero; thus,

$$LsI(s) + RI(s) + \frac{1}{C}\frac{I(s)}{s} = E_i(s) \tag{P8.1.3}$$

$$\frac{1}{C}\frac{I(s)}{s} = E_o(s) \tag{P8.1.4}$$

The transfer function of the system is

$$\frac{E_o(s)}{E_i(s)} = \frac{1}{LCs^2 + RCs + 1} \tag{P8.1.5}$$

From the relationship (P8.1.5) we get the differential equation for the system

$$\ddot{e}_o + \frac{R}{L}\dot{e}_o + \frac{1}{LC}e_o = \frac{1}{LC}e_i \tag{P8.1.6}$$

We consider the following state variables:

$$\left.\begin{array}{l} x_1 = e_o \\ x_2 = \dot{e}_o \end{array}\right\} \tag{P8.1.7}$$

The input and output variables are, respectively,

$$u = e_i \tag{P8.1.8}$$

$$y = e_o = x_1 \tag{P8.1.9}$$

The state-space representation of the system is

$$\begin{bmatrix} \dot{x}_1 \\ \dot{x}_2 \end{bmatrix} = \begin{bmatrix} 0 & 1 \\ -\dfrac{1}{LC} & -\dfrac{R}{L} \end{bmatrix} \begin{bmatrix} x_1 \\ x_2 \end{bmatrix} + \begin{bmatrix} 0 \\ \dfrac{1}{LC} \end{bmatrix} u \tag{P8.1.10}$$

$$y = \begin{bmatrix} 1 & 0 \end{bmatrix} \begin{bmatrix} x_1 \\ x_2 \end{bmatrix} \tag{P8.1.11}$$

8.2 Write down the state equations for the circuit depicted in the figure below. Consider as outputs the currents i_1 and i_2 and as inputs the voltages e_1 and e_2.

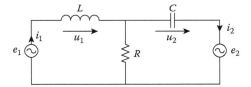

Solution
We choose as state variables the coil current i_1 and the voltage u_2 at the ends of the capacitor. The differential equations that describe the circuit are

$$L\frac{di_1}{dt} = e_1 - e_2 - u_2 \tag{P8.2.1}$$

$$C\frac{du_2}{dt} = i_1 - \frac{u_2 + e_2}{R} \tag{P8.2.2}$$

The outputs i_1 and i_2 of the circuit are

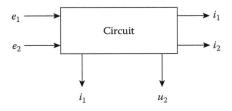

$$i_1 = i_1 \tag{P8.2.3}$$

$$i_2 = i_1 - \frac{u_2}{R} - \frac{e_2}{R} \tag{P8.2.4}$$

The state equations of the system are

$$\begin{bmatrix} \dot{i_1} \\ \dot{u_2} \end{bmatrix} = \begin{bmatrix} 0 & -\dfrac{1}{L} \\ \dfrac{1}{C} & -\dfrac{1}{RC} \end{bmatrix} \begin{bmatrix} i_1 \\ u_2 \end{bmatrix} + \begin{bmatrix} \dfrac{1}{L} & -\dfrac{1}{L} \\ 0 & -\dfrac{1}{RC} \end{bmatrix} \begin{bmatrix} e_1 \\ e_2 \end{bmatrix} \qquad \text{(P8.2.5)}$$

$$\begin{bmatrix} i_1 \\ i_2 \end{bmatrix} = \begin{bmatrix} 1 & 0 \\ 1 & -\dfrac{1}{R} \end{bmatrix} \begin{bmatrix} i_1 \\ u_2 \end{bmatrix} + \begin{bmatrix} 0 & 0 \\ 0 & -\dfrac{1}{R} \end{bmatrix} \begin{bmatrix} e_1 \\ e_2 \end{bmatrix} \qquad \text{(P8.2.6)}$$

8.3 Write the state-space representation for the mechanical system shown in the figure below.

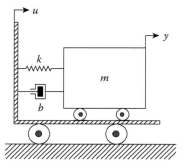

Solution

We apply Newton's second law to the system and get

$$m\frac{d^2y}{dt^2} = -b\left(\frac{dy}{dt} - \frac{du}{dt}\right) - k(y - u)$$

or

$$m\frac{d^2y}{dt^2} + b\frac{dy}{dt} + ky = b\frac{du}{dt} + ku \qquad \text{(P8.3.1)}$$

$$\text{(P8.3.1)} \Rightarrow \ddot{y} + \frac{b}{m}\dot{y} + \frac{k}{m}y = \frac{b}{m}\dot{u} + \frac{k}{m}u \qquad \text{(P8.3.2)}$$

The differential equation of the relationship (P8.3.2) has the form

$$\ddot{y} + a_1\dot{y} + a_2y = b_0\ddot{u} + b_1\dot{u} + b_2u \qquad \text{(P8.3.3)}$$

where

$$a_1 = \frac{b}{m}, \quad a_2 = \frac{k}{m}, \quad b_0 = 0, \quad b_1 = \frac{b}{m}, \quad b_2 = \frac{k}{m}$$

We regard as state variables the quantities

$$x_1 = y - \beta_0 u \qquad \text{(P8.3.4)}$$

$$x_2 = \dot{x}_1 - \beta_1 u \qquad \text{(P8.3.5)}$$

where β_0 and β_1 are computed as

$$\beta_0 = b_0 = 0 \qquad \text{(P8.3.6)}$$

$$\beta_1 = b_1 - a_1\beta_0 = \frac{b}{m} \qquad \text{(P8.3.7)}$$

Hence,

$$x_1 = y \qquad \text{(P8.3.8)}$$

$$x_2 = \dot{x}_1 - \frac{b}{m}u \qquad \text{(P8.3.9)}$$

The derivatives of the state variables are

$$\dot{x}_1 = x_2 + \beta_1 u = x_2 + \frac{b}{m}u \qquad \text{(P8.3.10)}$$

$$\dot{x}_2 = -a_2 x_1 - a_1 x_2 + \beta_2 u = -\frac{k}{m}x_1 - \frac{b}{m}x_2 + \left[\frac{k}{m} - \left(\frac{b}{m}\right)^2\right]u \qquad \text{(P8.3.11)}$$

The output is now

$$y = x_1 \qquad \text{(P8.3.12)}$$

The desired state equations are

$$\begin{bmatrix} \dot{x}_1 \\ \dot{x}_2 \end{bmatrix} = \begin{bmatrix} 0 & 1 \\ -\dfrac{k}{m} & -\dfrac{b}{m} \end{bmatrix} \begin{bmatrix} x_1 \\ x_2 \end{bmatrix} + \begin{bmatrix} \dfrac{b}{m} \\ \dfrac{k}{m} - \left(\dfrac{b}{m}\right)^2 \end{bmatrix} u \qquad \text{(P8.3.13)}$$

$$y = \begin{bmatrix} 1 & 0 \end{bmatrix} \begin{bmatrix} x_1 \\ x_2 \end{bmatrix} \qquad \text{(P8.3.14)}$$

8.4 Express in state-space form the mechanical system depicted in the next figure and draw its block diagram (using integrator blocks).

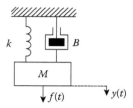

Solution

The differential equation that describes the system is

$$M\frac{d^2y(t)}{dt^2} = f(t) - B\frac{dy(t)}{dt} - Ky(t) \qquad \text{(P8.4.1)}$$

The transfer function of the system is

$$G(s) = \frac{Y(s)}{F(s)} = \frac{1}{Ms^2 + Bs + K} \qquad \text{(P8.4.2)}$$

We consider as state variables the following:

$$x_1 = y \qquad \text{(P8.4.3)}$$

$$x_2 = \dot{y} = \dot{x}_1 \qquad \text{(P8.4.4)}$$

where $x_1(t)$ is the position, and $x_2(t)$ the velocity:

$$(\text{P8.4.1}) \overset{\underset{(\text{P8.4.3})}{}}{\underset{(\text{P8.4.4})}{\Rightarrow}} \frac{d^2y}{dt^2} = \dot{x}_2 = -\left(\frac{B}{M}\right)x_2 - \left(\frac{K}{M}\right)x_1 + \left(\frac{1}{M}\right)f(t) \qquad \text{(P8.4.5)}$$

Therefore, the state equations are

$$\begin{bmatrix} \dot{x}_1 \\ \dot{x}_2 \end{bmatrix} = \begin{bmatrix} 0 & 1 \\ -\dfrac{K}{M} & -\dfrac{B}{M} \end{bmatrix} \begin{bmatrix} x_1 \\ x_2 \end{bmatrix} + \begin{bmatrix} 0 \\ \dfrac{1}{M} \end{bmatrix} f(t) \qquad \text{(P8.4.6)}$$

$$y = \begin{bmatrix} 1 & 0 \end{bmatrix} \begin{bmatrix} x_1 \\ x_2 \end{bmatrix} \qquad \text{(P8.4.7)}$$

The block diagram of the mechanical system is

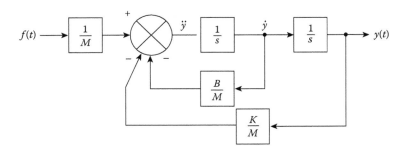

8.5 Write the state-space representation for the DC field motor, shown in the figure below.

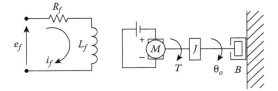

Solution

The equations that describe the system are

$$e_f = R_f i_f + L_f \frac{di_f}{dt} \overset{LT}{\Rightarrow} E_f(s) = (R_f + sL_f)I_f(s) \tag{P8.5.1}$$

$$T(t) = k_t i_f(t) \overset{LT}{\Rightarrow} T(s) = k_t I_f(s) \tag{P8.5.2}$$

$$T(t) = J\frac{d^2\theta_o(t)}{dt^2} + B\frac{d\theta_o(t)}{dt} \overset{LT}{\Rightarrow} T(s) = s(Js + B)\Theta_o(s) \tag{P8.5.3}$$

The transfer function of the system is

$$\frac{\Theta_o(s)}{E_f(s)} = \frac{k_t/R_f B}{s(T_m s + 1)(T_f s + 1)} \tag{P8.5.4}$$

where

$$\left. \begin{array}{l} T_f = \dfrac{L_f}{R_f} : \text{time constant of the field} \\[3mm] T_m = \dfrac{J}{M} : \text{time constant of the motor} \end{array} \right\} \tag{P8.5.5}$$

We regard as state variables the following:

$$x_1 = \theta_o \tag{P8.5.6}$$

$$x_2 = \dot{x}_1 = \dot{\theta}_o \tag{P8.5.7}$$

$$x_3 = i_f \tag{P8.5.8}$$

The input and output variables are

$$u = e_f \tag{P8.5.9}$$

$$y = \theta_o \tag{P8.5.10}$$

The derivatives of the state variables are

$$(\text{P8.5.7}) \Rightarrow \dot{x}_1 = x_2$$

$$\overset{(\text{P8.5.2}),(\text{P8.5.3})}{\Rightarrow} \quad k_t i_f(t) = J\ddot{\theta}_o + B\dot{\theta}_o \tag{P8.5.11}$$

$$(\text{P8.5.11}) \Rightarrow \dot{x}_2 = -\frac{B}{J}x_2 + \frac{k_t}{J}x_3 \tag{P8.5.12}$$

The model of the system in the state space is

$$\begin{bmatrix} \dot{x}_1 \\ \dot{x}_2 \\ \dot{x}_3 \end{bmatrix} = \begin{bmatrix} 0 & 1 & 0 \\ 0 & -\dfrac{B}{J} & \dfrac{k_t}{J} \\ 0 & 0 & -\dfrac{R_f}{L_f} \end{bmatrix} \begin{bmatrix} x_1 \\ x_2 \\ x_3 \end{bmatrix} + \begin{bmatrix} 0 \\ 0 \\ \dfrac{1}{L_f} \end{bmatrix} u \tag{P8.5.13}$$

$$y = \begin{bmatrix} 1 & 0 & 0 \end{bmatrix} \begin{bmatrix} x_1 \\ x_2 \\ x_3 \end{bmatrix} \tag{P8.5.14}$$

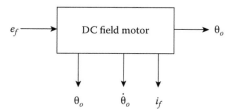

8.6 Write the state-space representation of the hydraulic system shown below, if
- a. The input and output variables are q and q_2, respectively
- b. The input and output variables are q and h_2, respectively

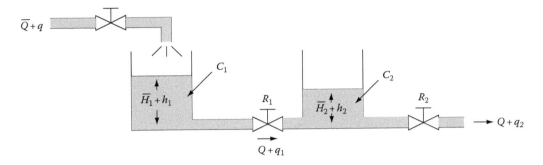

Solution

a. The hydraulic level control system has been discussed in Chapter 2. Its transfer function is

$$\frac{Q_2(s)}{Q(s)} = \frac{1}{R_1C_1R_2C_2s^2 + (R_1C_1 + R_2C_1 + R_2C_2)s + 1} \tag{P8.6.1}$$

The differential equation of the system is

$$R_1C_1R_2C_2\ddot{q}_2 + (R_1C_1 + R_2C_2 + R_2C_1)\dot{q}_2 + q_2 = q \tag{P8.6.2}$$

or

$$\ddot{q}_2 + \left(\frac{1}{R_2C_2} + \frac{1}{R_1C_1} + \frac{1}{R_1C_2}\right)\dot{q}_2 + \frac{1}{R_1C_1R_2C_2}q_2 = \frac{1}{R_1C_1R_2C_2}q \tag{P8.6.3}$$

We regard as state variables the following quantities:

$$x_1 = q_2 \tag{P8.6.4}$$

$$x_2 = \dot{q}_2 \tag{P8.6.5}$$

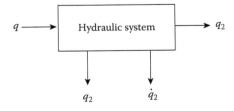

The state equations that describe the system are

$$\begin{bmatrix} \dot{x}_1 \\ \dot{x}_2 \end{bmatrix} = \begin{bmatrix} 0 & 1 \\ -\dfrac{1}{R_1C_1R_2C_2} & -\left(\dfrac{1}{R_2C_2}+\dfrac{1}{R_1C_1}+\dfrac{1}{R_1C_2}\right) \end{bmatrix}\begin{bmatrix} x_1 \\ x_2 \end{bmatrix} + \begin{bmatrix} 0 \\ \dfrac{1}{R_1C_1R_2C_2} \end{bmatrix} u \qquad (P8.6.6)$$

$$y = \begin{bmatrix} 1 & 0 \end{bmatrix}\begin{bmatrix} x_1 \\ x_2 \end{bmatrix} \qquad (P8.6.7)$$

b. In case the input and output variables are the quantities q and h_2, respectively, we consider as state variables the following:

$$x_1 = h_2 \qquad (P8.6.8)$$

$$x_2 = h_1 \qquad (P8.6.9)$$

From the equations that describe the operation of the system we have

$$C_2\frac{dh_2}{dt} = \frac{h_1 - h_2}{R_1} - \frac{h_2}{R_2} \qquad (P8.6.10)$$

$$C_1\frac{dh_1}{dt} = q - \frac{h_1 - h_2}{R} \qquad (P8.6.11)$$

or

$$\frac{dh_2}{dt} = -\left(\frac{1}{R_1C_2}+\frac{1}{R_2C_2}\right)h_2 + \frac{1}{R_1C_2}h_1 \qquad (P8.6.12)$$

$$\frac{dh_1}{dt} = \frac{1}{R_1C_1}h_2 - \frac{1}{R_1C_1}h_1 + \frac{1}{C_1}q \qquad (P8.6.13)$$

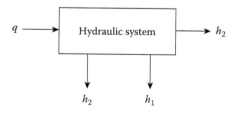

The state equations of the system are

$$\begin{bmatrix} \dot{x}_1 \\ \dot{x}_2 \end{bmatrix} = \begin{bmatrix} -\left(\dfrac{1}{R_1C_2}+\dfrac{1}{R_2C_2}\right) & \dfrac{1}{R_1C_2} \\ \dfrac{1}{R_1C_1} & -\dfrac{1}{R_1C_1} \end{bmatrix}\begin{bmatrix} x_1 \\ x_2 \end{bmatrix} + \begin{bmatrix} 0 \\ \dfrac{1}{C_1} \end{bmatrix} u \qquad (P8.6.14)$$

$$y = \begin{bmatrix} 1 & 0 \end{bmatrix}\begin{bmatrix} x_1 \\ x_2 \end{bmatrix} \qquad (P8.6.15)$$

We observe that many different representations in the state space are possible.

8.7 Represent in the state space, the double input, double output level control system, shown in the figure below.

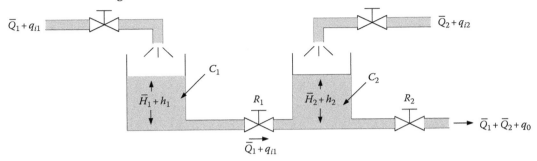

Solution
The equations that describe the depicted hydraulic system are

$$C_1 \frac{dh_1}{dt} = q_{i1} - q_1 \tag{P8.7.1}$$

$$\frac{h_1 - h_2}{R_1} = q_1 \tag{P8.7.2}$$

$$C_2 \frac{dh_2}{dt} = q_1 + q_{i2} - q_0 \tag{P8.7.3}$$

$$\frac{h_2}{R_2} = q_0 \tag{P8.7.4}$$

We consider as state variables the following quantities:

$$(P8.7.1), (P8.7.2) \Rightarrow \frac{dh_1}{dt} = \frac{1}{C_1} \left(q_{i1} - \frac{h_1 - h_2}{R_1} \right) \tag{P8.7.5}$$

$$(P8.7.3) \underset{(P8.7.4)}{\overset{(P8.7.2)}{\Rightarrow}} \frac{dh_2}{dt} = \frac{1}{C_2} \left(\frac{h_1 - h_2}{R_1} + q_{i2} - \frac{h_2}{R_2} \right) \tag{P8.7.6}$$

$$x_1 = h_1 \tag{P8.7.7}$$

$$x_2 = h_2 \tag{P8.7.8}$$

The system has the following form.

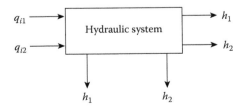

Equations P8.7.5 and P8.7.6 can be written as

$$\dot{x}_1 = -\frac{1}{R_1C_1}x_1 + \frac{1}{R_1C_1}x_2 + \frac{1}{C_1}u_1 \tag{P8.7.9}$$

$$\dot{x}_2 = -\frac{1}{R_1C_2}x_1 - \left(\frac{1}{R_1C_2}+\frac{1}{R_2C_2}\right)x_2 + \frac{1}{C_2}u_2 \tag{P8.7.10}$$

The state-space representation is

$$\begin{bmatrix}\dot{x}_1\\\dot{x}_2\end{bmatrix} = \begin{bmatrix}-\dfrac{1}{R_1C_1} & \dfrac{1}{R_1C_1}\\[2mm]\dfrac{1}{R_1C_2} & -\left(\dfrac{1}{R_1C_2}+\dfrac{1}{R_2C_2}\right)\end{bmatrix}\begin{bmatrix}x_1\\x_2\end{bmatrix}+\begin{bmatrix}\dfrac{1}{C_1} & 0\\[2mm]0 & \dfrac{1}{C_2}\end{bmatrix}\begin{bmatrix}u_1\\u_2\end{bmatrix} \tag{P8.7.11}$$

$$\begin{bmatrix}y_1\\y_2\end{bmatrix} = \begin{bmatrix}1 & 0\\0 & 1\end{bmatrix}\begin{bmatrix}x_1\\x_2\end{bmatrix} \tag{P8.7.12}$$

8.8 The block diagrams of two control systems are illustrated in the two following figures. Write their state-space representation.

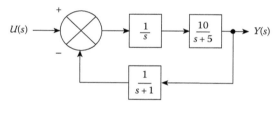

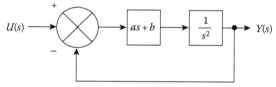

Solution

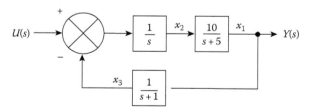

a. The system includes an integrator and two delay integrators. We consider as state variables (see figure above) the quantities x_1, x_2, x_3, for which

$$\frac{X_1(s)}{X_2(s)} = \frac{10}{s+5} \tag{P8.8.1}$$

$$\frac{X_2(s)}{U_2(s) - X_3(s)} = \frac{1}{s} \tag{P8.8.2}$$

$$\frac{X_3(s)}{X_1(s)} = \frac{1}{s+1} \tag{P8.8.3}$$

and

$$Y(s) = X_1(s) \tag{P8.8.4}$$

The relationships (P8.8.1), (P8.8.2), and (P8.8.3) can be rewritten as

$$sX_1(s) = -5X_1(s) + 10X_2(s) \tag{P8.8.5}$$

$$sX_2(s) = -X_3(s) + U(s) \tag{P8.8.6}$$

$$sX_3(s) = X_1(s) - X_3(s) \tag{P8.8.7}$$

We apply inverse Laplace transform to Equations P8.8.5 through P8.8.7, and P8.8.4 and get

$$\dot{x}_1 = -5x_1 + 10x_2 \tag{P8.8.8}$$

$$\dot{x}_2 = -x_3 + u \tag{P8.8.9}$$

$$\dot{x}_3 = x_1 - x_3 \tag{P8.8.10}$$

$$y = x_1 \tag{P8.8.11}$$

Hence, one representation of the system in the state space is

$$\begin{bmatrix} \dot{x}_1 \\ \dot{x}_2 \\ \dot{x}_3 \end{bmatrix} = \begin{bmatrix} -5 & 10 & 0 \\ 0 & 0 & -1 \\ 1 & 0 & -1 \end{bmatrix} \begin{bmatrix} x_1 \\ x_2 \\ x_3 \end{bmatrix} + \begin{bmatrix} 0 \\ 1 \\ 0 \end{bmatrix} u \tag{P8.8.12}$$

$$y = \begin{bmatrix} 1 & 0 & 0 \end{bmatrix} \begin{bmatrix} x_1 \\ x_2 \\ x_3 \end{bmatrix} \tag{P8.8.13}$$

b. The loop transfer function of the system can be written as

$$\frac{as+b}{s^2} = \left(a + \frac{b}{s} \right) \cdot \frac{1}{s} \tag{P8.8.14}$$

The block diagram of the second system can be transformed as

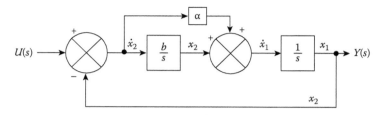

We regard as state variables the outputs of the integrators:

$$\dot{x}_1 = -ax_1 + x_2 + au \qquad\qquad (P8.8.15)$$

$$\dot{x}_2 = -bx_1 + bu \qquad\qquad (P8.8.16)$$

$$y = x_1 \qquad\qquad (P8.8.17)$$

The representation of the system in the state space is

$$\begin{bmatrix} \dot{x}_1 \\ \dot{x}_2 \end{bmatrix} = \begin{bmatrix} -a & 1 \\ -b & 0 \end{bmatrix}\begin{bmatrix} x_1 \\ x_2 \end{bmatrix} + \begin{bmatrix} a \\ b \end{bmatrix}u \qquad\qquad (P8.8.18)$$

$$y = \begin{bmatrix} 1 & 0 \end{bmatrix}\begin{bmatrix} x_1 \\ x_2 \end{bmatrix} \qquad\qquad (P8.8.19)$$

8.9 Suppose that a control system is described by the differential equation $\ddot{y} + 6\ddot{y} + 11\dot{y} + 6y = u$, where u denotes the input and y the output.
Write the state-space representation of the system
a. In phase-variable canonical form
b. In diagonal canonical form
Plot the block diagram for each case.

Solution
a. The differential equation of the system is

$$\dddot{y} + 6\ddot{y} + 11\dot{y} + 6y = 6u \qquad\qquad (P8.9.1)$$

We consider as state variables the following:

$$\left.\begin{array}{l} x_1 = y \\ x_2 = \dot{y} \\ x_3 = \ddot{y} \end{array}\right\} \qquad\qquad (P8.9.2)$$

From relationships (P8.9.1) and (P8.9.2), we get

$$\left.\begin{aligned}
\dot{x}_1 &= \dot{y} = x_2 \\
\dot{x}_2 &= \ddot{y} = x_3 \\
\dot{x}_3 &= \dddot{y} = -6x_1 - 11x_2 - 6x_3 + 6u
\end{aligned}\right\} \tag{P8.9.3}$$

The phase-variable canonical form representation is

$$\begin{bmatrix} \dot{x}_1 \\ \dot{x}_2 \\ \dot{x}_3 \end{bmatrix} = \begin{bmatrix} 0 & 1 & 0 \\ 0 & 0 & 1 \\ -6 & -11 & -6 \end{bmatrix} \begin{bmatrix} x_1 \\ x_2 \\ x_3 \end{bmatrix} + \begin{bmatrix} 0 \\ 0 \\ 6 \end{bmatrix} u \tag{P8.9.4}$$

$$y = \begin{bmatrix} 1 & 0 & 0 \end{bmatrix} \begin{bmatrix} x_1 \\ x_2 \\ x_3 \end{bmatrix} \tag{P8.9.5}$$

The implementation block diagram of the system, which corresponds to the representation of Equations P8.9.4 and P8.9.5, is

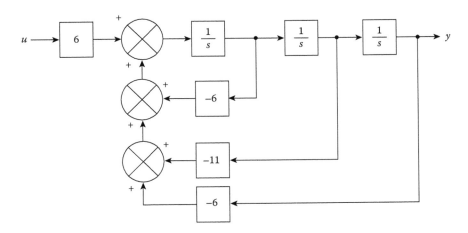

b. The transfer function of the system is

$$G(s) = \frac{Y(s)}{X(s)} = \frac{6}{s^3 + 6s^2 + 11s + 6} = \frac{6}{(s+1)(s+2)(s+3)} \tag{P8.9.6}$$

We express the transfer function as a sum of partial fractions

$$\frac{Y(s)}{U(s)} = \frac{3}{s+1} + \frac{-6}{s+2} + \frac{3}{s+3} \tag{P8.9.7}$$

thus,

$$Y(s) = \frac{3}{s+1}U(s) + \frac{-6}{s+2}U(s) + \frac{3}{s+3}U(s) \qquad \text{(P8.9.8)}$$

We consider the following state variables:

$$X_1(s) = \frac{3}{s+1}U(s) \qquad \text{(P8.9.9)}$$

$$X_2(s) = \frac{-6}{s+2}U(s) \qquad \text{(P8.9.10)}$$

$$X_3(s) = \frac{3}{s+3}U(s) \qquad \text{(P8.9.11)}$$

From the inverse Laplace transform of Equations P8.9.9 through P8.9.11 we get the relationships that provide the derivatives of the state variables:

$$\dot{x}_1 = -x_1 + 3u \qquad \text{(P8.9.12)}$$

$$\dot{x}_2 = -2x_2 - 6u \qquad \text{(P8.9.13)}$$

$$\dot{x}_3 = -3x_3 + 3u \qquad \text{(P8.9.14)}$$

Equations P8.9.12 through P8.9.14 are written in matrix form as

$$\begin{bmatrix} \dot{x}_1 \\ \dot{x}_2 \\ \dot{x}_3 \end{bmatrix} = \begin{bmatrix} -1 & 0 & 0 \\ 0 & -2 & 0 \\ 0 & 0 & -3 \end{bmatrix} \begin{bmatrix} x_1 \\ x_2 \\ x_3 \end{bmatrix} + \begin{bmatrix} 3 \\ -6 \\ 3 \end{bmatrix} u \qquad \text{(P8.9.15)}$$

Equation P8.9.8 can be written as

$$Y(s) = X_1(s) + X_2(s) + X_3(s) \qquad \text{(P8.9.16)}$$

$$\Rightarrow y = x_1 + x_2 + x_3 \qquad \text{(P8.9.17)}$$

or

$$y = \begin{bmatrix} 1 & 1 & 1 \end{bmatrix} \begin{bmatrix} x_1 \\ x_2 \\ x_3 \end{bmatrix} \qquad \text{(P8.9.18)}$$

The block diagram that implements the system according to the representation of Equations P8.9.15 and P8.9.18 is

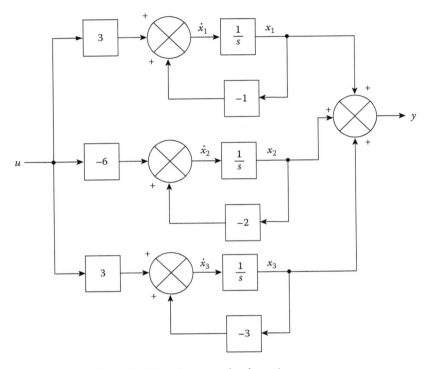

8.10 Consider a system described by the transfer function

$$G(s) = \frac{Y(s)}{U(s)} = \frac{s+3}{s^2 + 3s + 2}$$

Represent the system in the state space:
a. In the controllable canonical form
b. In the observable canonical form
c. In diagonal canonical form

Solution

a. The transfer function of the system is

$$G(s) = \frac{Y(s)}{U(s)} = \frac{s+3}{s^2 + 3s + 2} \tag{P8.10.1}$$

The differential equation has the form

$$y^{(2)} + 3y^{(1)} + 2y = u^{(1)} + 3u \tag{P8.10.2}$$

Based on Table F8.2, from the case 2(b), the state equations of the system in the controllable canonical form are

$$\begin{bmatrix} \dot{x}_1 \\ \dot{x}_2 \end{bmatrix} = \begin{bmatrix} 0 & 1 \\ -2 & -3 \end{bmatrix} \begin{bmatrix} x_1 \\ x_2 \end{bmatrix} + \begin{bmatrix} 0 \\ 1 \end{bmatrix} u \qquad\qquad \text{(P8.10.3)}$$

$$y = \begin{bmatrix} 3 & 1 \end{bmatrix} \begin{bmatrix} x_1 \\ x_2 \end{bmatrix} \qquad\qquad \text{(P8.10.4)}$$

b. Similarly, from Table F8.2, case 2(c), the state equations in the observable canonical form are

$$\begin{bmatrix} \dot{x}_1 \\ \dot{x}_2 \end{bmatrix} = \begin{bmatrix} 0 & -2 \\ 1 & -3 \end{bmatrix} \begin{bmatrix} x_1 \\ x_2 \end{bmatrix} + \begin{bmatrix} 3 \\ 1 \end{bmatrix} u \qquad\qquad \text{(P8.10.5)}$$

$$y = \begin{bmatrix} 0 & 1 \end{bmatrix} \begin{bmatrix} x_1 \\ x_2 \end{bmatrix} \qquad\qquad \text{(P8.10.6)}$$

c. We express the transfer function of the system as a sum of partial fractions:

$$\text{(P8.10.1)} \Rightarrow G(s) = \frac{2}{s+1} - \frac{1}{s+2} \qquad\qquad \text{(P8.10.7)}$$

From Table F8.2, case 3(a), the state equations of the system in the diagonal canonical form are

$$\begin{bmatrix} \dot{x}_1 \\ \dot{x}_2 \end{bmatrix} = \begin{bmatrix} -1 & 0 \\ 0 & -2 \end{bmatrix} \begin{bmatrix} x_1 \\ x_2 \end{bmatrix} + \begin{bmatrix} 1 \\ 1 \end{bmatrix} u \qquad\qquad \text{(P8.10.8)}$$

$$y = \begin{bmatrix} 2 & -1 \end{bmatrix} \begin{bmatrix} x_1 \\ x_2 \end{bmatrix} \qquad\qquad \text{(P8.10.9)}$$

8.11 Write a state-space representation for the systems shown in Figures (a) through (c).

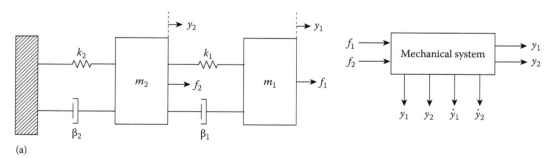

(a)

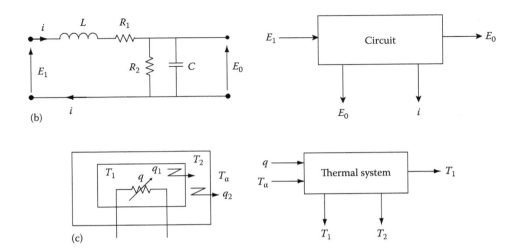

(b)

(c)

Solution

a. The equations of the mechanical system shown in Figure (a) are

$$f_1 - \beta_1(\dot{y}_1 - \dot{y}_2) - k_1(y_1 - y_2) = m_1\ddot{y}_1 \tag{P8.11.1}$$

$$f_2 + \beta_1(\dot{y}_1 - \dot{y}_2) + k_1(y_1 - y_2) - \beta_2\dot{y}_2 - k_2y_2 = m_2\ddot{y}_2 \tag{P8.11.2}$$

The state variables are

$$\left.\begin{array}{l} x_1 = y_1 \\ x_2 = y_2 \\ x_3 = \dot{y}_1 \\ x_4 = \dot{y}_2 \end{array}\right\} \tag{P8.11.3}$$

The derivatives of the state variables are

$$\left.\begin{array}{l} \dot{x}_1 = \dot{y}_1 = x_3 \\[4pt] \dot{x}_2 = \dot{y}_2 = x_4 \\[4pt] \dot{x}_3 = \ddot{y}_1 \overset{(P8.11.1)}{=} -\dfrac{k_1}{m_1}x_1 + \dfrac{k_1}{m_1}x_2 - \dfrac{\beta_1}{m_1}x_3 + \dfrac{\beta_1}{m_1}x_4 + \dfrac{1}{m_1}f_1 \\[10pt] \dot{x}_4 = \ddot{y}_2 \overset{(P8.11.2)}{=} \left(\dfrac{k_1}{m_2}\right)x_1 - \left(\dfrac{k_1 + k_2}{m_2}\right)x_2 + \left(\dfrac{\beta_1}{m_2}\right)x_3 - \left(\dfrac{\beta_1 + \beta_2}{m_2}\right)x_4 + \dfrac{1}{m_2}f_2 \end{array}\right\} \tag{P8.11.4}$$

The representation of the system in the state space is

$$
\begin{bmatrix} \dot{x}_1 \\ \dot{x}_2 \\ \dot{x}_3 \\ \dot{x}_4 \end{bmatrix} = \begin{bmatrix} 0 & 0 & 1 & 0 \\ 0 & 0 & 0 & 1 \\ -\dfrac{k_1}{m_1} & \dfrac{k_1}{m_1} & -\dfrac{\beta_1}{m_1} & \dfrac{\beta_1}{m_1} \\ \dfrac{k_1}{m_2} & -\dfrac{k_1+k_2}{m_2} & \dfrac{\beta_1}{m_2} & -\dfrac{\beta_1+\beta_2}{m_2} \end{bmatrix} \begin{bmatrix} x_1 \\ x_2 \\ x_3 \\ x_4 \end{bmatrix} + \begin{bmatrix} 0 & 0 \\ 0 & 0 \\ \dfrac{1}{m_1} & 0 \\ 0 & \dfrac{1}{m_2} \end{bmatrix} \begin{bmatrix} f_1 \\ f_2 \end{bmatrix} \qquad \text{(P8.11.5)}
$$

$$
\begin{bmatrix} y_1 \\ y_2 \end{bmatrix} = \begin{bmatrix} 1 & 0 & 0 & 0 \\ 0 & 1 & 0 & 0 \end{bmatrix} \begin{bmatrix} x_1 \\ x_2 \\ x_3 \\ x_4 \end{bmatrix} \qquad \text{(P8.11.6)}
$$

b. The electric circuit shown in Figure (b) is described by the following equations:

$$
E_o(s) = \frac{R_2 \cdot (1/sC)}{R_2 + (1/sC)} \cdot I(s) = \frac{R_2}{sR_2C+1} I(s) \qquad \text{(P8.11.7)}
$$

$$
E_1(s) = (sL+R_1)I(s) + E_o(s) \qquad \text{(P8.11.8)}
$$

We consider as state variables the following:

$$
\left. \begin{aligned} x_1 &= E_o \\ x_2 &= I \end{aligned} \right\} \qquad \text{(P8.11.9)}
$$

From relationships (P8.11.7) through (P8.11.9) we get the derivatives of the state variables:

$$
\left. \begin{aligned} \dot{x}_1 &= -\left(\frac{1}{R_2C}\right)x_1 + \left(\frac{1}{C}\right)x_2 \\ \dot{x}_2 &= -\left(\frac{1}{L}\right)x_1 - \left(\frac{R_1}{L}\right)x_2 + \left(\frac{1}{L}\right)E_1 \end{aligned} \right\} \qquad \text{(P8.11.10)}
$$

The representation of the system in the state space is

$$
\begin{bmatrix} \dot{x}_1 \\ \dot{x}_2 \end{bmatrix} = \begin{bmatrix} -\dfrac{1}{R_2C} & \dfrac{1}{C} \\ -\dfrac{1}{L} & -\dfrac{R_1}{L} \end{bmatrix} \begin{bmatrix} x_1 \\ x_2 \end{bmatrix} + \begin{bmatrix} 0 \\ \dfrac{1}{L} \end{bmatrix} E_1 \qquad \text{(P8.11.11)}
$$

$$
y = \begin{bmatrix} 1 & 0 \end{bmatrix} \begin{bmatrix} x_1 \\ x_2 \end{bmatrix} \qquad \text{(P8.11.12)}
$$

c. The equations that describe the thermal system shown in Figure (c) are

$$q_1 = h_1 A_1 (T_1 - T_2) = \frac{T_1 - T_2}{R_1} \qquad \text{(P8.11.13)}$$

$$q_2 = h_2 A_2 (T_2 - T_a) = \frac{T_2 - T_a}{R_2} \qquad \text{(P8.11.14)}$$

$$C_1 \frac{dT_1}{dt} = q - \left(\frac{T_1 - T_2}{R_1} \right) \qquad \text{(P8.11.15)}$$

$$C_2 \frac{dT_2}{dt} = \left(\frac{T_1 - T_2}{R_1} \right) - \left(\frac{T_2 - T_a}{R_2} \right) \qquad \text{(P8.11.16)}$$

The state variables are the following:

$$\left. \begin{aligned} x_1 &= T_1 \\ x_2 &= T_2 \end{aligned} \right\} \qquad \text{(P8.11.17)}$$

The derivatives of the state variables are

$$\left. \begin{aligned} \dot{x}_1 &= -\left(\frac{1}{R_1 C_1} \right) x_1 + \left(\frac{1}{R_1 C_1} \right) x_2 + \left(\frac{1}{C_1} \right) q \\ \dot{x}_2 &= \left(\frac{1}{R_1 C_2} \right) x_1 - \left(\frac{1}{R_1} + \frac{1}{R_2} \right) \frac{1}{C_2} x_2 + \left(\frac{1}{R_2 C_2} \right) T_a \end{aligned} \right\} \qquad \text{(P8.11.18)}$$

The state-space representation of the system is

$$\begin{bmatrix} \dot{x}_1 \\ \dot{x}_2 \end{bmatrix} = \begin{bmatrix} -\dfrac{1}{R_1 C_1} & \dfrac{1}{R_1 C_1} \\ \dfrac{1}{R_1 C_2} & -\left(\dfrac{1}{R_1} + \dfrac{1}{R_2} \right) \dfrac{1}{C_2} \end{bmatrix} \begin{bmatrix} x_1 \\ x_2 \end{bmatrix} + \begin{bmatrix} \dfrac{1}{C_1} & 0 \\ 0 & \dfrac{1}{R_2 C_2} \end{bmatrix} \begin{bmatrix} q \\ T_a \end{bmatrix} \qquad \text{(P8.11.19)}$$

$$y = \begin{bmatrix} 1 & 0 \end{bmatrix} \begin{bmatrix} x_1 \\ x_2 \end{bmatrix} \qquad \text{(P8.11.20)}$$

8.12 a. Find the state-transition matrix of the system

$$\begin{bmatrix} \dot{x}_1 \\ \dot{x}_2 \end{bmatrix} = \begin{bmatrix} 0 & 1 \\ -2 & -3 \end{bmatrix} \begin{bmatrix} x_1 \\ x_2 \end{bmatrix} + \begin{bmatrix} 0 \\ 1 \end{bmatrix} u$$

b. Provide the general solution of the state equations for initial conditions $\begin{bmatrix} x_1(0) \\ x_2(0) \end{bmatrix} = \begin{bmatrix} 0 \\ 0 \end{bmatrix}$, and input $u(t) = 1$.

Solution

a. The state-transition matrix is provided by the relationship

$$\Phi(t) = e^{At} = L^{-1}\left\{(sI - A)^{-1}\right\} \tag{P8.12.1}$$

In our case $A = \begin{vmatrix} 0 & 1 \\ -2 & -3 \end{vmatrix}$; thus,

$$sI - A = \begin{bmatrix} s & 0 \\ 0 & s \end{bmatrix} - \begin{bmatrix} 0 & 1 \\ -2 & -3 \end{bmatrix} = \begin{bmatrix} s & -1 \\ 2 & s+3 \end{bmatrix} \tag{P8.12.2}$$

$$(sI - A)^{-1} = \frac{\text{Adj}(sI - A)}{\det(sI - A)} = \frac{1}{(s+1)(s+2)} \begin{bmatrix} s+3 & 1 \\ -2 & s \end{bmatrix} \Rightarrow$$

$$(sI - A)^{-1} = \begin{bmatrix} \dfrac{s+3}{(s+1)(s+2)} & \dfrac{1}{(s+1)(s+2)} \\ \dfrac{-2}{(s+1)(s+2)} & \dfrac{s}{(s+1)(s+2)} \end{bmatrix} \tag{P8.12.3}$$

Therefore,

$$\Phi(t) = L^{-1}\left\{(sI - A)^{-1}\right\} = \begin{bmatrix} L^{-1}\left\{\dfrac{s+3}{(s+1)(s+2)}\right\} & L^{-1}\left\{\dfrac{1}{(s+1)(s+2)}\right\} \\ L^{-1}\left\{\dfrac{-2}{(s+1)(s+2)}\right\} & L^{-1}\left\{\dfrac{s}{(s+1)(s+2)}\right\} \end{bmatrix} \Rightarrow$$

$$\tag{P8.12.4}$$

$$\Phi(t) = \begin{bmatrix} 2e^{t} - e^{-2t} & e^{t} - e^{-2t} \\ -2e^{-t} + 2e^{-2t} & -e^{-t} + 2e^{-2t} \end{bmatrix}$$

b. The response to a unit-step input is

$$x(t) = e^{At}x(0) + \int_0^t \Phi(t - \lambda)bu(\lambda)\,d\lambda \tag{P8.12.5}$$

$$(\text{P8.12.5}) \Rightarrow x(t) = \begin{bmatrix} x_1(t) \\ x_2(t) \end{bmatrix} \int_0^t \begin{bmatrix} 2e^{-(t-\lambda)} - e^{-2(t-\lambda)} & e^{-(t-\lambda)} - e^{-2(t-\lambda)} \\ -2e^{-(t-\lambda)} + 2e^{-2(t-\lambda)} & -e^{-(t-\lambda)} + 2e^{-2(t-\lambda)} \end{bmatrix} \begin{bmatrix} 0 \\ 1 \end{bmatrix} [1]\,d\lambda$$

$$\Rightarrow \begin{bmatrix} x_1(t) \\ x_2(t) \end{bmatrix} = \begin{bmatrix} \dfrac{1}{2} - e^{-t} + \dfrac{1}{2}e^{-2t} \\ e^{-t} - e^{-2t} \end{bmatrix} \tag{P8.12.6}$$

8.13 A system is represented in the state space by the following matrices:

$$A = \begin{bmatrix} -1 & 0 \\ 0 & -2 \end{bmatrix} \quad B = \begin{bmatrix} 0 \\ 1 \end{bmatrix} \quad C = \begin{bmatrix} 1 & 0 \\ 0 & 1 \end{bmatrix} \quad x_0 = \begin{bmatrix} 2 \\ -3 \end{bmatrix}$$

Compute the system response to a unit-step input ($U(s) = 1/s$).

Solution

The output of the system is

$$y = Cx = C(sI - A)^{-1}(Bu + x_o) \qquad \text{(P8.13.1)}$$

Here

$$sI - A = \begin{bmatrix} s & 0 \\ 0 & s \end{bmatrix} - \begin{bmatrix} -1 & 0 \\ 0 & -2 \end{bmatrix} = \begin{bmatrix} s+1 & 0 \\ 0 & s+2 \end{bmatrix} \qquad \text{(P8.13.2)}$$

$$(sI - A)^{-1} = \frac{\text{Adj}(sI - A)}{\det(sI - A)} = \frac{1}{(s+1)(s+2)}\begin{bmatrix} s+2 & 0 \\ 0 & s+1 \end{bmatrix} \qquad \text{(P8.13.3)}$$

Hence,

$$y = \begin{bmatrix} 1 & 0 \\ 0 & 1 \end{bmatrix}\begin{bmatrix} \dfrac{1}{s+1} & 0 \\ 0 & \dfrac{1}{s+2} \end{bmatrix}\left(\begin{bmatrix} 0 \\ 1 \end{bmatrix}\dfrac{1}{s} + \begin{bmatrix} 2 \\ -3 \end{bmatrix}\right) \Rightarrow$$

$$Y(s) = \begin{bmatrix} \dfrac{1}{s+1} & 0 \\ 0 & \dfrac{1}{s+2} \end{bmatrix}\begin{bmatrix} 2 \\ -3 + \dfrac{1}{s} \end{bmatrix} = \begin{bmatrix} \dfrac{2}{s+1} \\ \dfrac{1-3s}{s(s+2)} \end{bmatrix} \qquad \text{(P8.13.4)}$$

Taking inverse Laplace transform we get

$$y(t) = L^{-1}\{Y(s)\} = \begin{bmatrix} L^{-1}\left\{\dfrac{2}{s+1}\right\} \\ L^{-1}\left\{\dfrac{1-3s}{s(s+2)}\right\} \end{bmatrix} = \begin{bmatrix} 2e^{-t} \\ 0.5(1-7e^{-2t}) \end{bmatrix} \qquad \text{(P8.13.5)}$$

8.14 Compute the state-transition matrix of the system $dx/dt = Ax$, where $A = \begin{bmatrix} -1 & 0 & 0 \\ 0 & -4 & 4 \\ 0 & -1 & 0 \end{bmatrix}$,

   with the use of

   a. Power series
   b. The Laplace transform
   c. The similarity transformation

Solution

   a. The system matrix is

$$A = \begin{bmatrix} -1 & 0 & 0 \\ 0 & -4 & 4 \\ 0 & -1 & 0 \end{bmatrix} \qquad (P8.14.1)$$

   With the use of power series the state-transition matrix $\Phi(t)$ is given by

$$\Phi(t) = e^{At} = I + At + \frac{A^2 t^2}{2!} + \cdots \qquad (P8.14.2)$$

   We have

$$\Phi(t) = \begin{bmatrix} 1 & 0 & 0 \\ 0 & 1 & 0 \\ 0 & 0 & 1 \end{bmatrix} + \begin{bmatrix} -t & 0 & 0 \\ 0 & -4t & 4t \\ 0 & -t & 0 \end{bmatrix} + \begin{bmatrix} t^2/2 & 0 & 0 \\ 0 & 6t^2 & -8t^2 \\ 0 & 2t^2 & -2t^2 \end{bmatrix} + \cdots \Rightarrow$$

$$\Phi(t) = \begin{bmatrix} 1 - t + \dfrac{t^2}{2} \cdots & 0 & 0 \\[2mm] 0 & 1 - 2t + \dfrac{4t^2}{2} - 2t(1-2t) + \cdots & 4t\left(1 - 2t + \dfrac{4t^2}{2} + \cdots\right) \\[2mm] 0 & -t\left(1 - 2t + \dfrac{4t^2}{2} + \cdots\right) & 1 - 2t + \dfrac{4t^2}{2} + 2t(1-2t) + \cdots \end{bmatrix}$$

$$\Rightarrow \Phi(t) = \begin{bmatrix} e^{-t} & 0 & 0 \\ 0 & (1-2t)e^{-2t} & 4te^{-2t} \\ 0 & -te^{-2t} & (1+2t)e^{-2t} \end{bmatrix} \qquad (P8.14.3)$$

   b. With the use of the Laplace transform the state-transition matrix is

$$\Phi(t) = e^{At} = L^{-1}\left\{(sI - A)^{-1}\right\} \qquad (P8.14.4)$$

We have

$$sI - A = \begin{bmatrix} s & 0 & 0 \\ 0 & s & 0 \\ 0 & 0 & s \end{bmatrix} - \begin{bmatrix} -1 & 0 & 0 \\ 0 & -4 & 4 \\ 0 & -1 & 0 \end{bmatrix} = \begin{bmatrix} s+1 & 0 & 0 \\ 0 & s+4 & -4 \\ 0 & 1 & s \end{bmatrix} \qquad \text{(P8.14.5)}$$

$$(P8.14.4) \overset{(P8.14.5)}{\Longrightarrow} \Phi(t) = L^{-1} \left\{ \frac{1}{(s+1)(s+2)^2} \begin{bmatrix} (s+2)^2 & 0 & 0 \\ 0 & s(s+1) & 4(s+1) \\ 0 & -(s+1) & (s+1)(s+4) \end{bmatrix} \right\} \Longrightarrow$$

$$\Phi(t) = \begin{bmatrix} L^{-1}\left\{\dfrac{1}{s+1}\right\} & 0 & 0 \\[2mm] 0 & L^{-1}\left\{\dfrac{s}{(s+2)^2}\right\} & L^{-1}\left\{\dfrac{4}{(s+2)^2}\right\} \\[2mm] 0 & L^{-1}\left\{\dfrac{-1}{(s+2)^2}\right\} & L^{-1}\left\{\dfrac{s+4}{(s+2)^2}\right\} \end{bmatrix} \Longrightarrow \qquad \text{(P8.14.6)}$$

$$\Phi(t) = \begin{bmatrix} e^{-t} & 0 & 0 \\ 0 & (1-2t)e^{-2t} & 4te^{-2t} \\ 0 & 0 & (1+2t)e^{-2t} \end{bmatrix}$$

c. With the use of the similarity transformation the state-transition matrix $\Phi(t)$ is

$$\Phi(t) = e^{At} = Te^{Jt}T^{-1} \qquad \text{(P8.14.7)}$$

The eigenvalues of matrix A are

$$\lambda_1 = -1 \quad \text{and} \quad \lambda_2 = \lambda_3 = -2$$

The eigenvectors can be found as follows:

$$\begin{bmatrix} -1 & 0 & 0 \\ 0 & -4 & 4 \\ 0 & -1 & 0 \end{bmatrix} \begin{bmatrix} u_{11} \\ u_{12} \\ u_{13} \end{bmatrix} = - \begin{bmatrix} u_{11} \\ u_{12} \\ u_{13} \end{bmatrix} \Longrightarrow u_1^T = [1 \quad 0 \quad 0] \qquad \text{(P8.14.8)}$$

$$\begin{bmatrix} -1 & 0 & 0 \\ 0 & -4 & 4 \\ 0 & -1 & 0 \end{bmatrix} \begin{bmatrix} u_{21} \\ u_{22} \\ u_{23} \end{bmatrix} = -2 \begin{bmatrix} u_{21} \\ u_{22} \\ u_{23} \end{bmatrix} \Longrightarrow u_2^T = [0 \quad 2 \quad 1] \qquad \text{(P8.14.9)}$$

The third vector is

$$u_3^T = [0 \quad 1 \quad 1] \qquad \text{(P8.14.10)}$$

The Jordan matrix is

$$J = \begin{bmatrix} -1 & 0 & 0 \\ 0 & -2 & 1 \\ 0 & 0 & -2 \end{bmatrix} \tag{P8.14.11}$$

It holds that

$$A = T \cdot J \cdot T^{-1} \tag{P8.14.12}$$

where

$$T = [u_1 \quad \vdots \quad u_2 \quad \vdots \quad u_3] \tag{P8.14.13}$$

Thus,

$$A = \begin{bmatrix} 1 & 0 & 0 \\ 0 & 2 & 1 \\ 0 & 1 & 1 \end{bmatrix} \begin{bmatrix} -1 & 0 & 0 \\ 0 & -2 & 1 \\ 0 & 0 & -2 \end{bmatrix} \begin{bmatrix} 1 & 0 & 0 \\ 0 & 1 & -1 \\ 0 & -1 & 2 \end{bmatrix} \tag{P8.14.14}$$

The state-transition matrix is

$$\Phi(t) = Te^{Jt}T^{-1} = \begin{bmatrix} 1 & 0 & 0 \\ 0 & 2 & 1 \\ 0 & 1 & 1 \end{bmatrix} \begin{bmatrix} e^{-t} & 0 & 0 \\ 0 & e^{-2t} & te^{-2t} \\ 0 & 0 & e^{-2t} \end{bmatrix} \begin{bmatrix} 1 & 0 & 0 \\ 0 & 1 & -1 \\ 0 & -1 & 2 \end{bmatrix} \Rightarrow$$

$$\Phi(t) = \begin{bmatrix} e^{-t} & 0 & 0 \\ 0 & (1-2t)e^{-2t} & 4te^{-2t} \\ 0 & 0 & (1+2t)e^{-2t} \end{bmatrix} \tag{P8.14.15}$$

8.15 A system is described by the following state equations:

$$\dot{x}(t) = Ax(t) + bu(t), \quad x(0) = x_0$$

where

$$A = \begin{bmatrix} 2 & 2 & -2 \\ 0 & 2 & 0 \\ 0 & 2 & 0 \end{bmatrix}, \quad b = \begin{bmatrix} 1 \\ 0 \\ 0 \end{bmatrix}, \quad x_0 = \begin{bmatrix} 1 \\ 0 \\ 0 \end{bmatrix}$$

a. Find the similarity transformation that makes the matrix A diagonal. Based on this diagonalization, find the state-transition matrix of the system.

b. Based on the previous query, find $x(t)$, $t > 0$ if $u(t)$ is the unit-step function.

Solution

a. The eigenvalues and the eigenvectors of matrix A are computed as follows:

$$|A - \lambda I| = (2 - \lambda)^2 (-\lambda) = 0 \tag{P8.15.1}$$

The eigenvalues are

$$\lambda_1 = 0, \quad \lambda_2 = \lambda_3 = 2$$

For the eigenvectors we have

For $\lambda_1 = 0$:

$$(A - 0I)u_1 = \begin{bmatrix} 2 & 2 & -2 \\ 0 & 2 & 0 \\ 0 & 2 & 0 \end{bmatrix} u_1 = \begin{bmatrix} 0 \\ 0 \\ 0 \end{bmatrix} \Rightarrow \tag{P8.15.2}$$

$$u_1 = \begin{bmatrix} 1 \\ 0 \\ 1 \end{bmatrix}$$

For $\lambda_2 = \lambda_3 = 2$

$$(A - 2I)u = \begin{bmatrix} 0 & 2 & -2 \\ 0 & 0 & 0 \\ 0 & 2 & -2 \end{bmatrix} u = \begin{bmatrix} 0 \\ 0 \\ 0 \end{bmatrix} \Rightarrow \tag{P8.15.3}$$

$$u_2 = \begin{bmatrix} 0 \\ 1 \\ 1 \end{bmatrix}, \quad u_3 = \begin{bmatrix} 1 \\ 0 \\ 0 \end{bmatrix}$$

The matrix of the eigenvectors is equal to

$$T = \begin{bmatrix} 1 & 0 & 1 \\ 0 & 1 & 0 \\ 1 & 1 & 0 \end{bmatrix} \tag{P8.15.4}$$

It holds that

$$A = TAT^{-1} = \begin{bmatrix} 1 & 0 & 1 \\ 0 & 1 & 0 \\ 1 & 1 & 0 \end{bmatrix} \begin{bmatrix} 0 & 0 & 0 \\ 0 & 2 & 0 \\ 0 & 0 & 2 \end{bmatrix} \begin{bmatrix} 0 & -1 & 1 \\ 0 & 1 & 0 \\ 1 & 1 & -1 \end{bmatrix}$$ (P8.15.5)

The desired similarity transformation is $x = Tz$.
The state-transition matrix is

$$\Phi(t) = e^{At} = e^{(TAT^{-1})t} = Te^{At}T^{-1} \Rightarrow$$

$$\Phi(t) = \begin{bmatrix} 1 & 0 & 1 \\ 0 & 1 & 0 \\ 1 & 1 & 0 \end{bmatrix} \begin{bmatrix} 1 & 0 & 0 \\ 0 & e^{2t} & 0 \\ 0 & 0 & e^{2t} \end{bmatrix} \begin{bmatrix} 0 & -1 & 1 \\ 0 & 1 & 0 \\ 1 & 1 & -1 \end{bmatrix} \Rightarrow$$ (P8.15.6)

$$\Phi(t) = \begin{bmatrix} e^{2t} & e^{2t}-1 & 1-e^{2t} \\ 0 & e^{2t} & 0 \\ 0 & e^{2t}-1 & 1 \end{bmatrix}$$

b. For $u(t) = 1, t > 0, x(t)$ is

$$x(t) = e^{At}x_o + \int_0^t e^{A(t-\lambda)}bu(\lambda)d\lambda =$$

$$e^{At}\begin{bmatrix} 1 \\ 0 \\ 0 \end{bmatrix} + \int_0^t \begin{bmatrix} e^{2(t-\lambda)} \\ 0 \\ 0 \end{bmatrix} d\lambda \Rightarrow$$ (P8.15.7)

$$x(t) = \begin{bmatrix} e^{2t} \\ 0 \\ 0 \end{bmatrix} + \begin{bmatrix} \frac{1}{2}e^{2t} - \frac{1}{2} \\ 0 \\ 0 \end{bmatrix} = \begin{bmatrix} \frac{3}{2}e^{2t} - \frac{1}{2} \\ 0 \\ 0 \end{bmatrix}$$

8.16 Assume that a system is described by the equations

$$\frac{dx}{dt} = Ax(t), \quad x(0) = x_0$$

where $x(t)$ is the 3 × 1 state vector and $A = \begin{bmatrix} 1 & 0 & 0 \\ 1 & 1 & 1 \\ -1 & 0 & 0 \end{bmatrix}$

a. Find the similarity transformation that diagonalizes the matrix A.
b. Compute the state-transition matrix.
c. Repeat the previous queries if matrix A is given by $A = \begin{bmatrix} 0 & 1 \\ -\omega^2 & 0 \end{bmatrix}$, where ω is a positive constant.

Solution

a. First we compute the eigenvalues and the eigenvectors of the matrix A. We have

$$|A - \lambda I| = -\lambda(1-\lambda)^2 = 0 \qquad \text{(P8.16.1)}$$

The eigenvalues are $\lambda_1 = 0$, $\lambda_2 = \lambda_3 = 1$
For the eigenvectors we have

$$\text{For } \lambda_1 = 0: \quad (A - 0I)u_1 = \begin{bmatrix} 1 & 0 & 0 \\ 1 & 1 & 1 \\ -1 & 0 & 0 \end{bmatrix} u_1 = \begin{bmatrix} 0 \\ 0 \\ 0 \end{bmatrix} \Rightarrow$$

$$\qquad \text{(P8.16.2)}$$

$$u_1 = \begin{bmatrix} 0 \\ 1 \\ -1 \end{bmatrix}$$

$$\text{For } \lambda_2 = \lambda_3 = 1: \quad (A - I)u = \begin{bmatrix} 0 & 0 & 0 \\ 1 & 0 & 1 \\ -1 & 0 & -1 \end{bmatrix} u = \begin{bmatrix} 0 \\ 0 \\ 0 \end{bmatrix} \Rightarrow$$

$$\qquad \text{(P8.16.3)}$$

$$u_2 = \begin{bmatrix} 0 \\ 1 \\ 0 \end{bmatrix}, \quad u_3 = \begin{bmatrix} 1 \\ 0 \\ -1 \end{bmatrix}$$

The matrix of the eigenvectors is

$$T = \begin{bmatrix} 0 & 0 & 1 \\ 1 & 1 & 0 \\ -1 & 0 & -1 \end{bmatrix} \qquad \text{(P8.16.4)}$$

It holds that

$$A = TAT^{-1} = \begin{bmatrix} 0 & 0 & 1 \\ 1 & 1 & 0 \\ -1 & 0 & -1 \end{bmatrix} \begin{bmatrix} 0 & 0 & 0 \\ 0 & 1 & 0 \\ 0 & 0 & 1 \end{bmatrix} \begin{bmatrix} -1 & 0 & -1 \\ 1 & 1 & 1 \\ 1 & 0 & 0 \end{bmatrix} \qquad \text{(P8.16.5)}$$

The desired similarity transformation is $x = Tz$.

b. The state-transition matrix is

$$\Phi(t) = e^{At} = e^{(TAT^{-1})t} = Te^{At}T^{-1} \Rightarrow$$

$$\Phi(t) = \begin{bmatrix} 0 & 0 & 1 \\ 1 & 1 & 0 \\ -1 & 0 & -1 \end{bmatrix} \begin{bmatrix} 1 & 0 & 0 \\ 0 & e^t & 0 \\ 0 & 0 & e^t \end{bmatrix} \begin{bmatrix} -1 & 0 & -1 \\ 1 & 1 & 1 \\ 1 & 0 & 0 \end{bmatrix} \Rightarrow \qquad \text{(P8.16.6)}$$

$$\Phi(t) = \begin{bmatrix} e^t & 0 & 0 \\ e^t - 1 & e^t & e^t - 1 \\ 1 - e^t & 0 & 1 \end{bmatrix}$$

c. The eigenvalues of matrix $A = \begin{bmatrix} 0 & 1 \\ -\omega^2 & 0 \end{bmatrix}$ are

$$[A - \lambda I] = \begin{bmatrix} -\lambda & 1 \\ -\omega^2 & -\lambda \end{bmatrix} = \lambda^2 + \omega^2 = 0 \Rightarrow \lambda_{1,2} = \pm j\omega \qquad \text{(P8.16.7)}$$

The eigenvectors are

$$\text{For } \lambda = j\omega: \qquad \begin{bmatrix} -j\omega & 1 \\ -\omega^2 & -j\omega \end{bmatrix} u_1 = 0 \Rightarrow u_1 = \begin{bmatrix} 1 \\ j\omega \end{bmatrix} \text{ and } u_2 = \begin{bmatrix} 1 \\ -j\omega \end{bmatrix} \qquad \text{(P8.16.8)}$$

The matrix of eigenvectors is equal to

$$T = \begin{bmatrix} 1 & 1 \\ j\omega & -j\omega \end{bmatrix} \qquad \text{(P8.16.9)}$$

It holds that

$$\Lambda = T^{-1}AT = \begin{bmatrix} j\omega & 0 \\ 0 & -j\omega \end{bmatrix} \qquad \text{(P8.16.10)}$$

The state-transition matrix is

$$\Phi(t) = e^{At} = e^{(T\Lambda T^{-1})t} = Te^{\Lambda t}T^{-1} \Rightarrow$$

$$\Phi(t) = \frac{1}{2j\omega}\begin{bmatrix} 1 & 1 \\ j\omega & -j\omega \end{bmatrix}\begin{bmatrix} e^{j\omega t} & 0 \\ 0 & e^{-j\omega t} \end{bmatrix}\begin{bmatrix} j\omega & 1 \\ j\omega & -1 \end{bmatrix} \Rightarrow$$

$$\Phi(t) = \begin{bmatrix} \frac{1}{2}(e^{j\omega t} + e^{-j\omega t}) & \frac{1}{2j\omega}(e^{j\omega t} - e^{-j\omega t}) \\ -\frac{\omega}{2j}(e^{j\omega t} - e^{-j\omega t}) & \frac{1}{2}(e^{j\omega t} + e^{-j\omega t}) \end{bmatrix} \Rightarrow \qquad \text{(P8.16.11)}$$

$$\Phi(t) = \begin{bmatrix} \cos\omega t & \frac{1}{\omega}\sin\omega t \\ -\omega\sin\omega t & \cos\omega t \end{bmatrix}$$

8.17 Find the transfer function of the systems described by the following state equations:

System 1:

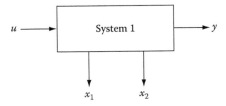

$$\begin{bmatrix} \dot{x}_1 \\ \dot{x}_2 \end{bmatrix} = \begin{bmatrix} -3 & 1 \\ -2 & 0 \end{bmatrix}\begin{bmatrix} x_1 \\ x_2 \end{bmatrix} + \begin{bmatrix} 0 \\ 1 \end{bmatrix}u$$

$$y = \begin{bmatrix} 1 & 0 \end{bmatrix}\begin{bmatrix} x_1 \\ x_2 \end{bmatrix}$$

System 2:

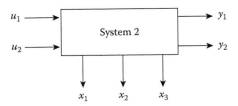

$$\begin{bmatrix} \dot{x}_1 \\ \dot{x}_2 \\ \dot{x}_3 \end{bmatrix} = \begin{bmatrix} -1 & 1 & -1 \\ 0 & -2 & 1 \\ 0 & 0 & -3 \end{bmatrix} \begin{bmatrix} x_1 \\ x_2 \\ x_3 \end{bmatrix} + \begin{bmatrix} 1 & 0 \\ 0 & 1 \\ 1 & 0 \end{bmatrix} \begin{bmatrix} u_1 \\ u_2 \end{bmatrix}$$

$$\begin{bmatrix} y_1 \\ y_2 \end{bmatrix} = \begin{bmatrix} 1 & 1 & 1 \\ 0 & 1 & 1 \end{bmatrix} \begin{bmatrix} x_1 \\ x_2 \\ x_3 \end{bmatrix}$$

Solution

a. For system 1 we have

$$A = \begin{bmatrix} -3 & 1 \\ -2 & 0 \end{bmatrix}, \quad B = \begin{bmatrix} 0 \\ 1 \end{bmatrix}, \quad C = [1 \quad 0] \qquad \text{(P8.17.1)}$$

The transfer function of system 1 is

$$G(s) = C(sI - A)^{-1}B + D = [1 \quad 0] \begin{bmatrix} s+3 & -1 \\ 2 & s \end{bmatrix} \begin{bmatrix} 0 \\ 1 \end{bmatrix} \Rightarrow$$

$$G(s) = [1 \quad 0] \begin{bmatrix} \dfrac{s}{(s+1)(s+2)} & \dfrac{1}{(s+1)(s+2)} \\ \dfrac{-2}{(s+1)(s+2)} & \dfrac{s+3}{(s+1)(s+2)} \end{bmatrix} \begin{bmatrix} 0 \\ 1 \end{bmatrix} \Rightarrow \qquad \text{(P8.17.2)}$$

$$G(s) = [1 \quad 0] \begin{bmatrix} \dfrac{1}{(s+1)(s+2)} \\ \dfrac{s+3}{(s+1)(s+2)} \end{bmatrix} = \dfrac{1}{(s+1)(s+2)}$$

b. For the MIMO system 2, we determine the transfer matrix as

$$G(s) = C(sI - A)^{-1}B + D \qquad \text{(P8.17.3)}$$

Here

$$A = \begin{bmatrix} -1 & 1 & -1 \\ 0 & -2 & 1 \\ 0 & 0 & -3 \end{bmatrix}, \quad B = \begin{bmatrix} 1 & 0 \\ 0 & 1 \\ 1 & 0 \end{bmatrix}$$

$$C = \begin{bmatrix} 1 & 1 & 1 \\ 0 & 1 & 1 \end{bmatrix}, \quad D = 0 \qquad \text{(P8.17.4)}$$

Hence,

$$G(s) = \begin{bmatrix} 1 & 1 & 1 \\ 0 & 1 & 1 \end{bmatrix} \begin{bmatrix} s+1 & -1 & 1 \\ 0 & s+2 & -1 \\ 0 & 0 & s+3 \end{bmatrix} \begin{bmatrix} 1 & 0 \\ 0 & 1 \\ 1 & 0 \end{bmatrix} \Rightarrow$$

$$G(s) = \begin{bmatrix} 1 & 1 & 1 \\ 0 & 1 & 1 \end{bmatrix} \begin{bmatrix} \dfrac{1}{s+1} & \dfrac{1}{(s+1)(s+2)} & \dfrac{-1}{(s+2)(s+3)} \\ 0 & \dfrac{1}{s+2} & \dfrac{1}{(s+2)(s+3)} \\ 0 & 0 & \dfrac{1}{s+3} \end{bmatrix} \begin{bmatrix} 1 & 0 \\ 0 & 1 \\ 1 & 0 \end{bmatrix} \Rightarrow \qquad (P8.17.5)$$

$$G(s) = \begin{bmatrix} G_{11}(s) & G_{12}(s) \\ G_{21}(s) & G_{22}(s) \end{bmatrix} = \begin{bmatrix} \dfrac{2(s+2)}{(s+1)(s+3)} & \dfrac{1}{s+1} \\ \dfrac{1}{s+2} & \dfrac{1}{s+2} \end{bmatrix}$$

8.18 Suppose that a control system is described by the following equations:

$$\begin{bmatrix} \dot{x}_1 \\ \dot{x}_2 \end{bmatrix} = \begin{bmatrix} -2 & -1 \\ -2 & -3 \end{bmatrix} \begin{bmatrix} x_1 \\ x_2 \end{bmatrix} + \begin{bmatrix} 0 \\ 1 \end{bmatrix} u \qquad y = \begin{bmatrix} 1 & 0 \end{bmatrix} \begin{bmatrix} x_1 \\ x_2 \end{bmatrix}$$

Compute the zero-input response of the system.

Solution

Denoting by $x(0) = [x_{10} \quad x_{20}]$ the initial conditions of the system's, its zero-input response is given by $x(t) = e^{At} x(0)$.

In order to compute e^{At} we have

$$(sI - A)^{-1} = \begin{bmatrix} s+2 & 1 \\ 2 & s+3 \end{bmatrix}^{-1} = \begin{bmatrix} \dfrac{s+3}{(s+1)(s+4)} & \dfrac{-1}{(s+1)(s+4)} \\ \dfrac{-2}{(s+1)(s+4)} & \dfrac{s+2}{(s+1)(s+4)} \end{bmatrix} \qquad (P8.18.1)$$

Next we compute the inverse Laplace transforms of the elements of the matrix $(sI - A)^{-1}$:

$$L^{-1}\left\{ \dfrac{s+3}{(s+1)(s+4)} \right\} = L^{-1}\left\{ \dfrac{2/3}{(s+1)} + \dfrac{1/3}{(s+4)} \right\} = \left\{ \dfrac{2}{3}e^{-t} + \dfrac{1}{3}e^{-4t} \right\} u(t) \qquad (P8.18.2)$$

$$L^{-1}\left\{ \dfrac{1}{(s+1)(s+4)} \right\} = L^{-1}\left\{ \dfrac{1/3}{(s+1)} + \dfrac{-1/3}{(s+4)} \right\} = \left\{ \dfrac{1}{3}e^{-t} + \dfrac{1}{3}e^{-4t} \right\} u(t) \qquad (P8.18.3)$$

$$L^{-1}\left\{ \dfrac{s+2}{(s+1)(s+4)} \right\} = L^{-1}\left\{ \dfrac{1/3}{(s+1)} + \dfrac{2/3}{(s+4)} \right\} = \left\{ \dfrac{1}{3}e^{-t} + \dfrac{2}{3}e^{-4t} \right\} u(t) \qquad (P8.18.4)$$

where $u(t)$ is the unit-step function. Thus,

$$e^{At} = \begin{bmatrix} \left(\dfrac{2}{3}e^{-t} + \dfrac{1}{3}e^{-4t}\right)U(t) & -\left(\dfrac{1}{3}e^{-t} + \dfrac{1}{3}e^{-4t}\right)u(t) \\[3mm] -2\left(\dfrac{1}{3}e^{-t} - \dfrac{1}{3}e^{-4t}\right)U(t) & \left(\dfrac{1}{3}e^{-t} + \dfrac{2}{3}e^{-4t}\right)u(t) \end{bmatrix}$$

(P8.18.5)

and

$$x(t) = \begin{bmatrix} x_1(t) \\ x_2(t) \end{bmatrix} = \begin{bmatrix} \left(\dfrac{2}{3}x_{10} - \dfrac{1}{3}x_{20}\right)e^{-t} + \left(\dfrac{1}{3}x_{10} + \dfrac{1}{3}x_{20}\right)e^{-4t} \\[3mm] \left(\dfrac{-2}{3}x_{10} + \dfrac{1}{3}x_{20}\right)e^{-t} + \left(\dfrac{2}{3}x_{10} + \dfrac{2}{3}x_{20}\right)e^{-4t} \end{bmatrix} u(t)$$

(P8.18.6)

8.19 For the systems described by the following state equations:

$$\dot{x}_1 = \begin{bmatrix} -\alpha & 1 & 0 \\ 0 & -\alpha & 1 \\ 0 & 0 & -\alpha \end{bmatrix} x_1 + B_1 u$$

$$\dot{x}_2 = \begin{bmatrix} -\alpha & 0 & 0 \\ 0 & -\alpha & 0 \\ 0 & 0 & -\alpha \end{bmatrix} x_2 + B_2 u$$

$$y_1 = C_1 x_1, \quad y_2 = C_2 x_2,$$

a. Find the zero-input responses of the state vectors
b. If $\alpha = 0$, examine whether the state vectors of the first query remain bounded

Solution

a. The matrix $A_1 = \begin{bmatrix} -\alpha & 1 & 0 \\ 0 & -\alpha & 1 \\ 0 & 0 & -\alpha \end{bmatrix}$ gives the following characteristic polynomial:

$Q(s) = (s + a)^3$.

In order to calculate $f(t) = e^{At}$ we have

$$F(s) = e^{st} = (s+a)^3 P(s) + b_0 + b_1 s + b_2 s^2$$

(P8.19.1)

It is

$$F^{(1)}(s) = te^{st} = 3(s+\alpha)^2 P(s) + (s+\alpha)^3 P^{(1)}(s) + b_1 + 2b_2 s$$

(P8.19.2)

$$F^{(2)}(s) = t^2 e^{st} = 6(s+\alpha)P(s) + 3(s+\alpha)^2 P^{(1)}(s) + 3(s+\alpha)^2 P^{(1)}(s)$$

$$+ (s+\alpha)^3 P^{(2)}(s) + 2b_2$$

(P8.19.3)

By substituting $s = -a$, we get

$$\begin{bmatrix} e^{-\alpha t} \\ te^{-\alpha t} \\ t^2 e^{-\alpha t} \end{bmatrix} = \begin{bmatrix} 1 & -\alpha & \alpha^2 \\ 0 & 1 & -2\alpha \\ 0 & 0 & 2 \end{bmatrix} \begin{bmatrix} b_0 \\ b_1 \\ b_2 \end{bmatrix} \tag{P8.19.4}$$

Consequently,

$$\begin{bmatrix} b_0 \\ b_1 \\ b_2 \end{bmatrix} = \begin{bmatrix} 1 & \alpha & 0.5\alpha^2 \\ 0 & 1 & \alpha \\ 0 & 0 & 0.5 \end{bmatrix} \begin{bmatrix} e^{-\alpha t} \\ te^{-\alpha t} \\ t^2 e^{-\alpha t} \end{bmatrix} = \begin{bmatrix} e^{-\alpha t} + \alpha t e^{-\alpha t} + 0.5\alpha^2 t^2 e^{-\alpha t} \\ te^{-\alpha t} + \alpha t^2 e^{-\alpha t} \\ 0.5 t^2 e^{-\alpha t} \end{bmatrix} \tag{P8.19.5}$$

From

$$A_1^2 = \begin{bmatrix} \alpha^2 & -2\alpha & 1 \\ 0 & \alpha^2 & -2\alpha \\ 0 & 0 & \alpha^2 \end{bmatrix} \tag{P8.19.6}$$

we get

$$e^{A_1 t} = \begin{bmatrix} 1 & 0 & 0 \\ 0 & 1 & 0 \\ 0 & 0 & 1 \end{bmatrix} b_0 + \begin{bmatrix} -\alpha & 1 & 0 \\ 0 & -\alpha & 1 \\ 0 & 0 & -\alpha \end{bmatrix} b_1 + \begin{bmatrix} \alpha^2 & -2\alpha & 1 \\ 0 & \alpha^2 & -2\alpha \\ 0 & 0 & \alpha^2 \end{bmatrix} b_2$$

$$= e^{-\alpha t} \begin{bmatrix} 1 & t & 0.5t^2 \\ 0 & 1 & t \\ 0 & 0 & 1 \end{bmatrix} \tag{P8.19.7}$$

For

$$A_2 = \begin{bmatrix} -\alpha & 0 & 0 \\ 0 & -\alpha & 0 \\ 0 & 0 & -\alpha \end{bmatrix}$$

it holds that

$$e^{A_2 t} = e^{-\alpha t} \begin{bmatrix} 1 & 0 & 0 \\ 0 & 1 & 0 \\ 0 & 0 & 1 \end{bmatrix} \tag{P8.19.8}$$

The matrices A_1 and A_2 have the same characteristic polynomial.

If the two systems are described by A_1 and A_2, respectively, and the initial conditions are $x_1(0) = x_2(0) = [1 \quad 1 \quad 1]^T$, then the state vectors are

$$x_1(t) = e^{-at}[1+t+0.5t^2 \quad 1+t \quad 1]^T \tag{P8.19.9}$$

and

$$x_2(t) = e^{-at}[1 \quad 1 \quad 1]^T \tag{P8.19.10}$$

b. If $\alpha = 0$, then as t approaches infinity, $x_1(t)$ diverges but $x_2(t)$ remains bounded. We conclude that the multiple eigenvalues of the matrix A on the imaginary axis do not result in the divergence of the state vector.

8.20 A system is represented in the state space as follows:

$$\begin{bmatrix} \dot{x}_1 \\ \dot{x}_2 \\ \dot{x}_3 \end{bmatrix} = \begin{bmatrix} 0 & 1 & 0 \\ 0 & 0 & 1 \\ -6 & -11 & -6 \end{bmatrix} \begin{bmatrix} x_1 \\ x_2 \\ x_3 \end{bmatrix} + \begin{bmatrix} 0 \\ 0 \\ 1 \end{bmatrix} u$$

$$x^T(0) = [1 \quad -1 \quad 2]$$

$$y = [20 \quad 9 \quad 1] \begin{bmatrix} x_1 \\ x_2 \\ x_3 \end{bmatrix}$$

Find
a. The state-transition matrix $\Phi(t)$
b. The solution of the homogeneous equation
c. The general solution, assuming a unit-step input signal
d. The output of the system

Solution
a. The state-transition matrix of the system is

$$\Phi(s) = (sI - A)^{-1} = \begin{bmatrix} s & -1 & 0 \\ 0 & s & -1 \\ 6 & 11 & s+6 \end{bmatrix}^{-1}$$

$$= \frac{1}{(s+1)(s+2)(s+3)} \begin{bmatrix} s^2+6s+11 & s+6 & 1 \\ -6 & s^2+6s & s \\ -6s & -11s-6 & s^2 \end{bmatrix} \tag{P8.20.1}$$

Hence,

$$\Phi(s) = L^{-1}\{(sI - A)^{-1}\}$$

$$= \begin{bmatrix} L^{-1}\left\{\dfrac{s^2+6s+11}{(s+1)(s+2)(s+3)}\right\} & L^{-1}\left\{\dfrac{s+6}{(s+1)(s+2)(s+3)}\right\} & L^{-1}\left\{\dfrac{1}{(s+1)(s+2)(s+3)}\right\} \\[3mm] L^{-1}\left\{\dfrac{-6}{(s+1)(s+2)(s+3)}\right\} & L^{-1}\left\{\dfrac{s^2+6s}{(s+1)(s+2)(s+3)}\right\} & L^{-1}\left\{\dfrac{s}{(s+1)(s+2)(s+3)}\right\} \\[3mm] L^{-1}\left\{\dfrac{-6s}{(s+1)(s+2)(s+3)}\right\} & L^{-1}\left\{\dfrac{-11s-6}{(s+1)(s+2)(s+3)}\right\} & L^{-1}\left\{\dfrac{s^2}{(s+1)(s+2)(s+3)}\right\} \end{bmatrix}$$

$$\Rightarrow \Phi(t) = \begin{bmatrix} 3e^{-t} - 3e^{-2t} + e^{-3t} & \dfrac{5}{2}e^{-t} - 4e^{-2t} + \dfrac{3}{2}e^{-3t} & \dfrac{1}{2}e^{-t} - e^{-2t} + \dfrac{1}{2}e^{-3t} \\[3mm] -3e^{-t} + 6e^{-2t} - 3e^{-3t} & -\dfrac{5}{2}e^{-t} + 8e^{-2t} - \dfrac{9}{2}e^{-3t} & -\dfrac{1}{2}e^{-t} + 2e^{-2t} - \dfrac{3}{2}e^{-3t} \\[3mm] 3e^{-t} - 12e^{-2t} + 9e^{-3t} & \dfrac{5}{2}e^{-t} - 16e^{-2t} + \dfrac{27}{2}e^{-3t} & \dfrac{1}{2}e^{-t} - 4e^{-2t} + \dfrac{9}{2}e^{-3t} \end{bmatrix}$$

$$(P8.20.2)$$

Note that the inverse Laplace transforms were computed from the relationship

$$L^{-1}\left\{\frac{as^2+bs+c}{(s+d)(s+m)(s+n)}\right\} = \frac{ad^2-bd+c}{(m-d)(n-d)}e^{-dt} + \frac{am^2-bm+c}{(d-m)(n-m)}e^{-mt} + \frac{an^2-bn+c}{(d-n)(m-n)}e^{-nt}$$

$$(P8.20.3)$$

b. The solution for the homogeneous equation is

$$x(t) = \Phi(t)x(0) \Rightarrow x(t) = \Phi(t)\begin{bmatrix} 1 \\ -1 \\ 2 \end{bmatrix} \Rightarrow$$

$$x(t) = \begin{bmatrix} \dfrac{3}{2}e^{-t} - e^{-2t} + \dfrac{1}{2}e^{-3t} \\[3mm] -\dfrac{3}{2}e^{-t} + 2e^{-2t} - \dfrac{3}{2}e^{-3t} \\[3mm] \dfrac{3}{2}e^{-t} - 4e^{-2t} + \dfrac{9}{2}e^{-3t} \end{bmatrix} \qquad (P8.20.4)$$

c. The general solution for the state vector is

$$x(t) = \Phi(t)x(0) + \int_0^t \Phi(t-\lambda)bu(\lambda)\,d\lambda \qquad (P8.20.5)$$

The convolution integral is computed as follows:

$$\int_0^t \Phi(t-\lambda)bu(\lambda)\,d\lambda = \int_0^t \begin{bmatrix} \frac{1}{2}e^{-(t-\lambda)} - e^{-2(t-\lambda)} + \frac{1}{2}e^{-3(t-\lambda)} \\ -\frac{1}{2}e^{-(t-\lambda)} + 2e^{-2(t-\lambda)} - \frac{3}{2}e^{-3(t-\lambda)} \\ \frac{1}{2}e^{-(t-\lambda)} - 4e^{-2(t-\lambda)} + \frac{9}{2}e^{-3(t-\lambda)} \end{bmatrix} u(\lambda)\,d\lambda \Rightarrow$$

$$\int_0^t \Phi(t-\lambda)bu(\lambda)\,d\lambda = \begin{bmatrix} \frac{1}{6} - \frac{1}{2}e^{-t} + \frac{1}{2}e^{-2t} - \frac{1}{6}e^{-3t} \\ \frac{1}{2}e^{-t} - e^{-2t} + \frac{1}{2}e^{-3t} \\ -\frac{1}{2}e^{-t} + 2e^{-2t} - \frac{3}{2}e^{-3t} \end{bmatrix} \qquad \text{(P8.20.6)}$$

$$\text{(P8.20.5)} \overset{\text{(P8.20.4)}}{\underset{\text{(P8.20.6)}}{\Rightarrow}} x(t) = \begin{bmatrix} x_1(t) \\ x_2(t) \\ x_3(t) \end{bmatrix} = \begin{bmatrix} \frac{1}{6} + e^{-t} - \frac{1}{2}e^{-2t} + \frac{1}{3}e^{-3t} \\ -e^{-t} + e^{-2t} - e^{-3t} \\ e^{-t} - 2e^{-2t} + 3e^{-3t} \end{bmatrix} \qquad \text{(P8.20.7)}$$

d. The output is

$$y(t) = [20 \quad 9 \quad 1] \begin{bmatrix} x_1(t) \\ x_2(t) \\ x_3(t) \end{bmatrix} = 20x_1(t) + 9x_2(t) + x_3(t) \Rightarrow$$

$$y(t) = \frac{20}{6} + 12e^{-t} - 3e^{-2t} + \frac{2}{3}e^{-3t}$$

8.21 Examine the controllability and observability of the system given at the previous problem.

Solution

The controllability matrix is

$$S = [B \quad \vdots \quad AB \quad \vdots \quad A^2B] \qquad \text{(P8.21.1)}$$

Here

$$B = \begin{bmatrix} 0 \\ 0 \\ 1 \end{bmatrix}, \quad AB = \begin{bmatrix} 0 \\ 1 \\ -6 \end{bmatrix}, \quad A^2B = \begin{bmatrix} 1 \\ -6 \\ 36 \end{bmatrix} \qquad \text{(P8.21.2)}$$

Thus,

$$S = \begin{bmatrix} 0 & 0 & 1 \\ 0 & 1 & -6 \\ 1 & -6 & 36 \end{bmatrix} \qquad \text{(P8.21.3)}$$

$$\det S = \begin{vmatrix} 0 & 0 & 1 \\ 0 & 1 & -6 \\ 1 & -6 & 36 \end{vmatrix} = -1 \neq 0 \qquad \text{(P8.21.4)}$$

The rank of matrix S is 3, hence the system is completely controllable.
  The observability matrix is

$$R^T = [c^T \quad \vdots \quad A^T c^T \quad \vdots \quad (A^T)^2 c^T] \qquad \text{(P8.21.5)}$$

Here,

$$c^T = \begin{bmatrix} 20 \\ 9 \\ 1 \end{bmatrix}, \quad A^T c^T = \begin{bmatrix} -6 \\ 9 \\ 3 \end{bmatrix}, \quad (A^T)^2 c^T = \begin{bmatrix} -18 \\ -39 \\ -9 \end{bmatrix} \qquad \text{(P8.21.6)}$$

Therefore,

$$R^T = \begin{bmatrix} 20 & -6 & -18 \\ 9 & 9 & -39 \\ 1 & 3 & -9 \end{bmatrix} \qquad \text{(P8.21.7)}$$

$$\det R^T = \begin{vmatrix} 20 & -6 & -18 \\ 9 & 9 & -39 \\ 1 & 3 & -9 \end{vmatrix} = 144 \neq 0 \qquad \text{(P8.21.8)}$$

The rank of matrix R^T is 3, thus the system is completely observable.

8.22 Compute the solution of the state equations for the system shown in the figure below, if the input signal is the unit-step function.

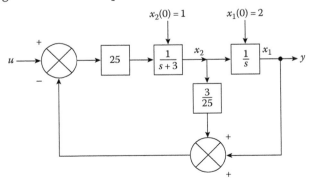

Solution

From the block diagram we derive the following state variables:

$$\dot{x}_1 = x_2 \tag{P8.22.1}$$

$$\dot{x}_2 = -3x_2 + 25\left(u - \frac{3}{25}x_2 - x_1\right) = -25x_1 - 6x_2 + 25u \tag{P8.22.2}$$

In matrix form it is written as $\dot{x} = Ax + Bu$, where

$$A = \begin{bmatrix} 0 & 1 \\ -25 & -6 \end{bmatrix} \quad \text{and} \quad B = \begin{bmatrix} 0 \\ 25 \end{bmatrix} \tag{P8.22.3}$$

The state-transition matrix is

$$\Phi(t) = (sI - A)^{-1} = \begin{bmatrix} s & -1 \\ 25 & s+6 \end{bmatrix}^{-1} = \frac{1}{(s+3)^2 + 4^2}\begin{bmatrix} s+6 & 1 \\ -25 & s \end{bmatrix} \tag{P8.22.4}$$

From the equation $X(s) = \Phi(s)x(0) + \Phi(s)BU(s)$ we have

$$X(s) = \frac{1}{(s+3)^2+4^2}\begin{bmatrix} s+6 & 1 \\ -25 & s \end{bmatrix}\begin{bmatrix} 2 \\ 1 \end{bmatrix} + \frac{1}{(s+3)^2+4^2}\begin{bmatrix} s+6 & 1 \\ -25 & s \end{bmatrix}\begin{bmatrix} 0 \\ 25 \end{bmatrix}\frac{1}{s} \Rightarrow$$

$$X(s) = \frac{1}{(s+3)^2+4^2}\begin{bmatrix} 2s+13 \\ s-5 \end{bmatrix} + \frac{1}{(s+3)^2+4^2}\begin{bmatrix} \frac{25}{s} \\ 25 \end{bmatrix} \Rightarrow$$

$$X(s) = \frac{1}{s\left[(s+3)^2+4^2\right]}\begin{bmatrix} 2s^2+13s+25 \\ s^2-25s \end{bmatrix} \tag{P8.22.5}$$

We write $X(s)$ in the following form:

$$X(s) = \frac{k_1}{s} + \frac{k_2 s + k_3}{(s+3)^2 + 4^2} \tag{P8.22.6}$$

where

$$k_1 = \lim_{s \to 0} \frac{1}{(s+3)^2+4^2}\begin{bmatrix} 2s^2+13s+25 \\ s^2-25s \end{bmatrix} = \begin{bmatrix} 1 \\ 0 \end{bmatrix} \tag{P8.22.7}$$

From Equation P8.22.5 through P8.22.7 we obtain

$$k_2 = \begin{bmatrix} 1 \\ 1 \end{bmatrix} \quad \text{and} \quad k_3 = \begin{bmatrix} 7 \\ -25 \end{bmatrix} \qquad \text{(P8.22.8)}$$

Hence,

$$X(s) = \frac{1}{s}\begin{bmatrix} 1 \\ 0 \end{bmatrix} + \frac{1}{(s+3)^2 + 4^2}\left(s\begin{bmatrix} 1 \\ 1 \end{bmatrix} + \begin{bmatrix} 7 \\ -25 \end{bmatrix} \right) \Rightarrow$$

$$x(t) = L^{-1}\{X(s)\} = L^{-1}\left\{ \frac{1}{s}\begin{bmatrix} 1 \\ 0 \end{bmatrix} \right\} + L^{-1}\left\{ \begin{bmatrix} \dfrac{s+7}{(s+3)^2 + 4^2} \\[2mm] \dfrac{s-25}{(s+3)^2 + 4^2} \end{bmatrix} \right\} \Rightarrow$$

$$x(t) = \begin{bmatrix} x_1(t) \\ x_2(t) \end{bmatrix} = \begin{bmatrix} 1 \\ 0 \end{bmatrix} + \begin{bmatrix} 1.414e^{-3t}\sin(4t+45°) \\ 7.08e^{-3t}\sin(4t+171.9°) \end{bmatrix} \qquad \text{(P8.22.9)}$$

8.23 At the diagram shown below, we consider as system S_1 the one with input u_1 and output y_1, as system S_2 the one with input u_2 and output y_2, and as system S_3 the one with input u_1 and output y_2.

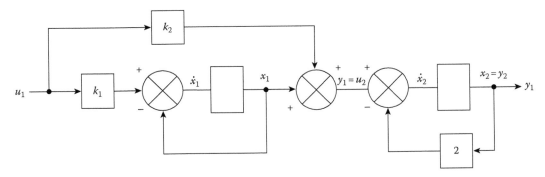

a. Write the state-space representation of S_1, S_2, and S_3.
b. Compute the transfer function of S_1, S_2, and S_3.
c. Determine the values of k_1 and k_2 for which the state vectors of the three systems are not controllable.
d. Compute the eigenvectors of the system matrix of S_3. Discuss their relationship with the input matrix of S_3 that is formulated by the k_1 and k_2 which make S_3 uncontrollable.

Solution

a. For the system S_1 we have

$$\left.\begin{aligned} \dot{x}_1 &= -x_1 + k_1 u_1, \quad x_1(0) = x_{10} \\ y_1 &= x_1 + k_2 u_1 \end{aligned}\right\} \qquad \text{(P8.23.1)}$$

For the system S_2 we have

$$\left.\begin{array}{l} \dot{x}_2 = -2x_2 + u_2, \qquad x_2(0) = x_{20} \\ y_2 = x_2 \end{array}\right\} \tag{P8.23.2}$$

For the system S_3 we have

$$\left.\begin{array}{ll} \dot{x}_1 = -x_1 + k_1 u_1 \\ \dot{x}_2 = -2x_2 + x_1 + k_2 u_1 \\ x_1(0) = x_{10}, & x_2(0) = x_{20} \\ y_2 = x_2 \end{array}\right\} \tag{P8.23.3}$$

b. The transfer function of S_1 is

$$G_1(s) = \frac{Y_1(s)}{U_1(s)} = 1 \cdot (s+1)^{-1} k_1 + k_2 = \frac{k_1 + k_2(s+1)}{s+1} \tag{P8.23.4}$$

The transfer function of S_2 is

$$G_2(s) = \frac{Y_2(s)}{U_2(s)} = \frac{1}{s+2} \tag{P8.23.5}$$

The transfer function of S_3 is computed as

$$G_3(s) = \frac{Y_2(s)}{U_1(s)} = C(sI - A)^{-1} B = \begin{bmatrix} 0 & 1 \end{bmatrix} \begin{bmatrix} s+1 & 0 \\ -1 & s+2 \end{bmatrix} \begin{bmatrix} k_1 \\ k_2 \end{bmatrix} \Rightarrow$$

$$G_3(s) = \begin{bmatrix} 0 & 1 \end{bmatrix} \begin{bmatrix} \dfrac{1}{s+1} & 0 \\ \dfrac{1}{(s+1)(s+2)} & \dfrac{1}{s+2} \end{bmatrix} \begin{bmatrix} k_1 \\ k_2 \end{bmatrix} = \begin{bmatrix} \dfrac{1}{(s+1)(s+2)} & \dfrac{1}{s+2} \end{bmatrix} \begin{bmatrix} k_1 \\ k_2 \end{bmatrix} \Rightarrow$$

$$G_3(s) = \frac{Y_2(s)}{U_1(s)} = \frac{k_1 + k_2(s+1)}{(s+1)(s+2)} \tag{P8.23.6}$$

c. The state vector of S_1 is not controllable for $k_1 = 0$, $k_2 \in R$, since there is pole-zero cancellation at $G_1(s)$. The system 2 is controllable for any k_1, k_2.
The controllability matrix for system 3 is

$$S = [B \;\; \vdots \;\; AB] = \begin{bmatrix} k_1 & -k_1 \\ k_2 & k_1 - 2k_2 \end{bmatrix} \tag{P8.23.7}$$

$$|S| = 0 \Rightarrow k_1(k_1 - 2k_2) + k_1 k_2 = k_1(k_1 - k_2) = 0 \tag{P8.23.8}$$

Hence, for $k_1 = 0, k_2 \in R$ and for $k_1 = k_2$ the state vector of S_3 is not controllable:

For $k_1 = 0, k_2 \in R \Rightarrow G_3(s) = k_2/(s + 2)$, there is pole-zero cancellation at $s = -1$.

For $k_1 = k_2 \Rightarrow G_3(s) = k_2/(s + 1)$, there is pole-zero cancellation at $s = -2$.

d. The system matrix of S_3 for $k_1 = 0, k_2 \in R$ is $A = \begin{bmatrix} -1 & 0 \\ 1 & -2 \end{bmatrix}$. Its eigenvalues are obtained from the relationship

$$|A - \lambda I| = 0 \Rightarrow$$

$$\left| \begin{bmatrix} -1 & 0 \\ 1 & -2 \end{bmatrix} - \begin{bmatrix} \lambda & 0 \\ 0 & \lambda \end{bmatrix} \right| = \begin{vmatrix} -1-\lambda & 0 \\ 0 & -2-\lambda \end{vmatrix} = (-1-\lambda)(-2-\lambda) = 0 \qquad \text{(P8.23.9)}$$

Therefore, the eigenvalues are $\lambda_1 = -1$ and $\lambda_2 = -2$.
We compute the eigenvectors from the relationship

$$Au_i = \lambda_i u_i \Rightarrow (A - \lambda_i I)u_i = 0 \qquad \text{(P8.23.10)}$$

For $\lambda_1 = -1$, $\quad Au_1 = \lambda_1 u_1 \Rightarrow \begin{bmatrix} -1 & 0 \\ 1 & -2 \end{bmatrix} \begin{bmatrix} u_{11} \\ u_{12} \end{bmatrix} = -\begin{bmatrix} u_{11} \\ u_{12} \end{bmatrix} \Rightarrow u_{11} = u_{12} \qquad \text{(P8.23.11)}$

and the eigenvector has the form $u = \begin{bmatrix} 1 \\ 1 \end{bmatrix}$.

For $\lambda_2 = -2$, $\quad Au_2 = \lambda_2 u_2 \Rightarrow \begin{bmatrix} -1 & 0 \\ 1 & -2 \end{bmatrix} \begin{bmatrix} u_{21} \\ u_{22} \end{bmatrix} = -2\begin{bmatrix} u_{21} \\ u_{22} \end{bmatrix} \Rightarrow u_{21} = 0 \qquad \text{(P8.23.12)}$

and the eigenvector has the form $u = \begin{bmatrix} 0 \\ 1 \end{bmatrix}$.

The input matrices of S_3 that result from the substitution of $k_1 = 0, k_2 \in R$, and $k_1 = k_2$ are $B = \begin{bmatrix} 0 \\ k_2 \end{bmatrix}$ and $B = \begin{bmatrix} k_2 \\ k_2 \end{bmatrix}$, respectively.

Consequently the system is not controllable if the input matrix lies on an (real-valued) eigenvector.

8.24 Consider a system represented in state space by the following equations:

$$\begin{bmatrix} \dot{x}_1 \\ \dot{x}_2 \end{bmatrix} = \begin{bmatrix} 0 & 6 \\ -1 & -5 \end{bmatrix} \begin{bmatrix} x_1 \\ x_2 \end{bmatrix} + \begin{bmatrix} 0 \\ 1 \end{bmatrix} u$$

$$y = \begin{bmatrix} 1 & k \end{bmatrix} \begin{bmatrix} x_1 \\ x_2 \end{bmatrix}$$

a. Find the state-transition matrix $\Phi(t)$ by diagonalizing the system.

b. Find $x(t)$ for $t > 0$ if the input is $u(t) = 1$, $t > 0$ and the initial conditions are $x(0) = \begin{bmatrix} -1 \\ 0 \end{bmatrix}$.

c. For which values of k is the state vector unobservable?

Solution

a. The eigenvalues of matrix A are computed as follows:

$$|A - \lambda I| = 0 \Rightarrow \begin{vmatrix} -\lambda & 6 \\ -1 & -5 - \lambda \end{vmatrix} = 0 \Rightarrow \lambda^2 + 5\lambda + 6 = 0 \qquad \text{(P8.24.1)}$$

Hence, $\lambda_1 = -3$ and $\lambda_2 = -2$.

The eigenvectors are calculated by the relationship

$$Au_i = \lambda_i u_i \Rightarrow (A - \lambda_i I)u_i = 0 \qquad \text{(P8.24.2)}$$

$$\text{For } \lambda_1 = -3, \quad (A - \lambda_1 I)u_1 = \begin{bmatrix} 3 & 6 \\ -1 & -2 \end{bmatrix}\begin{bmatrix} u_{11} \\ u_{12} \end{bmatrix} = \begin{bmatrix} 0 \\ 0 \end{bmatrix} \Rightarrow$$

$$3u_{11} + 6u_{12} = 0 \Rightarrow u_{11} = -2u_{12} \qquad \text{(P8.24.3)}$$

The eigenvector for $(\lambda_1 = -3)$ is of the form $\begin{bmatrix} -2 \\ 1 \end{bmatrix} u_{12}$.

Thus, the first eigenvector is

$$u_1 = \begin{bmatrix} -2 \\ 1 \end{bmatrix} \qquad \text{(P8.24.4)}$$

$$\text{For } \lambda_2 = -2, \quad (A - \lambda_2 I)u_2 = \begin{bmatrix} 2 & 6 \\ -1 & -3 \end{bmatrix}\begin{bmatrix} u_{21} \\ u_{22} \end{bmatrix} = \begin{bmatrix} 0 \\ 0 \end{bmatrix} \Rightarrow \qquad \text{(P8.24.5)}$$

$$2u_{21} + 6u_{22} = 0 \Rightarrow u_{21} = -3u_{22}$$

The eigenvector for $\lambda_2 = -2$ has the form

$$u_2 = \begin{bmatrix} -3 \\ 1 \end{bmatrix} u_{22} \qquad \text{(P8.24.6)}$$

Thus the second eigenvector (for $\lambda_2 = -2$) is

$$u_2 = \begin{bmatrix} -3 \\ 1 \end{bmatrix} \qquad \text{(P8.24.7)}$$

The matrix of the eigenvectors is

$$T = \begin{bmatrix} -2 & -3 \\ 1 & 1 \end{bmatrix} \quad \text{(P8.24.8)}$$

With the transformation

$$x = Tz \quad \text{(P8.24.9)}$$

we get the diagonal system:

$$\left. \begin{aligned} \dot{z} &= T^{-1}ATz + T^{-1}bu = \Lambda z + T^{-1}bu, \quad z_0 = T^{-1}x_0 \\ y &= c^T Tz \end{aligned} \right\} \quad \text{(P8.24.10)}$$

where

$$\Lambda = \begin{bmatrix} -3 & 0 \\ 0 & -2 \end{bmatrix} \quad \text{(P8.24.11)}$$

The state-transition matrix is

$$\Phi(t) = e^{At} = e^{(T\Lambda T^{-1})t} = Te^{\Lambda t}T^{-1} = T \begin{bmatrix} e^{-3t} & 0 \\ 0 & e^{-2t} \end{bmatrix} T^{-1} \Rightarrow$$

$$\text{(P8.24.12)}$$

$$\Phi(t) = \begin{bmatrix} 3e^{-2t} - 2e^{-3t} & 6e^{-2t} - 6e^{-3t} \\ -e^{-2t} + e^{-3t} & -2e^{-2t} + 3e^{-3t} \end{bmatrix}$$

b. For $x_0 = x(0) = [-1 \quad 0]^T$, $b = [0 \quad 1]^T$ and $u(t) = 1$ the state vector $x(t)$ is given by

$$x(t) = e^{At}x(0) + \int_0^t e^{A(t-\lambda)}bu(\lambda)\,d\lambda = e^{At}\begin{bmatrix} -1 \\ 0 \end{bmatrix} + \int_0^t \begin{bmatrix} 6e^{-2(t-\lambda)} - 6e^{-3(t-\lambda)} \\ -2e^{-2(t-\lambda)} + 3e^{-3(t-\lambda)} \end{bmatrix} \Rightarrow$$

$$\text{(P8.24.13)}$$

$$x(t) = \begin{bmatrix} -3e^{-2t} + 2e^{-3t} \\ e^{-2t} - e^{-3t} \end{bmatrix} + \begin{bmatrix} 1 - 3e^{-2t} + 2e^{-3t} \\ e^{-2t} - e^{-3t} \end{bmatrix} = \begin{bmatrix} 1 - 6e^{-2t} + 4e^{-3t} \\ 2e^{-2t} - 2e^{-3t} \end{bmatrix}$$

c. The observability matrix is

$$\begin{bmatrix} c^T \\ c^T A \end{bmatrix} = \begin{bmatrix} 1 & k \\ -k & 6 - 5k \end{bmatrix} \quad \text{(P8.24.14)}$$

We set the determinant equal to zero, and we get the values of k, for which the state vector is not observable:

$$\begin{vmatrix} 1 & k \\ -k & 6 - 5k \end{vmatrix} = 6 - 5k + k^2 = 0 \Rightarrow k_1 = 2 \quad \text{and} \quad k_2 = 3 \quad \text{(P8.24.15)}$$

The transfer function of the system is

$$G(s) = c(sI - A)^{-1}b = \begin{bmatrix} 1 & k \end{bmatrix} \begin{bmatrix} s & -6 \\ 1 & s+5 \end{bmatrix}^{-1} \begin{bmatrix} 0 \\ 1 \end{bmatrix} \Rightarrow$$

$$G(s) = \begin{bmatrix} 1 & k \end{bmatrix} \begin{bmatrix} \dfrac{s+5}{(s+2)(s+3)} & \dfrac{6}{(s+2)(s+3)} \\ \dfrac{-1}{(s+2)(s+3)} & \dfrac{s}{(s+2)(s+3)} \end{bmatrix} \begin{bmatrix} 0 \\ 1 \end{bmatrix} \Rightarrow$$

$$G(s) = \frac{6+ks}{(s+2)(s+3)} \tag{P8.24.16}$$

Thus, for $k = 2$ and for $k = 3$ we have pole-zero cancellations at the points $s = -3$ and $s = -2$, respectively.

8.25 The block diagram of a control system is illustrated in the figure below.

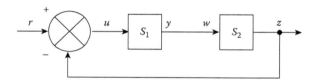

The dynamic systems S_1 and S_2 are represented in the state space as follows:

S_1	S_2
$\dot{x}_1 = x_2 + u$	$\dot{x}_3 = x_3 + w$
$\dot{x}_2 = -2x_1 - 3u_2$	$z = x_3$
$y = ax_1 + x_2$	

a. Examine if the state vector of the open-loop system is controllable and observable.

b. Repeat the previous query for the closed-loop system.

Solution

a. We observe from the block diagram that the two, connected in series, systems formulate a new open-loop system S_3.

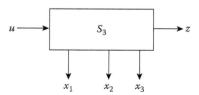

The state-space representation of S_3 is

$$
\left.\begin{aligned}
\dot{x}_1 &= x_2 + u \\
\dot{x}_2 &= -2x_1 - 3x_2 \\
\dot{x}_3 &= x_3 + w = x_3 + y = ax_1 + x_2 + x_3 \\
z &= x_3
\end{aligned}\right\} \tag{P8.25.1}
$$

Hence, the matrices A, b, and c are

$$
A = \begin{bmatrix} 0 & 1 & 0 \\ -2 & -3 & 0 \\ a & 1 & 1 \end{bmatrix}, \quad b = \begin{bmatrix} 1 \\ 0 \\ 0 \end{bmatrix}, \quad c = [0 \quad 0 \quad 1] \tag{P8.25.2}
$$

The controllability matrix for S_3 is

$$
S = [b \quad \vdots \quad Ab \quad \vdots \quad A^2b] = \begin{bmatrix} 1 & 0 & -2 \\ 0 & -2 & 6 \\ 0 & a & a-2 \end{bmatrix} \tag{P8.25.3}
$$

Thus,

$$
\det S = \begin{vmatrix} 1 & 0 & -2 \\ 0 & -2 & 6 \\ 0 & a & a-2 \end{vmatrix} = 4(1 - 2a) \tag{P8.25.4}
$$

The rank of matrix S is 3, hence the state vector is observable if $a \neq 1/2$.
The observability matrix is

$$
R = \begin{bmatrix} c \\ cA \\ cA^2 \end{bmatrix} = \begin{bmatrix} 0 & 0 & 1 \\ a & 1 & 1 \\ a-2 & a-2 & 1 \end{bmatrix} \tag{P8.25.5}
$$

Thus,

$$
\det R = \begin{vmatrix} 0 & 0 & 1 \\ a & 1 & 1 \\ a-2 & a-2 & 1 \end{vmatrix} = (a-2)(a-1) \tag{P8.25.6}
$$

The rank of matrix R is 3, hence the state vector is observable if $a \neq 1$ and $a \neq 2$.

b. For the closed-loop system S_4 it holds that

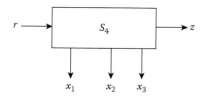

$$u = r - z = r - x_3 \tag{P8.25.7}$$

It is represented in the state space as follows:

$$
\left.
\begin{aligned}
\dot{x}_1 &= x_2 - x_3 + r \\
\dot{x}_2 &= -2x_1 - 3x_2 \\
\dot{x}_3 &= x_3 + w = x_3 + y = ax_1 + x_2 + x_3 \\
z &= x_3
\end{aligned}
\right\} \tag{P8.25.8}
$$

The controllability matrix of S_4 is

$$
S = [b \quad \vdots \quad Ab \quad \vdots \quad A^2b] =
\begin{bmatrix}
1 & 0 & -a-2 \\
0 & -2 & 6 \\
0 & a & a-2
\end{bmatrix}
\tag{P8.25.9}
$$

Thus,

$$
\det S =
\begin{vmatrix}
1 & 0 & -a-2 \\
0 & -2 & 6 \\
0 & a & a-2
\end{vmatrix}
= 4(1-2a)
\tag{P8.25.10}
$$

The state vector is controllable if $a \neq 1/2$.
The observability matrix of S_4 is

$$
R =
\begin{bmatrix}
c \\
cA \\
cA^2
\end{bmatrix}
=
\begin{bmatrix}
0 & 0 & 1 \\
a & 1 & 1 \\
a-2 & a-2 & 1-a
\end{bmatrix}
\tag{P8.25.11}
$$

Thus,

$$
\det R =
\begin{vmatrix}
0 & 0 & 1 \\
a & 1 & 1 \\
a-2 & a-2 & 1-a
\end{vmatrix}
= (a-1)(a-2)
\tag{P8.25.12}
$$

The state vector is observable if $a \neq 1$ and $a \neq 2$.

8.26 Examine if the state vectors of the electric circuits shown in Figure (a) and (b) are controllable.

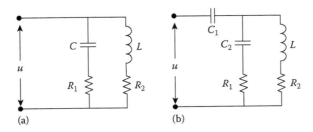

(a)   (b)

Solution

a. For the electric circuit shown in Figure (a) we consider the following state variables:

$$\left.\begin{aligned} x_1(t) = x_1 = u_c(t) \\ x_2(t) = x_2 = i_L(t) \end{aligned}\right\} \qquad \text{(P8.26.1)}$$

The equations of the circuit are

$$\left.\begin{aligned} u = x_1 + R_1\dot{x}_1 \\ u = R_2 x_2 + L\dot{x}_2 \end{aligned}\right\} \qquad \text{(P8.26.2)}$$

or

$$\begin{bmatrix} \dot{x}_1 \\ \dot{x}_2 \end{bmatrix} = \begin{bmatrix} -\dfrac{1}{R_1 C} & 0 \\ 0 & -\dfrac{R_2}{L} \end{bmatrix} \begin{bmatrix} x_1 \\ x_2 \end{bmatrix} + \begin{bmatrix} \dfrac{1}{R_1 C} \\ \dfrac{1}{L} \end{bmatrix} u \qquad \text{(P8.26.3)}$$

The state vector is controllable, if the determinant of the controllability matrix is not zero. The controllability matrix is

$$S = [B \quad \vdots \quad AB] = \begin{bmatrix} \dfrac{1}{R_1 C} & -\dfrac{1}{R_1^2 C^2} \\ \dfrac{1}{L} & -\dfrac{R_2}{L^2} \end{bmatrix} \qquad \text{(P8.26.4)}$$

$$\det S = -\dfrac{R_2}{R_1 C L^2} + \dfrac{1}{R_1^2 C^2 L} = \dfrac{1}{R_1 C L}\left(\dfrac{1}{R_1 C} - \dfrac{R_2}{L} \right) \qquad \text{(P8.26.5)}$$

The system is controllable if

$$\dfrac{1}{R_1 C} \neq \dfrac{R_2}{L} \quad \text{or} \quad R_1 R_2 \neq \dfrac{L}{C} \qquad \text{(P8.26.6)}$$

b. The state variables of the electric circuit shown in the Figure (b) are

$$\left. \begin{array}{l} x_1(t) = x_1 = u_{c_1}(t) \\ x_2(t) = x_2 = u_{c_2}(t) \\ x_3(t) = x_3 = i_{L_3}(t) \end{array} \right\} \qquad \text{(P8.26.7)}$$

The equations of the system are

$$\left. \begin{array}{l} u = x_1 + L\dot{x}_3 + Rx_3 \\ u = x_1 + x_2 + R_1 C_2 \dot{x}_2 \\ C_1 \dot{x}_1 = C_2 \dot{x}_2 + x_3 \end{array} \right\} \qquad \text{(P8.26.8)}$$

or

$$\begin{bmatrix} \dot{x}_1 \\ \dot{x}_2 \\ \dot{x}_3 \end{bmatrix} = \begin{bmatrix} -1 & -1 & 1 \\ -1 & -1 & 0 \\ -1 & 0 & -R \end{bmatrix} \begin{bmatrix} x_1 \\ x_2 \\ x_3 \end{bmatrix} + \begin{bmatrix} 1 \\ 1 \\ 1 \end{bmatrix} u \qquad \text{(P8.26.9)}$$

The controllability matrix of the system is

$$S = [B \quad \vdots \quad AB \quad \vdots \quad A^2 B] = \begin{bmatrix} 1 & -1 & 2-R \\ 1 & -2 & 3 \\ 1 & -1-R & R^2+R+1 \end{bmatrix} \qquad \text{(P8.26.10)}$$

$$\det S = \begin{vmatrix} 1 & -1 & 2-R \\ 1 & -2 & 3 \\ 1 & -1-R & R^2+R+1 \end{vmatrix}$$

$$= \begin{vmatrix} -2 & 3 \\ -1-R & R^2+R+1 \end{vmatrix} - \begin{vmatrix} -1 & 2-R \\ -1-R & R^2+R+1 \end{vmatrix} + \begin{vmatrix} -1 & 2-R \\ -2 & 3 \end{vmatrix}$$

$$= 1-R \qquad \text{(P8.26.11)}$$

For $R \neq 1$ the system is controllable.

8.27 Consider the system $\dot{x}(t) = Ax(t) + bu(t)$ where

$$A = \begin{bmatrix} 1 & 0 & 0 \\ 0 & 2 & 0 \\ 0 & 0 & 3 \end{bmatrix}, \quad b = \begin{bmatrix} 1 \\ 1 \\ 2 \end{bmatrix}$$

Express the system in phase-variable canonical form.

Solution

The observability matrix of the system is

$$S = [B \;\vdots\; Ab \;\vdots\; A^2b] = \begin{bmatrix} 1 & 1 & 1 \\ 1 & 2 & 4 \\ 2 & 6 & 18 \end{bmatrix} \tag{P8.27.1}$$

We have

$$\det S = \begin{vmatrix} 1 & 1 & 1 \\ 1 & 2 & 4 \\ 2 & 6 & 18 \end{vmatrix} = 4 \neq 0$$

Hence, the system can be transformed in the phase-variable canonical form. We compute the matrix S^{-1}:

$$S^{-1} = \frac{\text{Adj}\,S}{|S|} = \frac{1}{4}\begin{vmatrix} 12 & -12 & 2 \\ -10 & 16 & -3 \\ 2 & -4 & 1 \end{vmatrix} \tag{P8.27.2}$$

The last row S^{-1} is

$$q = \frac{1}{4}\begin{pmatrix} 2 & -4 & 1 \end{pmatrix} \tag{P8.27.3}$$

We form the matrix P:

$$P = \begin{bmatrix} q \\ qA \\ qA^2 \end{bmatrix} = \frac{1}{4}\begin{bmatrix} 2 & -4 & 1 \\ 2 & -8 & 3 \\ 2 & -16 & 9 \end{bmatrix} \tag{P8.27.4}$$

$$\det P = -16\left(\frac{1}{4}\right) = -4 \tag{P8.27.5}$$

The transformation matrix T is

$$T = P^{-1}\begin{bmatrix} 6 & -5 & 1 \\ 3 & -4 & 1 \\ 4 & -6 & 2 \end{bmatrix} \tag{P8.27.6}$$

We consider the linear transformation $x = Tz$. The system in the phase-variable canonical form is

$$\dot{z} = T^{-1}ATz + T^{-1}bu = A^*z + b^*u \tag{P8.27.7}$$

where

$$A^* = T^{-1}AT = \begin{bmatrix} 0 & 1 & 0 \\ 0 & 0 & 1 \\ 6 & -11 & 6 \end{bmatrix} \tag{P8.27.8}$$

$$b^* = T^{-1}b = \begin{bmatrix} 0 \\ 0 \\ 1 \end{bmatrix} \tag{P8.27.9}$$

8.28 Consider a system described by the state equation $\dot{x}(t) = Ax(t) + bu(t)$, where $A = \begin{bmatrix} 0 & 1 \\ 20.6 & 0 \end{bmatrix}$, $B = \begin{bmatrix} 0 \\ 1 \end{bmatrix}$.

Based on the state-feedback control law $u = -Kx$, determine the state-feedback matrix, so that the poles of the closed-loop system are at the points $s = -1.8 \pm j2.4$.

Solution

First solution:

First of all, we examine if the system is controllable. The controllability matrix is

$$S = [B \;\vdots\; AB] = \begin{bmatrix} 0 & 1 \\ 1 & 0 \end{bmatrix} \tag{P8.28.1}$$

$\det S = \begin{vmatrix} 0 & 1 \\ 1 & 0 \end{vmatrix} = -1 \neq 0$, hence rank $S = 2$. The system is controllable and it is possible to place the poles in a desired position.

The characteristic polynomial of the desired system is

$$|sI - A + BK| = \left| \begin{bmatrix} s & 0 \\ 0 & s \end{bmatrix} - \begin{bmatrix} 0 & 1 \\ 20.6 & 0 \end{bmatrix} + \begin{bmatrix} 0 \\ 1 \end{bmatrix}[k_1 \quad k_2] \right|$$

$$= \begin{vmatrix} s & -1 \\ -20.6 + k_1 & s + k_2 \end{vmatrix} = s^2 + k_2 s + k_1 - 20.6 \tag{P8.28.2}$$

The characteristic polynomial must be equal to

$$(s + \lambda_1)(s + \lambda_2) = (s + 1.8 - j2.4)(s + 1.8 + j2.4) = s^2 + 3.6s + 9 \tag{P8.28.3}$$

$$(P8.28.2),(P8.28.3) \Rightarrow k_1 = 29.6 \quad \text{and} \quad k_2 = 3.6$$

or

$$K = [29.6 \quad 3.6] \tag{P8.28.4}$$

The block diagram of the system with state feedback is

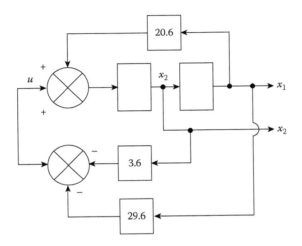

Second solution:
We will apply Ackermann's formula. The characteristic polynomial is

$$\left|sI-(A-BK)\right| = \left|sI-\tilde{A}\right| = s^2 + 3.6s + 9 = \Phi(s) \tag{P8.28.5}$$

Thus, we have

$$\Phi(A) = A^2 + 3.6A + 9I \Rightarrow$$

$$\Phi(A) = \begin{bmatrix} 0 & 1 \\ 20.6 & 0 \end{bmatrix}\begin{bmatrix} 0 & 1 \\ 20.6 & 0 \end{bmatrix} + 3.6\begin{bmatrix} 0 & 1 \\ 20.6 & 0 \end{bmatrix} + 9\begin{bmatrix} 1 & 0 \\ 0 & 1 \end{bmatrix} \Rightarrow$$

$$\Phi(A) = \begin{bmatrix} 29.6 & 3.6 \\ 74.16 & 29.6 \end{bmatrix} \tag{P8.28.6}$$

Therefore,

$$K = [0 \quad 1][B \quad \vdots \quad AB]^{-1}\Phi(A)$$

$$= [0 \quad 1]\begin{bmatrix} 0 & 1 \\ 1 & 0 \end{bmatrix}^{-1}\begin{bmatrix} 29.6 & 3.6 \\ 74.16 & 29.6 \end{bmatrix} = [29.6 \quad 3.6] \tag{P8.28.7}$$

Third solution:

The system is in controllable canonical form, hence the transformation matrix is $T = I$. From the characteristic equation of the system we have

$$|sI - A| = \begin{vmatrix} s & -1 \\ -20.6 & s \end{vmatrix} = s^2 - 20.6 = 0 \tag{P8.28.8}$$

Thus,

$$a_1 = 0, \quad a_2 = -20.6 \tag{P8.28.9}$$

The desired characteristic polynomial is

$$(s - \mu_1)(s - \mu_2) = (s + 1.8 - j2.4)(s + 1.8 + j2.4)$$

$$= s^2 + 3.6s + 9 = s^2 + b_1 s + b_2 \tag{P8.28.10}$$

Therefore,

$$b_1 = 3.6, \quad b_2 = 9 \tag{P8.28.11}$$

Hence,

$$K = [b_2 - a_2 \quad \vdots \quad b_1 - a_1]T^{-1} = [9 + 20.6 \quad \vdots \quad 3.6 - 0]I^{-1} \Rightarrow$$

$$k = [29.6 \quad 3.6] \tag{P8.28.12}$$

8.29 The linearized equations that describe the operation of the inverted pendulum system shown in the figure below are

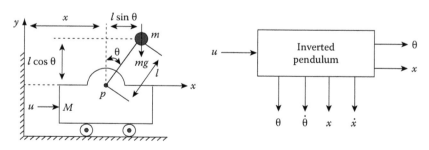

$$(M + m)\ddot{x} + ml\ddot{\theta} = u$$

$$m\ddot{x} + ml\ddot{\theta} = mg\theta$$

Given that $M = 2\,\mathrm{Kg}$, $m = 0.1\,\mathrm{Kg}$, and $l = 0.5\,\mathrm{m}$, determine the state-feedback matrix so that the poles of the closed-loop system are at the points $s_1 = s_2 = -10$ and $s_{3,4} = -2 \pm j3.464$.

Solution

The linearized equations that describe the operation of the inverse pendulum system are

$$(M+m)\ddot{x} + ml\ddot{\theta} = u \tag{P8.29.1}$$

$$m\ddot{x} + ml\ddot{\theta} = mg\theta \tag{P8.29.2}$$

$$Ml\ddot{\theta} = (M+m)g\theta - u \tag{P8.29.3}$$

$$M\ddot{x} = u - mg\theta \tag{P8.29.4}$$

We consider the following state variables:

$$\left.\begin{aligned} x_1 &= \theta \\ x_2 &= \dot{\theta} \\ x_3 &= x \\ x_4 &= \dot{x} \end{aligned}\right\} \tag{P8.29.5}$$

We have

$$\left.\begin{aligned} \dot{x}_1 &= x_2 \\ \dot{x}_2 &\overset{(P8.29.3)}{=} \frac{M+m}{Ml}gx_1 - \frac{1}{Ml}u \\ \dot{x}_3 &= x_4 \\ \dot{x}_4 &\overset{(P8.29.4)}{=} -\frac{m}{M}gx_1 + \frac{1}{M}u \end{aligned}\right\} \tag{P8.29.6}$$

The representation of the system in the state space is

$$\begin{bmatrix} \dot{x}_1 \\ \dot{x}_2 \\ \dot{x}_3 \\ \dot{x}_4 \end{bmatrix} = \begin{bmatrix} 0 & 1 & 0 & 0 \\ \dfrac{M+m}{Ml}g & 0 & 0 & 0 \\ 0 & 0 & 0 & 1 \\ -\dfrac{m}{M}g & 0 & 0 & 0 \end{bmatrix} \begin{bmatrix} x_1 \\ x_2 \\ x_3 \\ x_4 \end{bmatrix} + \begin{bmatrix} 0 \\ -\dfrac{1}{Ml} \\ 0 \\ \dfrac{1}{M} \end{bmatrix} u$$

$$\begin{bmatrix} y_1 \\ y_2 \end{bmatrix} = \begin{bmatrix} 1 & 0 & 0 & 0 \\ 0 & 0 & 1 & 0 \end{bmatrix} \begin{bmatrix} x_1 \\ x_2 \\ x_3 \\ x_4 \end{bmatrix} \tag{P8.29.7}$$

From the Equation (3) we get the transfer function

$$\frac{\Theta(s)}{U(s)} = \frac{-1}{Mls^2 - (M+m)g} = \frac{-1}{s^2 - 20.601} = \frac{-1}{s^2 - (4.539)^2} \tag{P8.29.8}$$

We observe that the system is unstable, since one pole ($s = 4.539$) is in the right-half s-plane.
We will use the state-feedback control law $u = -Kx$.
By substituting the given values, the matrices A and B become

$$A = \begin{bmatrix} 0 & 1 & 0 & 0 \\ 20.601 & 0 & 0 & 0 \\ 0 & 0 & 0 & 1 \\ -0.4905 & 0 & 0 & 0 \end{bmatrix}, \quad B = \begin{bmatrix} 0 \\ -1 \\ 0 \\ 0.5 \end{bmatrix} \tag{P8.29.9}$$

We examine first if the system is controllable. The controllability matrix is

$$S = [B \quad \vdots \quad AB \quad \vdots \quad A^2B \quad \vdots \quad A^3B] = \begin{bmatrix} 0 & -1 & 0 & -20.601 \\ -1 & 0 & -20.601 & 0 \\ 0 & 0.5 & 0 & 0.4905 \\ 0.5 & 0 & 0.4905 & 0 \end{bmatrix} \tag{P8.29.10}$$

The rank of the matrix S is 4, hence the system is completely controllable.
The characteristic equation of the system is

$$|sI - A| = \begin{vmatrix} s & -1 & 0 & 0 \\ -20.601 & s & 0 & 0 \\ 0 & 0 & s & -1 \\ 0.4905 & 0 & 0 & s \end{vmatrix} = s^4 - 20.601s^2 = 0 \tag{P8.29.11}$$

Therefore,

$$a_1 = 0, \quad a_2 = -20.601, \quad a_3 = 0, \quad a_4 = 0 \tag{P8.29.12}$$

The desired characteristic polynomial is

$$(s - \mu_1)(s - \mu_2)(s - \mu_3)(s - \mu_4)$$
$$= (s+10)(s+10)(s+2-j3.464)(s+2+j3.464)$$
$$= (s^2 + 4s + 16)(s^2 + 20s + 100)$$
$$= s^4 + 24s^3 + 196s^2 + 720s + 1600 \tag{P8.29.13}$$

It follows that

$$b_1 = 24, \quad b_2 = 196, \quad b_3 = 720, \quad b_4 = 1600 \tag{P8.29.14}$$

We calculate the state-feedback gain matrix K.

It is

$$K = [b_4 - a_4 \quad b_3 - a_3 \quad b_2 - a_2 \quad b_1 - a_1]T^{-1} \tag{P8.29.15}$$

The matrix T is

$$T = SW \tag{P8.29.16}$$

where

$$W = \begin{bmatrix} a_3 & a_2 & a_1 & 0 \\ a_2 & a_1 & 1 & 0 \\ a_1 & 1 & 0 & 0 \\ 1 & 0 & 0 & 0 \end{bmatrix} = \begin{bmatrix} 0 & -20.601 & 0 & 1 \\ -20.601 & 0 & 1 & 0 \\ 0 & 1 & 0 & 0 \\ 1 & 0 & 0 & 0 \end{bmatrix} \tag{P8.29.17}$$

Hence,

$$T \underset{(P8.29.17)}{\overset{(P8.29.10)}{=}} \begin{bmatrix} 0 & 0 & -1 & 0 \\ 0 & 0 & 0 & -1 \\ -9.81 & 0 & 0.5 & 0 \\ 0 & -9.81 & 0 & 0.5 \end{bmatrix} \tag{P8.29.18}$$

and

$$T^{-1} = \begin{bmatrix} -\dfrac{0.5}{9.81} & 0 & -\dfrac{1}{9.81} & 0 \\ 0 & -\dfrac{0.5}{9.81} & 0 & -\dfrac{1}{9.81} \\ -1 & 0 & 0 & 0 \\ 0 & -1 & 0 & 0 \end{bmatrix} \tag{P8.29.19}$$

The state-feedback gain matrix K is

$$K = [1600 \quad 720 \quad 216.601 \quad 24] \begin{bmatrix} -\dfrac{0.5}{9.81} & 0 & -\dfrac{1}{9.81} & 0 \\ 0 & -\dfrac{0.5}{9.81} & 0 & -\dfrac{1}{9.81} \\ -1 & 0 & 0 & 0 \\ 0 & -1 & 0 & 0 \end{bmatrix} \Rightarrow$$

$$K = [-298.15 \quad -60.697 \quad -163.099 \quad -73.394] \tag{P8.29.20}$$

The next figure represents the inverted pendulum system with state feedback.

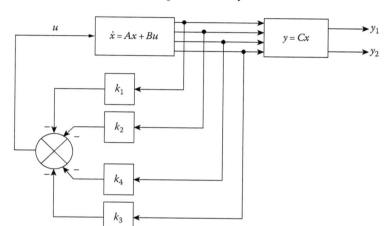

8.30 The state equations of a system are

$$\dot{x} = Ax + Bu$$

$$y = Cx$$

where

$$A = \begin{bmatrix} 0 & 1 & 0 & 0 \\ 0 & 0 & 1 & 0 \\ 0 & 0 & 0 & 0 \\ 0 & 0 & 0 & 1 \end{bmatrix}, \quad B = \begin{bmatrix} 0 & 0 \\ 0 & 0 \\ 1 & 0 \\ 0 & 1 \end{bmatrix}, \quad C = \begin{bmatrix} 1 & 0 & 0 & 0 \\ 0 & 0 & 0 & 1 \end{bmatrix}$$

Decouple the system with state feedback and examine the stability of the decoupled system.

Solution

We determine the integer numbers d_1 and d_2 as follows:

$$c_1 A^0 B = [0 \quad 0] \tag{P8.30.1}$$

$$c_1 A B = [0 \quad 0] \tag{P8.30.2}$$

$$c_1 A^2 B = [1 \quad 0 \quad 0 \quad 0] \begin{bmatrix} 0 & 0 & 1 & 0 \\ 0 & 0 & 0 & 0 \\ 0 & 0 & 0 & 0 \\ 0 & 0 & 0 & 1 \end{bmatrix} \begin{bmatrix} 0 & 0 \\ 0 & 0 \\ 1 & 0 \\ 0 & 1 \end{bmatrix} = [1 \quad 0] \tag{P8.30.3}$$

Thus $d_1 = 2$
  Moreover,

$$c_2 A^0 B = \begin{bmatrix} 0 & 0 & 0 & 1 \end{bmatrix} \begin{bmatrix} 0 & 0 \\ 0 & 0 \\ 1 & 0 \\ 0 & 1 \end{bmatrix} = \begin{bmatrix} 0 & 1 \end{bmatrix} \qquad \text{(P8.30.4)}$$

Hence $d_2 = 0$
  Then

$$B^* = \begin{bmatrix} c_1 A^{d_1} B \\ \cdots \\ c_2 A^{d_2} B \end{bmatrix} = \begin{bmatrix} 1 & 0 \\ 0 & 1 \end{bmatrix} \qquad \text{(P8.30.5)}$$

Since $\det(B^*) \neq 0$, the given system can be decoupled as follows:

$$A^* = \begin{bmatrix} c_1 A^{d_1+1} \\ \cdots \\ c_2 A^{d_2+1} \end{bmatrix} = \begin{bmatrix} 0 & 0 & 0 & 0 \\ 0 & 0 & 0 & 1 \end{bmatrix} \qquad \text{(P8.30.6)}$$

The state-feedback matrix is

$$K = -(B^*)^{-1} \cdot A^* = -\begin{bmatrix} 1 & 0 \\ 0 & 1 \end{bmatrix} \begin{bmatrix} 0 & 0 & 0 & 0 \\ 0 & 0 & 0 & 1 \end{bmatrix} = -\begin{bmatrix} 0 & 0 & 0 & 0 \\ 0 & 0 & 0 & 1 \end{bmatrix} \qquad \text{(P8.30.7)}$$

The decoupled system is written as

$$\left. \begin{aligned} \dot{x} &= (A + BK)x + B(B^*)^{-1}\omega \\ y &= Cx \end{aligned} \right\} \qquad \text{(P8.30.8)}$$

The transfer-function matrix of the closed-loop system is

$$G(s) = C(sI - A - BK)^{-1} B(B^*)^{-1} = \begin{bmatrix} \dfrac{1}{s^3} & 0 \\ 0 & \dfrac{1}{s} \end{bmatrix} \qquad \text{(P8.30.9)}$$

To determine if the system is stable we compute

$$|sI - (A + BK)| = \begin{vmatrix} s & 1 & 0 & 0 \\ 0 & s & 1 & 0 \\ 0 & 0 & s & 0 \\ 0 & 0 & 0 & s \end{vmatrix} = s^4 \qquad \text{(P8.30.10)}$$

Hence, the system is unstable.

8.31 Design a state observer for the observable system

$$(A,C) = \left(\begin{bmatrix} 0 & 1 & 0 \\ 0 & 0 & 1 \\ 0 & 2 & -1 \end{bmatrix}, \; [1 \quad 0 \quad 0] \right)$$

Solution

The characteristic polynomial of the system is

$$|sI - A| = \begin{vmatrix} s & -1 & 0 \\ 0 & s & -1 \\ 0 & -2 & s+1 \end{vmatrix} = s(s^2 + s - 2) \tag{P8.31.1}$$

We observe that not all eigenvalues (0, 1, and –2) are stable.
   Suppose that the desired polynomial is of the form

$$a_d(s) = s^3 + \gamma_2 s^2 + \gamma_1 s + c_0 \tag{P8.31.2}$$

We need a matrix K such that

$$|sI - (A - KC)| = a_d(s) \tag{P8.31.3}$$

Therefore,

$$|sI - (A - KC)| = \begin{vmatrix} s+k_0 & -1 & 0 \\ k_1 & s & -1 \\ k_2 & -2 & s+1 \end{vmatrix} \Rightarrow$$

$$|sI - (A - KC)| = s^3 + (1 + k_0)s^2 + (k_0 + k_1 - 2)s + (k_1 + k_2 - 2k_0) \tag{P8.31.4}$$

where

$$K = [k_0, \, k_1, \, k_2]^T \tag{P8.31.5}$$

Thus,

$$k_0 = \gamma_2 - 1$$
$$k_1 = \gamma_1 - \gamma_2 + 3 \tag{P8.31.6}$$
$$k_2 = \gamma_0 - \gamma_1 + 3\gamma_2 - 5$$

The state observer is

$$\dot{\tilde{x}} = A\tilde{x} + Bu + [\gamma_2 - 1 \quad \gamma_1 - \gamma_2 + 3 \quad \gamma_0 - \gamma_1 + 3\gamma_2 - 5]^T (y - \tilde{y})$$
$$\tilde{y} = C\tilde{x} \tag{P8.31.7}$$

8.32 Consider a system described by the following state-space equations:

$$\begin{bmatrix} \dot{x}_1 \\ \dot{x}_2 \end{bmatrix} = \begin{bmatrix} 0 & 20.6 \\ 1 & 0 \end{bmatrix} \begin{bmatrix} x_1 \\ x_2 \end{bmatrix} + \begin{bmatrix} 0 \\ 1 \end{bmatrix} u$$

$$y = \begin{bmatrix} 0 & 1 \end{bmatrix} \begin{bmatrix} x_1 \\ x_2 \end{bmatrix}$$

Design a full state observer with eigenvalues $\lambda_1 = -1.8 + j2.4$, $\lambda_2 = -1.8 - j2.4$.

Solution

First way: Direct substitution

We have

$$\tilde{e} = (A - KC)e \tag{P8.32.1}$$

where

$$K = [k_1 \quad k_2]^T \tag{P8.32.2}$$

We substitute K at the characteristic polynomial $|sI - A + KC|$ and we equal it with the desired polynomial:

$$(s - \lambda_1)(s - \lambda_2) = s^2 + \gamma_1 s + \gamma_0 \tag{P8.32.3}$$

We have

$$|sI - A + KC| = \begin{vmatrix} \begin{bmatrix} s & 0 \\ 0 & s \end{bmatrix} - \begin{bmatrix} 0 & 20.6 \\ 1 & 0 \end{bmatrix} + \begin{bmatrix} k_1 \\ k_2 \end{bmatrix} \begin{bmatrix} 0 & 1 \end{bmatrix} \end{vmatrix} = \begin{bmatrix} s & -20.6 + k_1 \\ -1 & s + k_2 \end{bmatrix} \Rightarrow$$

$$|sI - A + KC| = s^2 + k_2 s + k_1 - 20.6 \tag{P8.32.4}$$

The desired characteristic equation is

$$(s + 1.8 - j2.4)(s + 1.8 + j2.4) = s^2 + 3.6s + 9 = 0 \tag{P8.32.5}$$

From (P8.32.4) and (P8.32.5) we have

$$\begin{matrix} k_1 = 29.6 \\ k_2 = 3.6 \end{matrix} \quad \text{hence} \quad K = \begin{bmatrix} 29.6 \\ 3.6 \end{bmatrix} \tag{P8.32.6}$$

Second way: Ackermann's formula

We have

$$\Phi(s) = s^2 + 3.6s + 9 \tag{P8.32.7}$$

Thus,

$$\Phi(A) = A^2 + 3.6A + 9I = \begin{bmatrix} 29.6 & 74.16 \\ 3.6 & 29.6 \end{bmatrix} \tag{P8.32.8}$$

From relationship (8.82) we get

$$K = \Phi(A) \begin{bmatrix} 0 & 1 \\ 1 & 0 \end{bmatrix}^{-1} \begin{bmatrix} 0 \\ 1 \end{bmatrix} = \begin{bmatrix} 29.6 \\ 3.6 \end{bmatrix} \tag{P8.32.9}$$

Third way: Similarity transformation

The given system is in observable canonical form, thus $Q = I$.

The characteristic polynomial of the initial system is

$$|sI - A| = \begin{vmatrix} s & -20.6 \\ -1 & s \end{vmatrix} = s^2 - 20.6s = s^2 + a_1 s + a_0 = 0 \tag{P8.32.10}$$

Thus, $\alpha_1 = 0$, $\alpha_0 = -20.6$.

From the desired characteristic equation we have $\gamma_1 = 3.6$, $\gamma_0 = 9$. From relationship (8.86) we get

$$K = Q \begin{bmatrix} \gamma_0 - a_0 \\ \gamma_1 - a_1 \end{bmatrix} = \begin{bmatrix} 1 & 0 \\ 0 & 1 \end{bmatrix} \begin{bmatrix} 9 + 20.6 \\ 3.6 - 0 \end{bmatrix} = \begin{bmatrix} 29.6 \\ 3.6 \end{bmatrix} \tag{P8.32.11}$$

8.33 Consider a system described by the state-space equations

$$\begin{bmatrix} \dot{x}_1 \\ \dot{x}_2 \\ \dot{x}_3 \end{bmatrix} = \begin{bmatrix} 0 & 1 & 0 \\ 0 & 0 & 1 \\ -6 & -11 & -6 \end{bmatrix} \begin{bmatrix} x_1 \\ x_2 \\ x_3 \end{bmatrix} + \begin{bmatrix} 0 \\ 0 \\ 1 \end{bmatrix} u$$

$$y = \begin{bmatrix} 1 & 0 & 0 \end{bmatrix} \begin{bmatrix} x_1 \\ x_2 \\ x_3 \end{bmatrix}$$

Design a full state observer with eigenvalues $\lambda_1 = -2 + j3.464$, $\lambda_2 = -2 - j3.464$, $\lambda_3 = -5$. Use the similarity transformation.

Solution

The system is observable, thus we only have to compute vector K.

The characteristic polynomial of the given system is

$$|sI - A| = \begin{vmatrix} s & -1 & 0 \\ 0 & s & -1 \\ 6 & 11 & s+6 \end{vmatrix} \Rightarrow$$

$$|sI - A| = s^3 + 6s^2 + 11s + 6 = s^3 + a_2 s^2 + a_1 s + a_0 \tag{P8.33.1}$$

The desired characteristic equation is

$$(s-\lambda_1)(s-\lambda_2)(s-\lambda_3) = (s+2-j3.464)(s+2+j3.464)(s+5) = 0$$

$$\Rightarrow s^3 + 9s^2 + 36s + 80 = s^3 + \gamma_2 s^2 + \gamma_1 s + \gamma_0 = 0 \tag{P8.33.2}$$

Thus,

$$K = Q\begin{bmatrix} \gamma_0 - a_0 \\ \gamma_1 - a_1 \\ \gamma_2 - a_2 \end{bmatrix}, \tag{P8.33.3}$$

where

$$Q = (WN^*)^{-1} \tag{P8.33.4}$$

and

$$N^* = \begin{bmatrix} 1 & 0 & 0 \\ 0 & 1 & 0 \\ 0 & 0 & 1 \end{bmatrix}^* = \begin{bmatrix} 1 & 0 & 0 \\ 0 & 1 & 0 \\ 0 & 0 & 1 \end{bmatrix} \quad \text{and} \quad W = \begin{bmatrix} 11 & 6 & 1 \\ 6 & 1 & 0 \\ 1 & 0 & 0 \end{bmatrix} \tag{P8.33.5}$$

Therefore,

$$Q = \left\{ \begin{bmatrix} 1 & 0 & 0 \\ 0 & 1 & 0 \\ 0 & 0 & 1 \end{bmatrix} \begin{bmatrix} 11 & 6 & 1 \\ 6 & 1 & 0 \\ 1 & 0 & 0 \end{bmatrix} \right\}^{-1} = \begin{bmatrix} 0 & 0 & 1 \\ 0 & 1 & -6 \\ 1 & -6 & 25 \end{bmatrix} \tag{P8.33.6}$$

and

$$K = Q\begin{bmatrix} \gamma_0 - a_0 \\ \gamma_1 - a_1 \\ \gamma_2 - a_2 \end{bmatrix} = \begin{bmatrix} 0 & 0 & 1 \\ 0 & 1 & -6 \\ 1 & -6 & 25 \end{bmatrix} \begin{bmatrix} 80-6 \\ 36-11 \\ 9-6 \end{bmatrix} = \begin{bmatrix} 3 \\ 7 \\ 1 \end{bmatrix} \tag{P8.33.7}$$

The state observer is given by

$$\dot{\tilde{x}} = (A - KC)\tilde{x} + Bu + Ky \tag{P8.33.8}$$

More specifically,

$$\begin{bmatrix} \dot{\tilde{x}}_1 \\ \dot{\tilde{x}}_2 \\ \dot{\tilde{x}}_3 \end{bmatrix} = \begin{bmatrix} -3 & 1 & 0 \\ -7 & 0 & 1 \\ -5 & 11 & -6 \end{bmatrix} \begin{bmatrix} \tilde{x}_1 \\ \tilde{x}_2 \\ \tilde{x}_3 \end{bmatrix} + \begin{bmatrix} 0 \\ 0 \\ 1 \end{bmatrix} u + \begin{bmatrix} 3 \\ 7 \\ -1 \end{bmatrix} y \tag{P8.33.9}$$

9

Control-System Compensation

9.1 Introduction

System design has to follow certain system objectives so as to satisfy desired specifications. As sometimes a control system may not have the proper behavior, it is necessary to alternate or adjust the system design. This can occur by inserting a new component to the control system with a proper configuration. This procedure is known as **control-system compensation**. The compensating systems are called **compensators** or **controllers**. Some of the most usual compensation configurations are as follows:

Cascade compensation is the most common control-system compensation, in which the controller is connected in series with the control system.

Feedback compensation refers to a controller being connected with a feedback loop to the control system.

Mixed compensation is a controller connected both in series and in feedback with the control system.

The system design entails the determination of the proper input signal, so that the process under control (or plant) behaves as desired.

The proper choice of a controller depends on the structure of the plant and the design specifications. The latter are summarized in the following:

1. The system output must remain bounded for every bounded external input and for any initial condition of the system (bounded response).
2. The system output behaves as desired (command following).
3. The poles of the system are at the desired positions.
4. During the transient response, the rise time has to be small.
5. At the steady state, the error must be the smallest possible.
6. The closed-loop system must be robust, that is, it should not be sensitive to changes in the conditions and to the errors of the process.

The usual compensation methods are not capable of satisfying all specifications at the same time, because some are contradictory to each other. Two targets must be balanced: **performance** and **robustness**. A system performs satisfactorily if it provides a fast and smooth response to changes in the desired value and in the disturbances with small or no oscillations.

Robustness can be achieved by choosing a conservative calibration for the controller, but this usually results in low levels of performance. Hence, a conservative compensation reduces performance in favor of robustness.

Several methods have been developed for determining the proper compensating systems. The classic compensation methods are primarily graphic and simple to use, but usually they do not provide criteria for determining the necessary conditions that must be satisfied for a successful design.

The modern compensation methods are analytical and they provide the necessary conditions for solving the problem of designing an automatic control system. Several of the analytical methods rely on the minimization of a **cost criterion**.

The basic categories of controllers, depending on the control they provide, are

- *PID* controllers
- Lead-lag controllers

9.2 *PID* Controllers

A *PID* controller is basically a compensator connected in cascade with the closed-loop system. It controls the signal that governs the system by taking into account the error signal. Its name stems from proportional, integral, and derivative, which are the actions that each of its three parts performs. Depending on the properties of the system, several combinations of these actions can be used. In general, every closed-loop system with cascade compensation has the structure shown in Figure 9.1.

The transfer function of the plant is $G_p(s)$ and the transfer function of the *PID* controller is $G_c(s)$.

A *PID* controller is sometimes called **three-term controller and is** described by the following equation:

$$u(t) = K_p \left(e(t) + \frac{1}{T_i} \int_0^t e(t)d(t) + T_d \frac{de(t)}{dt} \right) \tag{9.1}$$

where

$u(t)$ is called control variable and is the output of the controller and input of the plant

$e(t) = y_{set\ point} - y(t) = y_{sp} - y(t)$ is the **error** between the real and the desired value of the output of the controlled process

The control variable is the sum of the three terms of relationship (9.1):

a. The **proportional term** P, which depends on the present error

b. The **integral term** I, which provides the accumulation of past errors

c. The **derivative term** D, which is prediction of future errors, based on current rate of change

Figure 9.2 represents the block diagram of a *PID* controller.

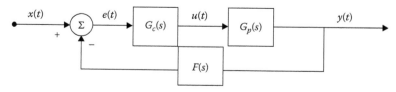

FIGURE 9.1
Closed-loop system with cascade compensation.

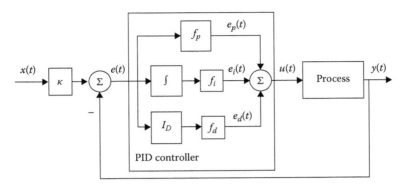

FIGURE 9.2
Block diagram of a *PID* controller.

The parameters of an industrial *PID* controller are

a. The proportional gain K_p
b. The integral action time T_i
c. The derivative action time T_d

The results of the action of every *P*, *I*, and *D* controller in a closed-loop system are summarized in Table 9.1.

9.2.1 *P* Controller

With a proper choice of the proportional gain K_p of the *P* controller, the steady-state oscillations are eliminated and the output signal of the closed-loop system is stabilized. However, the system tends to present steady-state error, which can be decreased but not eliminated.
The equations that describe the *P* controller are

$$u(t) = K_p e(t) \tag{9.2}$$

and

$$G_c(s) = \frac{U(s)}{E(s)} = K_p \tag{9.3}$$

TABLE 9.1

Effects of Controllers

Controller Type	Rise Time	Maximum Overshoot	Settling Time	Steady-State Error
P	Decrease	Increase	Small change	Decrease
I	Decrease	Increase	Increase	Elimination
D	Small change	Decrease	Decrease	Small change

If K_p increases,

- The oscillations increase, that is, the transient response behavior is not satisfactory
- The natural frequency ω_n of the system increases and the damping ratio J decreases

To better understand the action of the P controller, we can think of the mechanical equivalent of a spring. The gain K_p corresponds to the hardness of the spring. The greater the deviation from equilibrium, the greater is the return force. The return force increases with K_p and this shows that the proportional term of the controller makes the closed-loop system faster, while it decreases the steady-state error. However, for second- and higher-order systems (which are systems with time delay), large values of K_p result in oscillations of the response, and thus, the closed-loop system can become unstable. The proportional controller is not adequate to provide zero steady-state error as the system is usually subjected to external disturbances.

9.2.2 *I* Controller

With the addition of the integral term of the *I* controller, the steady-state error is eliminated but the settling time t_s of the system increases.

The mathematical expression of the *I* controller is

$$u(t) = K_i \int e(t)dt = \frac{K_p}{T_i} \int e(t)dt \tag{9.4}$$

and

$$G_c(s) = \frac{U(s)}{E(s)} = \frac{K_i}{s} \tag{9.5}$$

The use of an *I* controller inserts an additional pole to the system at $s = 0$. It follows that

- The system type is increased by one; thus, for every disturbance with a step-function form, the steady-state error becomes zero.
- The original root locus of the characteristic equation is pulled to the right; hence, the system is getting slower and the relative stability of the system is decreased.

9.2.3 *D* Controller

With the addition of the integral term of the *D* controller, the stability of the system increases and the behavior of the transient response improves.

The mathematical expression is

$$u(t) = K_d \frac{d(e(t))}{dt} = K_p T_d \frac{d(e(t))}{dt} \tag{9.6}$$

and

$$G_c(s) = \frac{U(s)}{E(s)} = K_d s \tag{9.7}$$

The use of a D controller adds one more zero to the system at $s = 0$. It follows that

- The system type decreases by one; thus, the possibility of nonzero steady-state error increases.
- The response time of the system response is getting smaller.

For the mechanical equivalent, the derivative term of the controller has the same effect as the linear (non-static) friction, which is proportional to velocity (the position's derivative) but with an opposite direction. In the same way that friction is used in suspensions for decreasing oscillations, the derivative term of the controller is used for increasing the damping of the closed-loop system.

The D controller has no influence on the steady-state error, since the derivative at the steady state is equal to zero. Nevertheless, in the case of regulatory control, the use of the derivative term can decrease substantially the maximum error, which emerges when the system is subjected to external disturbances. This is due to the fact that the derivative term recognizes the change in the error (from the inclination) long before the error is large enough to make the proportional action significant. The most important disadvantage in the use of the derivative term is the amplification of high-frequency measurement noise. In order to bypass this problem, a filter is often used in combination with the D controller.

9.2.4 *PI* Controller

A proportional-integral controller provides zero steady-state error and improves the response time of the system. The mathematical expression is

$$u(t) = K_p e(t) + \frac{K_p}{T_i} \int e(t) dt \tag{9.8}$$

$$G_c(s) = \frac{U(s)}{E(s)} = K_p + \frac{K_i}{s} = \frac{sK_p + K_i}{s} \overset{(9.8)}{=} K_p \left(1 + \frac{1}{T_i s}\right) \tag{9.9}$$

A *PI* controller inserts an additional zero to the system at $s = -K_i/K_p$ and a pole at $s = 0$. Hence,

- The order and the type of the system increase by one and the steady-state error is reduced.
- It is detrimental for the stability of the system because of the pole at $s = 0$. For some values of K_p and K_i the system may become unstable.

9.2.5 *PD* Controller

The proportional-derivative (*PD*) controller allows the operation of the plant with larger (compared to a P controller) gain values. It diminishes the oscillations and it decreases steady-state error.

The mathematical expression is

$$u(t) = K_p e(t) + K_d \frac{d(e(t))}{dt} \tag{9.10}$$

and

$$G_c(s) = \frac{U(s)}{E(s)} = K_p + K_d s \qquad (9.11)$$

A *PI* controller adds a zero to the system at $s = -K_p/K_d$. Therefore, the root locus is pulled to the left, and it is possible to decrease the oscillations and increase the proportional gain.

9.2.6 *PID* Controller

The proportional-integral-derivative controller with a proper choice of K_p, K_i and K_d combines a low rise time, a small maximum overshoot, a low settling time, and zero steady-state error.

The mathematical expression is

$$u(t) = K_p e(t) + K_i \int e(t)dt + K_d \frac{d(e(t))}{dt} \qquad (9.12)$$

and

$$G_c(s) = \frac{U(s)}{E(s)} = K_p + \frac{K_i}{s} + K_d s = \frac{K_d s^2 + K_p s + K_i}{s} = K_p\left(1 + \frac{1}{sT_i} + T_d s\right) \qquad (9.13)$$

9.2.7 *PID* Controller Forms

The algorithm of the industrial controller can be described by the following forms:

1. In the **parallel form,** the controller has the transfer function of the relationship (9.13), and its block diagram is depicted in the following figure:

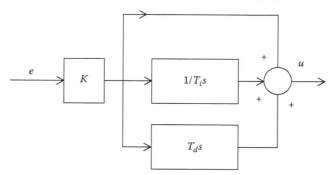

    The term K is the proportional gain K_p, which influences all terms of the controller, while the time constants T_i and T_d do not interact.

2. In the **serial form,** the controller has the transfer function of the relation (9.14). Its block diagram is shown in the following figure.

$$G_c(s) = \frac{U(s)}{E(s)} = K^*\left(1 + \frac{1}{T_i^* s}\right)(1 + T_d^* s) \qquad (9.14)$$

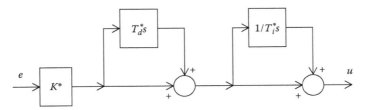

The disadvantage of the serial form is that the constants of the integral and the derivative terms interact with each other, so it is not often used.

3. In the analytic form, the transfer function of the controller is given by

$$G_c(s) = K_p + \frac{K_i}{s} + K_d s = \frac{k}{s}(s+z_1)(s+z_2)$$

(9.15)

This form is useful for analyzing the stability of the closed-loop system and has the advantage that the three terms are independent from each other.

9.3 Design of *PID* Controllers

Designing *PID* controllers means choosing the proper parameters (tuning), so that the plant operates within the desired specifications.

The methods for computing the parameters of a *PID* controller are empirical and computational. With the empirical methods we compute the parameters from the step response of the system, while most of the times the partial transfer functions of the closed-loop system are unknown. Designing a controller aims at a satisfactory maximum overshoot, a relatively fast response, and a small steady-state error.

9.3.1 First Ziegler–Nichols Method

Ziegler and Nichols (1940) developed two experimental methods. Their main advantage is that the analytical model of the system is not necessary.

The first Ziegler–Nichols method is based on the experimental determination of the step response of the system.

The method is characterized by two parameters that involve the response delay and the maximum speed of the step response, as shown in the following figure:

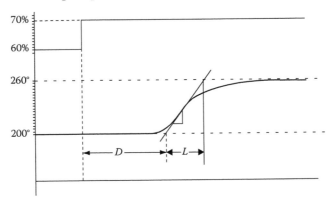

Control System Problems: Formulas, Solutions and Simulation Tools

TABLE 9.2

Proposed Values for K_p, K_i, and K_d in Terms of D and L

Controller	K_p	K_i	K_d
P	L/D	0	0
PI	$0.9\,L/D$	$3.33/D$	0
PID	$1.2\,L/D$	$0.5/D$	$0.5/D$

Table 9.2 provides the proposed values for the parameters K_p, K_i, and K_d of the PID controller based on the values of the dead time D and the time constant L determined by drawing a tangent line at the inflection point of the curve and then by finding the intersections of the tangent line with the time axis and the steady-state line.

This method is suitable for systems of type 0 with dead time. The transfer function is approximately of the form

$$G(s) = \frac{K}{L}\frac{e^{-Ds}}{s+1/L} \tag{9.16}$$

Based on the values of the Table 9.2, the transfer function of the controller is

$$G_c(s) = 0.6L\frac{(s+(1/D))^2}{s} \tag{9.17}$$

The relationship (9.17) results from the input–output relationship of the controller. It is

$$u(t) = 1.2\frac{L}{D}e(t) + 0.6\frac{L}{D^2}\int_0^t e(t)dt + 0.6L\frac{de(t)}{dt} \tag{9.18}$$

9.3.2 Second Ziegler–Nichols Method

The second Ziegler–Nichols method can be applied to any system type, in which the output can lead to undamped oscillations. The steps for determining the two parameters are as follows:

- We add only the proportional controller to the closed loop system. The gains K_i and K_d are set equal to zero.
- The external input of the process is set equal to a continuous time step signal.
- We increase the proportional gain K_p until the closed-loop system becomes critically stable, that is, until the step response presents undamped oscillations.
- We note the values of the period of oscillation T_{cr} and of the gain K_{cr} that provoked the oscillation.

Table 9.3 shows the suggested values of the parameters K_p, K_i, and K_d of the PID controller based on the values of K_{cr} and T_{cr}.

Based on the values of Table 9.3, the transfer function of the controller is

$$G_c(s) = 0.075K_{cr}T_{cr}\frac{(s+(4/T_{cr}))^2}{s} \tag{9.19}$$

TABLE 9.3

Proposed Values for K_p, K_i, and K_d in Terms of K_{cr} and T_{cr}

Controller	K_p	K_i	K_d
P	$0.5K_{cr}$	0	0
PI	$0.45K_{cr}$	$1.2/T_{cr}$	0
PID	$0.6K_{cr}$	$2/T_{cr}$	$0.125T_{cr}$

The input–output relationship of the controller is

$$u(t) = 0.6K_{cr}e(t) + 1.2\frac{K_{cr}}{T_{cr}}\int_0^t e(t)dt + 0.075K_{cr}T_{cr}\frac{de(t)}{dt} \tag{9.20}$$

The tuning methods proposed by Ziegler and Nichols provide closed-loop responses with a 50% damping. If proportional control is exclusively used, then the values calculated by the Table 9.3 give a gain margin equal to 2 for K_p, as $K_p = K_{cr}/2$. If integral control is added (i.e., the controller is PI), then K_p decreases from $0.5K_{cr}$ to $0.45K_{cr}$. The stabilizing effect of the derivative action (i.e., the controller is *PID*), allows an increase of K_p to $0.6K_{cr}$. However, this method has certain disadvantages:

1. The long duration of the experiments may lead to a detrimental productivity or to a low product quality.
2. In many applications the oscillation that emerges in the case of critical stability is not acceptable, since the system is pushed to the limits of its stable behavior. If external disturbances occur, then it is likely that the operation becomes unstable or even dangerous.
3. This method of finding the parameters cannot be used in systems with integral behavior because they are in general unstable for high and low values of K_p, despite the fact that they are stable for intermediate values.

9.3.3 Chien–Hrones–Reswick (CHR) Tuning Method

Chien, Hrones, and Reswick proposed a modification of the first Ziegler and Nichols method so that the final system has a greater damping ratio. This method achieves a faster response with a given maximum overshoot (0% or 20%). The method of Chien, Hrones, and Reswick differs from the Ziegler and Nichols method in that it uses three procedure parameters instead of two.

For tuning a controller based on the *CHR* method, the Ziegler–Nichols coefficients are selected from the Table 9.4.

The choice of the controller is based on Table 9.5.

The controller parameters for a nonperiodic response are tuned as shown in Table 9.6.

9.3.4 Cohen-Coon (CC) Tuning Method

This is another process reaction curve method. The response of the open-loop system to a unit-step input (without adding the controller to the system) is measured and approached with small straight lines for which we follow the same procedure followed in Ziegler–Nichols method. The main advantage is the simplicity of the procedure, but the disadvantage is that the responses present oscillations.

TABLE 9.4

Proposed Values for K_p, K_i, and K_d Based on the *CHR* Method

Controller	K_p	T_i	T_d
P	$\dfrac{1}{K}\left(0.35+\dfrac{1}{\mu}\right),\ \ \mu=\dfrac{D}{L}$	0	0
PI	$\dfrac{1}{K}\left(0.083+\dfrac{0.9}{\mu}\right)$	$\dfrac{3.33+0.31\mu}{1+2.2\mu}D$	0
PD	$\dfrac{1}{K}\left(0.16+\dfrac{1.24}{\mu}\right)$	0	$\dfrac{0.27-0.088\mu}{1+0.13\mu}D$
PID	$\dfrac{1}{K}\left(0.25+\dfrac{1.35}{\mu}\right)$	$\dfrac{2.5+0.46\mu}{1+0.61\mu}D$	$\dfrac{0.37}{1+0.19\mu}D$

TABLE 9.5

Controller Selection Based
on the *CHR* Method

Controller Type	$R = L/D$
P	$R > 10$
PI	$7.5 < R < 10$
PID	$3 < R < 7.5$

TABLE 9.6

Controller Parameters for a
Nonperiodic Response

Controller	K_p	T_i	T_d
P	$0.3R/K$	0	0
PI	$0.35R/K$	$1.2L$	0
PID	$0.6R/K$	L	$0.5D$

The Cohen-Coon method is based on the placement of dominant poles. Given that

$$G_p(s) = K\frac{e^{-t_d s}}{\tau s+1} \tag{9.21}$$

the main aim of this method is to restrain the load disturbances at the output of the system with the proper placement of the poles of the closed-loop system so as to have a damping ratio of 0.25. This method leads to the minimization of the steady-state error due to load disturbances for two- and three-term controllers.

The parameters of the *PID* controller are computed from the following formulas:

$$K_p = \frac{1}{K}\frac{\tau}{t_d}\left(\frac{4}{3}+\frac{t_d}{4\tau}\right)$$

$$T_i = t_d\frac{32+(6t_d/\tau)}{13+(8t_d/\tau)} \tag{9.22}$$

$$T_d = \frac{4t_d}{11+(2t_d/\tau)}$$

9.4 Compensation with Lead-Lag Controllers

9.4.1 Phase-Lead Compensation

Phase-lead compensation has the same behavior as the *PD* controller and it is applied to control systems with satisfactory steady-state characteristics, but whose transient response needs improvement.

The transfer function of a first-order phase-lead controller is

$$G_{lead}(s) = K \frac{(sT_1 + 1)}{(sT_2 + 1)} = \frac{1 + aTs}{a(1 + Ts)} = \frac{s + (1/aT)}{s + (1/T)} = \frac{s + z}{s + p}, \quad a > 1 \tag{9.23}$$

The phase-lead controller adds a zero to the transfer function of the system at $s = -1/aT$ and a pole at $s = -1/T$. It follows that the contribution of positive phase to the phase of the original system leads to an increase in the phase margin of the system, thus making it more stable.

In the frequency domain, the phase-lead controllers improve the transfer function of the system for high frequencies. In the time domain, they improve the transient response by decreasing the rise time and the maximum overshoot. Moreover, they improve the steady-state response by partially reducing the steady-state error. The disadvantages of compensating with the use of a phase-lead controller are (a) the increase of the bandwidth, which makes the compensated system vulnerable to noise, and (b) the need for additional amplifier gain.

The maximum value of the phase lead appears in frequency ω_m, which is between the frequency of the pole $s = -1/T$ and the zero $s = -1/aT$. For the frequency ω_m it holds that

$$\omega_m = \sqrt{zp} = \frac{1}{T\sqrt{a}} \tag{9.24}$$

The frequency response of the phase-lead circuit is

$$G_{lead}(j\omega) = \frac{j\omega + z}{j\omega + p} = \frac{1}{a} \frac{(1 + aTj\omega)}{(1 + Tj\omega)} \tag{9.25}$$

From the relationship (9.25) the phase of the frequency response is

$$\varphi(\omega) = \tan^{-1} a\omega T - \tan^{-1} \omega T \tag{9.26}$$

We substitute (9.26) to (9.24) and get the relationship for the maximum argument φ_m that corresponds to the phase lead. From this, the parameter a can be calculated. The relationship for the maximum argument is

$$\tan \varphi_m = \frac{a - 1}{2\sqrt{a}} \quad \text{or} \quad \sin \varphi_m = \frac{a - 1}{a + 1} \tag{9.27}$$

The design of the phase-lead controller is based on the root locus, but it can be also specified with the use of Bode diagrams.

The steps for designing a phase-lead controller are introduced, first with the use of the root locus, and then with the use of Bode plots.

9.4.1.1 Design with the Root Locus

In general, the zero of the controller is placed at the larger real pole of the system (excluding $s = 0$). In this way, the response is greatly improved. If the system is of type 0, then the zero of the controller is placed at the position of the second larger real pole. If there are certain specifications for the transient state, these can be determined by the poles of the second-order system (determination of the dominant poles). The root locus must traverse the desired poles.

The design steps are the following:

1. From the desired specifications we determine the position of the dominant poles of the closed-loop system.

2. We plot the root locus of the uncompensated system and we examine whether the positions of the desired roots belong to the locus.

3. If necessary, we connect a lead controller in series with the original system; hence, the zero of the controller is placed right below the desired roots or on the left side of the first two real poles.

4. We find the position of the poles so that the desired angle of the pole is on the root locus and thus equal to 180°.

5. The total gain of the system for the desired root is calculated, and the steady-state error is found.

6. After designing the controller, we examine if the specifications are met. If they are not met, then we repeat the design procedure by displacing the pole and the zero positions of the controller. If a large steady-state error emerges, then an additional phase-lag controller must be connected in series with the designed phase-lead controller.

9.4.1.2 Designing with the Use of Bode Plots

The frequency response of the phase-lead controller is added to the frequency response of the original system. The new transfer function is $G_{lead}(j\omega)G_p(j\omega)F(j\omega)$.

The steps for designing with the use of Bode diagrams are as follows:

1. We plot the Bode diagram of $G_p(j\omega)F(j\omega)$ for the uncompensated system and we compute the phase margin so that the specification for the steady-state error is satisfied.

2. We determine the phase lead φ_m, paying attention to the phase margins.

3. We apply the relationship $\sin \varphi_m = a - 1/a + 1$ in order to find a.

4. We compute the frequency for which the magnitude curve of the uncompensated system is equal to $-10 \log a$ dB. The phase-lead controller presents a gain of $10 \log a$ dB at the frequency ω_m. The new critical frequency that corresponds to 0 dB is also calculated.

5. We compute the pole from the relationship $p = \omega_m \sqrt{a}$ and the zero from the relationship $z = p/a$.

6. We draw the Bode plot of $G_{lead}(j\omega)G_p(j\omega)F(j\omega)$ for the compensated system and we examine the new phase margin.

7. We repeat the designing steps until we arrive at a satisfactory result.

9.4.2 Phase-Lag Compensation

Phase-lag compensation presents the same behavior as a *PI* controller. It is applied to control systems with satisfactory transient response characteristics, but with non-satisfactory steady-state response characteristics.

The transfer function of a first-order phase-lag controller is

$$G_{lag}(s) = \frac{1+Ts}{1+aTs} = \frac{1}{a}\frac{s+(1/T)}{s+(1/aT)} = \frac{1}{a}\frac{s+z}{s+p}, \quad a > 1 \tag{9.28}$$

The phase-lag controller adds a zero to the system's transfer function at $s = -1/T$ and a pole at $s = -1/aT$. From the relationship (9.28), we observe that the initial gain of the open loop is decreased by $1/a$. In order to leave unchanged the dominant poles, the static gain must increase by a times. This results in a decrease of the steady-state error by a times. The mathematical expression is

$$K' = a\frac{\displaystyle\prod_{i=1}^{n}(s_d + p_i)}{\displaystyle\prod_{j=1}^{m}(s_d + z_j)} \tag{9.29}$$

The positions of the pole and zero of the controller must be very close so that their contributions at the transient response are cancelled by each other. They must also be close to the origin of the complex *s*-plane.

The frequency response of the phase-lag network is

$$G_{lag}(j\omega) = \frac{j\omega+z}{j\omega+p} = \frac{1+Tj\omega}{1+aTj\omega} \tag{9.30}$$

The maximum value of phase delay appears at the frequency for which $\omega_m = \sqrt{zp}$. The lag network is not useful in providing lag angle, which results in decreased relative stability.

In the frequency domain, the phase-lag controllers improve the loop transfer function of the system for low frequencies. In the time domain, they reduce the steady-state error but they increase the rise time and thus the transient response delay.

A description of the steps needed for designing phase-lag controllers with the use of the root locus, and with the use of Bode plots, follows.

9.4.2.1 *Designing with the Use of the Root Locus*

1. We plot the root locus of the uncompensated system.
2. From the desired specifications, we determine the positions of the dominant poles of the closed-loop system.
3. We compute open-loop static gain, from the relationship

$$|GH(s)| = K\frac{\displaystyle\prod_{j=1}^{m}|s+z_j|}{\displaystyle\prod_{i=1}^{n}|s+p_i|} = 1 \tag{9.31}$$

4. We examine the steady-state error for the K that was computed previously and if necessary we add a phase-lag controller.

5. We arbitrarily choose the zero of the controller at $s = -1/T$, (usually $s = -0.1$). We then choose the pole of the controller at $s = -1/aT$, where a is calculated from the general relationship:

$$a = \frac{Steady\ state\ error}{Initial\ steady\ state\ error} \qquad (9.32)$$

6. The phase that corresponds to the dominant pole is added by the controller according to the relationship

$$Arg(GH(s)) = Arg\left(\prod_{j=1}^{m}|s+z_j|\right) - Arg\left(\prod_{i=1}^{n}|s+p_i|\right) = \begin{cases} \pm(2k+1)\pi, & k>0 \\ \pm2k\pi, & k<0 \end{cases} \qquad (9.33)$$

7. If the phase does not satisfy the specifications, we have to choose again the zero of the controller and repeat the procedure.

9.4.2.2 Designing with the Use of Bode Plot

The steps are the following:

1. We design the Bode plot of $G_p(j\omega)F(j\omega)$ of the uncompensated system and for such a gain value that fulfills the specification for the error constant.

2. We determine the phase margin of the original system and if it is not satisfactory then we proceed as follows.

3. We compute the frequency for which the phase margin satisfies the specifications. This is the frequency for which the magnitude is 0 dB. In the determination of the new critical frequency we have to consider that the contribution of the phase-lag network must be less than 5°.

4. In order to secure that the lag controller contributes only a 5° additional phase lag, we place the zero of the controller a decade below the new critical frequency.

5. We compute the needed attenuation in the magnitude such that the magnitude is 0 dB at the critical frequency.

6. We compute a from the relationship $-20\log a = 20\log \omega T - 20\log \omega aT = 20\log|G_{lag}(j\omega)|_{\omega \gg \frac{1}{T}}$.

7. We compute the pole at $s = -1/aT$ from the relationship $G_{lag}(s) = 1 + Ts/1 + aTs$.

9.4.3 Lag-Lead Compensation

The **lag-lead compensation** is applied to control systems, which need improvement in both transient and steady-state response. The name stems from the fact that the phase of the output signal presents a phase lag at low frequencies and a phase lead at high frequencies. The lag-lead controllers combine the characteristics of the lag and the lead controllers. They increase, however, the order of the system by two. This renders the study of the system rather complicated.

The design methodology of a phase-lead controller with the use of root locus is the following:

- From the desired specifications, we determine the desired position of the dominant poles of the closed-loop system.
- We compute the phase of the open-loop transfer function of the original system at one of the dominant poles s_d. Based on the condition $Arg(GH(s)) = Arg\left(\prod_{j=1}^{m}|s+z_j|\right) - Arg\left(\prod_{i=1}^{n}|s+p_i|\right) = \begin{cases} \pm(2k+1)\pi, & k>0 \\ \pm 2k\pi, & k<0 \end{cases}$, we calculate the necessary phase that the controller contributes.
- We determine the static gain K of the controlled process, so that the specification for the steady-state error is satisfied.
- We determine the pole and the zero of the phase-lead component, so that the magnitude and phase conditions for the partially compensated system are fulfilled for $s = s_d$. We have to ignore the phase-lag component of the controller. The mathematical expressions are

$$Arg(s_d + (1/T_1)) - Arg(s_d + (1/aT_1)) = \pi - Arg(GH(s_d)) \qquad (9.34)$$

$$|GH(s_d)| = \left|\frac{s_d + (1/T_1)}{s_d + (1/aT_1)}\right| = 1 \qquad (9.35)$$

- We choose the pole and the zero of the phase-lag component, so that the phase that corresponds to the dominant pole and is added to the phase-lag controller is as small as possible. It must usually be less than 3°. The magnitude condition must also hold:

$$|GH(s_d)| = \left|\frac{s_d + (1/T_2)}{s_d + (1/aT_2)}\right| = 1 \qquad (9.36)$$

The design methodology for the phase-lead controller with the use of Bode diagrams is based on the computation of the parameters of the phase-lag and phase-lead controllers described in the previous paragraphs:

1. First, we compute the phase-lag controller so that the steady-state requirements are satisfied.
2. We calculate the phase-lead controller, so that the dynamic characteristics of the plant are fulfilled.

9.4.4 Lead-Lag Compensation

The transfer function of a lead-lag controller has been analyzed in Chapter 6. It is of the form:

$$G_{lead-lag}(s) = \frac{(1+aT_1s)(1+bT_2s)}{(1+T_1s)(1+T_2s)}, \quad a>1, \ b<1 \tag{9.37}$$

The parameter a is defined in a way that in the new critical frequency (that corresponds to magnitude of $0\,$dB) we have an additional phase lead. The parameter b is determined so that there is attenuation for low frequencies.

Formulas

TABLE F9.1

PID Controller

Controller Type	Transfer Function	Parameters
P	$G_c(s) = \dfrac{U(s)}{E(s)} = K_p$	K_p: proportional gain
I	$G_c(s) = \dfrac{U(s)}{E(s)} = \dfrac{K_i}{s}$	K_i: integral gain
D	$G_c(s) = \dfrac{U(s)}{E(s)} = K_d s$	K_d: derivative gain
PI	$G_c(s) = \dfrac{U(s)}{E(s)} = K_p + \dfrac{K_i}{s} = \dfrac{sK_p + K_i}{s} = K_p\left(1 + \dfrac{1}{T_i s}\right)$	T_i: integration time
PD	$G_c(s) = \dfrac{U(s)}{E(s)} = K_p + K_d s$	
PID	$G_c(s) = \dfrac{U(s)}{E(s)} = K_p + \dfrac{K_i}{s} + K_d s = K_p\left(1 + \dfrac{1}{sT_i} + T_d s\right)$	T_d: differentiation time

TABLE F9.2

PID Controllers and Their Effect on the Behavior of a Closed-Loop System

Controller Type	Rise Time	Maximum Overshoot	Settling Time	Steady-State Error
P	Decrease	Increase	Small change	Decrease
I	Decrease	Increase	Increase	Elimination
D	Small change	Decrease	Decrease	Small change

TABLE F9.3

Methods for Designing *PID* Controllers (Tuning)

		Proper Controller Parameters		
Method	**Controller**	K_p	K_i	K_d
First Ziegler–Nichols method	P	L/D	0	0
	PI	$0.9\,L/D$	$3.33/D$	0
	PID	$1.2\,L/D$	$0.5/D$	$0.5/D$
Second Ziegler–Nichols method	P	$0.5K_{cr}$	0	0
	PI	$0.45K_{cr}$	$1.2/T_{cr}$	0
	PID	$0.6K_{cr}$	$2/T_{cr}$	$0.125T_{cr}$

TABLE F9.4

Methods for Designing *PID* Controllers (Tuning)

		Proper Controller Parameters		
Method	**Controller**	K_p	T_i	T_d
Chien–Hrones–Reswick tuning method	P	$0.3R/K$	0	0
	PI	$0.35R/K$	$1.2L$	0
	PID	$0.6R/K$	L	$0.5D$

Controller	$R = L/D$
P	$R > 10$
PI	$7.5 < R < 10$
PID	$3 < R < 7.5$

Cohen-Coon tuning method

$$K_p = \frac{1}{K}\frac{\tau}{t_d}\left(\frac{4}{3}+\frac{t_d}{4\tau}\right)$$

$$T_i = t_d \frac{32+(6t_d/\tau)}{13+(8t_d/\tau)}$$

$$T_d = \frac{4t_d}{11+(2t_d/\tau)}$$

$$G_p(s) = K\frac{e^{-t_d s}}{\tau s + 1}$$

TABLE F9.5

Lead–Lag Controllers

Lead: phase lead

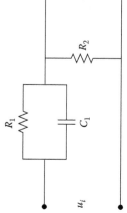

$$G_c(s) = K\frac{(sT_1+1)}{(sT_2+1)} = K_c\frac{1+aTs}{a(1+Ts)} = K_c\frac{s+(1/aT)}{s+(1/T)}, \quad a>1$$

$$G_c(s) = \frac{R_2}{R_2+(R_1/1+sR_1C_1)} = \frac{E_o(s)}{E_i(s)}$$

K_c: static gain of the amplifier

T: time constant

$$T = \frac{R_1R_2C_1}{R_1+R_2}$$

$$a = \frac{R_1+R_2}{R_2}$$

Equations for tuning with phase lead

$$\theta = \tan^{-1}\frac{\omega}{1/\alpha T} - \tan^{-1}\frac{\omega}{1/T}$$

$$\frac{d\theta}{d\omega} = 0 \Rightarrow \omega_m = \sqrt{\frac{1}{\alpha T}\cdot\frac{1}{T}}$$

$$\theta_m = \tan^{-1}\sqrt{\frac{1/T}{1/\alpha T}} - \tan^{-1}\sqrt{\frac{1/\alpha T}{1/T}}$$

$$\tan\theta_m = \frac{a-1}{2\sqrt{a}}, \quad \sin\theta_m = \frac{a-1}{a+1}$$

θ: phase lead

ω_m: maximum frequency

θ_m: maximum phase

Lag: phase lag

$$G_c(s) = \frac{1+Ts}{1+aTs} = \frac{1}{a}\frac{s+(1/T)}{s+(1/aT)}, \quad a>1$$

$$G_c(s) = \frac{R_2+(1/sC_1)}{R_1+R_2+(1/sC_1)} = \frac{E_o(s)}{E_i(s)}$$

K_c: static gain of the amplifier

T: time constant

$$T = R_2C_1$$

$$a = \frac{R_1+R_2}{R_2}$$

Lag lead: phase lag lead

$$G_c(s) = \left(\frac{s+1/T_1}{s+1/bT_1}\right)\left(\frac{s+1/T_2}{s+1/\alpha T_2}\right)$$

$$b > 1, \quad \alpha < 1.$$

$$G_c(s) = \frac{(s+(1/R_1C_1))(s+(1/R_2C_2))}{s^2+(1/R_1C_1+1/R_2C_2+1/R_2C_1)s+(1/R_1R_2C_1C_2)}$$

$$R_1C_1 = T_1, \quad R_2C_2 = T_2$$
$$R_1R_2C_1C_2 = \alpha b T_1 T_2$$
$$\alpha b = 1$$

Lead lag: phase lead lag

$$G_{lead-lag}(s) = \frac{(1+aT_1s)(1+bT_2s)}{(1+T_1s)(1+T_2s)}$$
$$a > 1, \quad b < 1$$

T_1: time constant of the lead network

T_2: time constant of the lag network

Problems

9.1 Consider a first-order system with the following transfer function:

$$G_p(s) = \frac{K}{1+sT_1}$$

Assuming unity feedback, express the parameters of the *PI* controller in terms of the system's characteristics.

Solution

The transfer function of the PI controller is given in relationship (9.9). The transfer function of a closed loop system connected in series with a PI controller is

$$H(s) = \frac{G_p(s)G_c(s)}{1+G_p(s)G_c(s)F(s)} \tag{P9.1.1}$$

The characteristic equation of the system is

$$1+G_p(s)G_c(s)F(s) = 0 \Rightarrow 1+\frac{K}{1+sT_1}K_p\left(1+\frac{1}{T_i s}\right)F(s) = 0 \tag{P9.1.2}$$

For unity feedback ($F(s) = 1$), relationship (P9.1.2) becomes

$$s^2 + \frac{1+KK_p}{T_1}s + \frac{KK_p}{T_1 T_i} = 0 \tag{P9.1.3}$$

The closed-loop system is of second order. The poles of the closed-loop system are the roots of the characteristic equation. They are determined by the coefficients of the characteristic polynomial. In terms of the natural frequency ω_n and the damping ratio J, the characteristic equation is written as

$$s^2 + 2J\omega_n s + \omega_n^2 = 0 \tag{P9.1.4}$$

By equaling the coefficients of the polynomials (P9.1.3) and (P9.1.4) it yields that

$$K_p = \frac{2J\omega_n T_1 - 1}{K}, \quad T_i = \frac{2J\omega_n T_1 - 1}{\omega_n^2 T_1} \tag{P9.1.5}$$

From (P9.1.5), we conclude that

- The gain K_p of the controller is positive if $\omega_n > 1/2JT_1$.
- If $\omega_n \gg 1/2JT_1$, the integration time constant becomes $T_i = 2J/\omega_n$, which means that it is independent from the time constant of the plant.

9.2 Consider a second-order control system with the following transfer function:

$$G_p(s) = \frac{K}{(1+sT_1)(1+sT_2)}$$

Assuming unity feedback, express the parameters of the *PID* controller in terms of the system's characteristics.

Solution

The characteristic equation of the unity feedback control system with a *PID* controller is

$$1+G_p(s)G_c(s)F(s) = 0 \Rightarrow 1+K_p\left(1+\frac{1}{sT_i}+T_ds\right)\frac{K}{(1+sT_1)(1+sT_2)} = 0 \qquad \text{(P9.2.1)}$$

or

$$s^3 + \left(\frac{1}{T_1}+\frac{1}{T_2}+\frac{KK_pT_d}{T_1T_2}\right)s^2 + \frac{1+KK_p}{T_1T_2}s + \frac{KK_p}{T_1T_2T_i} = 0 \qquad \text{(P9.2.2)}$$

The closed-loop system is of third order. The desired characteristic equation is

$$(s+T\omega_n)(s^2+2J\omega_ns+\omega_n^2) = s^3 +(T+2J)\omega_ns^2 +(1+2TJ)(\omega_n)^2s+(\omega_n)^3T = 0 \quad \text{(P9.2.3)}$$

By equaling the coefficients of the polynomials (P9.2.2) and (P9.2.3), we get

$$K_p = \frac{T_1T_2\omega_n^2(1+2TJ)-1}{K}$$

$$T_i = \frac{T_1T_2\omega_n^2(1+2TJ)-1}{\omega_n^3T_1T_2T}$$

$$T_d = \frac{T_1T_2\omega_n(T+2J)-T_1-T_2}{\omega_n^2T_1T_2(1+2TJ)-1}$$

9.3 Consider a system with the following transfer function:

$$G(s) = \frac{1}{(s-a)^2}, \quad a > 0$$

a. Demonstrate that the system cannot be stabilized with use of a *PI* controller.

b. Show that the system cannot be stabilized with use of a *PID* controller.

Solution

a. The *PI* controller can be written as

$$G_c(s) = K+\frac{K_i}{s} = \frac{Ks+K_i}{s} = \frac{K(s+b)}{s} \qquad \text{(P9.3.1)}$$

The characteristic polynomial of the compensated system is

$$P(s) = s(s-a)^2 + K(s+b) = s^3 - 2as^2 + (a^2+K)s + Kb \qquad \text{(P9.3.2)}$$

Because $a > 0$ and $K > 0$, the coefficients s^3 and s^2 are of the opposite sign and the closed-loop control system cannot be stable.

b. The *PID* controller can be written as

$$G_c(s) = K_a + \frac{K_i}{s} + K_d s = \frac{K_a s + K_i + K_d s^2}{s} = \frac{K(s^2 + bs + c)}{s} \qquad \text{(P9.3.3)}$$

The characteristic polynomial of the compensated system is

$$P(s) = s(s-a)^2 + K((s)^2 + bs + c) = s^3 + (K-2a)s^2 + (a^2+Kb)s + Kc \qquad \text{(P9.3.4a)}$$

In order to examine the stability of the compensated system we use Routh's tabulation:

$$
\begin{array}{ccc}
s^3 & 1 & a^2 + Kb \\[4pt]
s^2 & K - 2a & Kc \\[6pt]
s & \dfrac{(K-2a)(a^2+Kb) - Kc}{K-2a} & \\[10pt]
s^0 & Kc &
\end{array}
$$

The closed-loop system is stable if

$$K - 2a > 0 \qquad \text{(P9.3.4b)}$$

$$Kc > 0 \qquad \text{(P9.3.5)}$$

$$(K-2a)(a^2+Kb) - Kc = K^2 b + K(a^2 - 2ab - c) - 2a^3 > 0 \qquad \text{(P9.3.6)}$$

We suppose that $K > 0$, $c > 0$, and from (P9.3.4a and P9.3.4b), we get

$$K > 2a \qquad \text{(P9.3.7)}$$

Assuming that $b > 0$, the relationship (6) is fulfilled if K is located outside the interval between the roots of the polynomial

$$f(x) = x^2 b + x(a^2 - 2ab - c) - 2a^3 \qquad \text{(P9.3.8)}$$

The determinant of the polynomial is greater than zero, thus the polynomial has two real roots. As the product of the roots is negative, the roots are of opposite sign. For

$$K > \max\left\{ 2a, \frac{-(a^2 - 2ab - c) + \sqrt{(a^2 - 2ab - c)^2 + 8a^3 b}}{2b} \right\} \qquad \text{(P9.3.9)}$$

the compensated system is stable.

9.4 A system of three tanks with the same volume of liquid V is illustrated in the following figure. By q we denote the water supply, T_e is the input temperature, and h is the power of the thermal medium. The temperatures of the units are T_1, T_2, T_3 and they coincide with the relevant output temperatures, due to perfect stirring. The system is regulated by a P controller. Design the block diagram and compute the critical gain of the system.

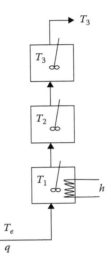

Solution
The differential equation that describes the dynamic behavior at the first unit is

$$V\rho c_p \frac{dT_1}{dt} = q\rho c_p(T_e - T_1) + h \tag{P9.4.1}$$

The equation is nonlinear; thus, we cannot apply Laplace transform. We linearize around the equilibrium point $(T_{1,s},\ q_s,\ T_{e,s},\ h_s)$ and we get the relevant linearized differential equation

$$\frac{V}{q_s}\frac{d\bar{T}_1}{dt} + \bar{T}_1 = \frac{T_{e,s} - T_{1,s}}{q_s}\bar{q} + \bar{T}_e + \frac{\bar{h}}{q_s\rho c_p} \tag{P9.4.2}$$

where the deviation variables $\bar{T}_1, \bar{q}, \bar{T}_e, \bar{h}$ are used.

If we set $\tau = V/q_s$, then the transfer functions between the input and the output variables of the first unit are

$$\frac{\bar{T}_1(s)}{\bar{q}(s)} = \frac{T_{e,s} - T_{1,s}}{q_s}\frac{1}{\tau s + 1}$$

$$\frac{\bar{T}_1(s)}{\bar{T}_e(s)} = \frac{1}{\tau s + 1} \tag{P9.4.3}$$

$$G_1(s) = \frac{\bar{T}_1(s)}{\bar{h}(s)} = \frac{1}{q_s\rho c_p}\frac{1}{\tau s + 1}$$

In the same way, the differential equations for the two other units are

$$V\rho c_p \frac{dT_2}{dt} = q\rho c_p(T_1 - T_2)$$

$$V\rho c_p \frac{dT_3}{dt} = q\rho c_p(T_2 - T_3)$$

(P9.4.4)

By linearization, we get

$$G_2(s) = \frac{\overline{T}_2(s)}{\overline{T}_1(s)} = \frac{1}{\tau s + 1}$$

$$G_3(s) = \frac{\overline{T}_3(s)}{\overline{T}_2(s)} = \frac{1}{\tau s + 1}$$

(P9.4.5)

We plot the block diagram of the closed-loop system:

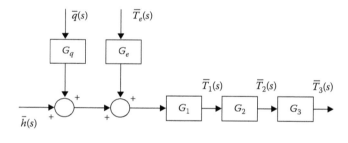

The block diagram must represent properly the transfer functions given in (P9.4.5). Hence,

$$\frac{\overline{T}_1(s)}{\overline{q}(s)} = G_1(s)G_q(s) = \frac{T_{e,s} - T_{1,s}}{q_s} \frac{1}{\tau s + 1}$$

(P9.4.6)

$$\frac{\overline{T}_1(s)}{\overline{T}_e(s)} = G_1(s)G_e(s) = \frac{1}{\tau s + 1}$$

(P9.4.7)

It yields that

$$G_q(s) = \rho c_p (T_{e,s} - T_{1,s})$$

(P9.4.8)

$$G_e(s) = q_s \rho c_p$$

(P9.4.9)

If we now use feedback and a P controller, then the closed-loop block diagram becomes

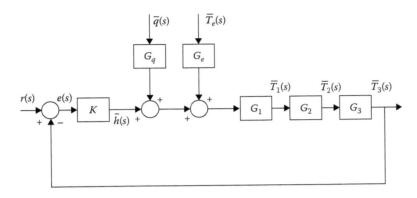

It holds that

$$\frac{\bar{T}_1(s)}{\bar{q}(s)} = \frac{G_1(s)G_q(s)}{1+KG_1(s)G_2(s)G_3(s)} \tag{P9.4.10}$$

and

$$\frac{\bar{T}_1(s)}{\bar{T}_e(s)} = \frac{G_1(s)G_e(s)}{1+KG_1(s)G_2(s)G_3(s)} \tag{P9.4.11}$$

These transfer functions are different than the ones used to describe the open-loop control system.

On the contrary, the transfer function $\bar{T}_1(s)/\bar{h}(s)$ does not change, as both of its signals are internal:

$$\frac{\bar{T}_1(s)}{\bar{h}(s)} = G_1(s) = \frac{1}{q_s\rho c_p}\frac{1}{\tau s+1} \tag{P9.4.12}$$

In order to compute the critical gain, we observe from relationships (P9.4.10) and (P9.4.11) that the characteristic equation of the closed-loop system is

$$1+KG_1(s)G_2(s)G_3(s) = 0 \tag{P9.4.13}$$

Hence,

$$\tau^3 s^3 + 3\tau^2 s^2 + 3\tau s + 1 + \frac{K}{q_s\rho c_p} = 0 \tag{P9.4.14}$$

By applying the Hurwitz criterion, we get

$$\Delta 3 = \begin{bmatrix} 3\tau^2 & 1+\dfrac{K}{q_s\rho c_p} & 0 \\ \tau^3 & 3\tau & 0 \\ 0 & 3\tau^2 & 1+\dfrac{K}{q_s\rho c_p} \end{bmatrix} \tag{P9.4.15}$$

where all the minor determinants must be positive.
   The determinant of the 1×1 matrix is obviously positive.
   For the 2×2 matrix, we have

$$9\tau^3 - \tau^3 - \frac{K}{q_s\rho c_p}\tau^3 > 0 \tag{P9.4.16}$$

The determinant of the 3×3 matrix is positive if the determinant of the 2×2 matrix is positive.
   From the inequality (P9.4.16), it yields that the closed-loop system is stable if

$$K < 8q_s\rho c_p \tag{P9.4.17}$$

Hence, the critical gain for which the system is critically stable is $K = 8q_s\rho c_p$. This depends on the position of the equilibrium point at which linearization was applied.

9.5 Consider a system with transfer function $G_p(s) = \dfrac{-s+2}{(s+1)^2}$

   Determine a controller of the form $G_c(s) = K(s - z/s - p)$ that if connected in cascade with the plant, the error of the closed-loop system is zero when the input signal is a step function.

Solution

The steady-state error for a step input is

$$e_{ss} = \frac{1}{1+K_p} \tag{P9.5.1}$$

where

$$K_p = \lim G_p(s)G_c(s) \tag{P9.5.2}$$

The error e_{ss} tends to zero, if K_p tends to infinity and, therefore, if $G_p(s)G_c(s)$ has a pole at 0. It follows that

$$p = 0 \tag{P9.5.3}$$

The final value theorem is applicable, if the closed-loop system is asymptotically stable. The characteristic polynomial of the compensated system is

$$P_c(s) = s(s+1)^2 + K(-s+2)(s-z) = s^3 + (2-K)s^2 + (1+2K+2Kz)s - 2Kz \tag{P9.5.4}$$

In order to examine the stability of the compensated system we use Routh's tabulation:

$$
\begin{array}{ccc}
s^3 & 1 & 1+2K+2Kz \\
s^2 & 2-K & -2Kz \\
s & \dfrac{(2-K)(1+2K+2Kz)+2Kz}{2-K} & \\
s^0 & -2Kz &
\end{array}
$$

For asymptotic stability, the following inequalities must hold:

$$2-K>0 \tag{P9.5.5}$$

$$-2Kz>0 \tag{P9.5.6}$$

$$(2-K)(1+2K+2Kz)+2Kz = K^2(-2z-2)+K(6z+3)+2>0 \tag{P9.5.7}$$

From relationship (P9.5.1) it follows that $K<2$.
From (P9.5.2), it yields that $Kz<0$.
We may distinguish between two cases.

First case: $K>0$; hence, $z<0$.
From (P9.5.7) the polynomial has to be positive.

i. If $2z+2>0 \Rightarrow z>-1$, the product of the polynomial's roots is negative. This means that the roots are real and of opposite sign. As the coefficient of the higher-order term is negative, the inequality holds between the roots of the polynomial. Thus,

$$0<K<\min\left\{2,\frac{6z+3+\sqrt{(6z+3)^2+8(2z+2)}}{2(2z+2)}\right\} \tag{P9.5.8}$$

ii. If $2z+2<0 \Rightarrow z<-1$, the coefficient of the higher-order term is positive. Hence, the inequality is true, if the polynomial has complex roots. In the case of real roots, the inequality is fulfilled outside of the interval between the roots. The discriminant of the polynomial is

$$D=(6z+3)^2+8(2z+2)=36z^2+52z+25 \tag{P9.5.9}$$

The discriminant is positive because the polynomial of z has complex roots. Consequently, the relationship (P9.5.7) holds outside the interval between the roots of the polynomial of K.

As $6z+3<6z+6=3(2z+2)<0$, the polynomial of K has a positive product and a negative sum of the roots. Thus, both roots are negative and stability is achieved for $0<K<2$.

Second case: $K<0$; hence, $z>0$.
The roots of the polynomial of K are real and of opposite sign because their sum is positive and their product is negative. As the coefficient of the higher-order term is negative, the inequality holds between the roots of the polynomial K, that is,

$$\frac{6z-3-\sqrt{(6z+3)^2+8(2z+2)}}{2(2z+2)}<K<\frac{6z+3+\sqrt{(6z+3)^2+8(2z+2)}}{2(2z+2)} \tag{P9.5.10}$$

Both inequalities are true, for

$$\frac{6z+3-\sqrt{(6z+3)^2+8(2z+2)}}{2(2z+2)} < K < 0 \tag{P9.5.11}$$

From the previous analysis, and for $z = -0.5$ and $K = 1$, the controller is

$$G_c(s) = \frac{s+0.5}{s} \tag{P9.5.12}$$

9.6 A system has the transfer function $G(s) = \dfrac{(1+10s)}{(1+5s)(1+s)s^2}$.

Find a controller of the form $G_c(s) = K\left(\dfrac{s+z}{s+p}\right)^{\mu}$ so that the closed-loop system has poles

at $-1 \pm j$. Find the positions of the other poles of the system.

Solution

The point $-1 - j$ is on the root locus of the system $G(s)G_c(s)$, if

$$Arg(G(-1+j)G_c(-1+j)) = Arg(G(-1+j) + Arg(G_c(-1+j)) = 180° \tag{P9.6.1}$$

We have

$$Arg\{G(-1+j)\} = Arg\{-1+j-(-0.1)\} - Arg\{-1+j-(-0.2)\} - Arg\{-1+j-(-1)\}$$
$$- 2Arg\{-1+j-(0)\} = 131.99° - 128.66° - 90° - 2\cdot135° = -356.67° \tag{P9.6.2}$$

The controller must satisfy the relationship

$$Arg\{G_c(-1+j)\} = \mu(Arg\{-1+j-(z)\} - Arg\{-1+j-(p)\}) = Arg\{K(-1+j)\}$$
$$= \mu(Arg\{-1+j-(z)\} - Arg\{-1+j-(p)\}) = -180° - (-356.67°) = 176.67° \tag{P9.6.3}$$

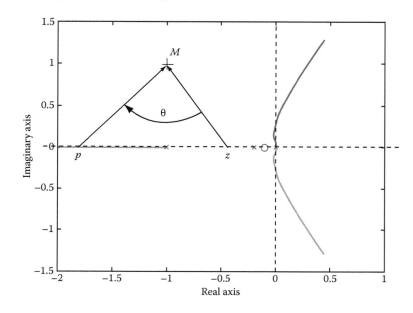

If μ = 1, then the angle θ of the figure must equal 176.67°. This means that the zero must be placed in the right-half plane and the pole in the left-half plane. In this case, the interval [0, z] is part of the root locus and the closed-loop system is unstable.

If μ = 2, the angle θ of the figure must equal 88.33°.

If we choose z = −0.2, then the pole position becomes

$$p = -0.2 - 1 \cdot \tan(38.66°) - 1 \cdot \tan(49.67°) = -0.2 - 0.8 - 1.17 = -2.17 \qquad \text{(P9.6.4)}$$

For the point −1 + j, we get

$$K = \frac{1}{\left|G(-1+j)\right|\left|(-1+j+0.2)^2/(-1+j+2.17)^2\right|} = 1.287 \qquad \text{(P9.6.5)}$$

For this value of K the rest of the poles of the compensated system will be at the positions −0.2 and −0.05 ± j0.071.

9.7 Consider the control system shown in the figure below:

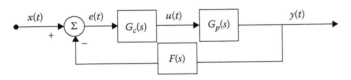

$$\text{where} \quad G_p(s) = \frac{1}{s(s^2 + 4s + 13)}, \quad \text{and} \quad F(s) = \frac{1}{s+1}$$

Find the values of the parameters of the *PID* controller by applying the Ziegler–Nichols methods.

Solution

The plant has the transfer function

$$G_p(s) = \frac{1}{s(s^2 + 4s + 13)} \qquad \text{(P9.7.1)}$$

From relationship (P9.7.1) we observe that the system is of type 1, hence we cannot use the first Zieger–Nichols method.

We compute the value of the critical gain for which the system is tuned, by applying Routh's tabulation at the characteristic equation of the system, and by assuming that we only have proportional control.

The characteristic equation is

$$1 + G_p(s)F(s) = s(s+1)(s^2 + 4s + 13) + K_p = 0 \Rightarrow$$

$$\Rightarrow s^4 + 5s^3 + 17s^2 + 13s + K_p = 0 \qquad \text{(P9.7.2)}$$

Routh's tabulation is

$$
\begin{array}{c|ccc}
s^4 & 1 & 17 & K_p \\
s^3 & 5 & 13 & \\
s^2 & 14.4 & K_p & \\
s^1 & \dfrac{187.2 - 5K_p}{14.4} & & \\
s^0 & K_p & &
\end{array}
$$

Based on Routh's stability criterion, the system is stable if

$$0 < K_p < 37.44 \tag{P9.7.3}$$

It yields that

$$K_{cr} = 37.44 \tag{P9.7.4}$$

Since we can compute the critical gain, the second Zieger–Nichols method can be applied. The critical period of oscillations is given by

$$T_{cr} = \frac{2\pi}{\omega_{cr}} \tag{P9.7.5}$$

The critical frequency of the system is computed by solving the auxiliary equation of row s^2 in Routh's tabulation. Thus,

$$14.4\,s^2 + K_{cr} = 0 \Rightarrow s = \pm j\omega_{cr} = \pm j1.61 \tag{P9.7.6}$$

$$(\text{P9.7.5}) \overset{(\text{P9.7.6})}{\Rightarrow} T_{cr} \simeq 3.9\,\text{s} \tag{P9.7.7}$$

Consequently, the demanded values of the controller's parameters are

$$K_p = 0.6K_{cr} = 22.5$$

$$K_i = \frac{2}{T_{cr}} = 0.513\,\text{s}^{-1} \tag{P9.7.8}$$

$$K_d = 0.125T_{cr} = 0.487\,\text{s}$$

9.8 The transfer function of a control system is $G_p(s) = 100/(s(s^2 + 10s + 100))$. Design a cascade compensation that satisfies the following specifications:
- The velocity error constant is $K_v = 9$.
- The closed-loop transfer function is $M(s) = K/((s + a)(s^2 + 20s + 200))$. where K and a are constants.

Solution

We have

$$G_p(s) = \frac{100}{s(s^2 + 10s + 100)} \tag{P9.8.1}$$

$$M(s) = \frac{G_c(s)G_p(s)}{1 + G_c(s)G_p(s)} \tag{P9.8.2}$$

From the relationship (P9.8.2), it follows that

$$G_c(s)G_p(s) = \frac{M(s)}{1 - M(s)} = \frac{K}{(s+a)(s^2 + 20s + 200) - K} \tag{P9.8.3}$$

Since the velocity error constant is 9, G_cG_p is of type one and

$$200a = K \tag{P9.8.4}$$

The velocity error constant equals to

$$K_v = \frac{K}{200 + 20a} = 9 \tag{P9.8.5}$$

Thus,

$$\frac{200a}{200 + 20a} = 9 \Rightarrow a = 90 \text{ and } K = 18,000 \tag{P9.8.6}$$

Hence, the controller in cascade connection is given by

$$G_c(s) = \frac{G_c(s)G_p(s)}{G_p(s)} = \frac{18,000/(s+90)(s^2 + 20s + 200) - 18,000}{100/s(s^2 + 10s + 100)} \Rightarrow$$

$$G_c(s) = \frac{180(s^2 + 10s + 100)}{s^2 + 110s + 2000} \tag{P9.8.7}$$

9.9 A system has the transfer function $G_p(s) = 100/(s^2 + 10s + 100)$. Suppose a cascade *PI* controller.

a. Find the value of K_i so that the velocity error constant is $K_v = 100$.

b. For the value of K_i computed in the first query, find the critical value of K_P so that the system is stable.

Solution

a. The loop transfer function of the system is

$$G_cG_p = \left(K_p + \frac{K_i}{s}\right)\frac{100}{s^2 + 10s + 100} \tag{P9.9.1}$$

The velocity error constant is

$$K_v = \frac{100K_i}{100} = 100 \Rightarrow K_i = 100 \qquad \text{(P9.9.2)}$$

b. The characteristic equation is

$$1 + \left(K_p + \frac{100}{s} \right) \frac{100}{s^2 + 10s + 100} = 0 \Rightarrow$$

$$s^3 + 10s^2 + 100s + 10,000 + 100K_p s = 0 \Rightarrow$$

$$s^3 + 10s^2 + 100(1 + K_p)s + 10,000 = 0 \qquad \text{(P9.9.3)}$$

We construct Routh's tabulation in order to find the critical value of K_p for stability:

$$
\begin{array}{c|cc}
s^3 & 1 & 100(1 + K_p) \\
s^2 & 10 & 10,000 \\
s^1 & \dfrac{1,000(1 + K_p) - 10,000}{10} & \\
s^0 & 10,000 &
\end{array}
$$

We equal row s^1 to zero; hence,

$$\frac{1,000(1 + K_p) - 10,000}{10} = 0 \Rightarrow K_{p_{cr}} = 9 \qquad \text{(P9.9.4)}$$

9.10 A system has the following transfer function:

$$G_p(s) = \frac{200}{s(s + 1)(s + 10)}$$

Design a cascade *PD* controller that fulfills the following specifications:
- The maximum overshoot is zero.
- The settling time is less than 2.5 s.

Solution

The controller has the following transfer function:

$$G_c(s) = K_p + K_d s = K_p \left(1 + \frac{K_d}{K_p} s \right) \qquad \text{(P9.10.1)}$$

If we place the zero of the controller to one of the poles of the system $G(s)$, we get a second-order system, which can be designed based on the specifications. Hence, we choose a zero at −10, as $K_d/K_p = 1$. The characteristic equation of the system is

$$1 + K_p \frac{200}{s(s + 10)} = 0 \qquad \text{(P9.10.2)}$$

If we choose $K_p = 0.125$, then the roots are at -5 and -5, and the maximum overshoot is zero. Also it follows that

$$t_s = \frac{4}{J\omega_n} = \frac{4}{5} = 0.8 \qquad \text{(P9.10.3)}$$

Therefore, the settling time is less than 2.5 s.
   Based on the previous, the *PD* controller is

$$G_c(s) = 0.125 + 0.125s \qquad \text{(P9.10.4)}$$

The root locus of the system is

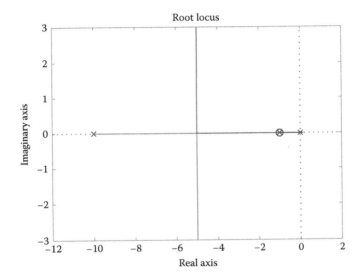

9.11 Consider a control system with transfer function $G_p(s) = 8/(s + 4)$ and a controller in cascade connection, with transfer function $G(s) = K/(s + a)$.
   a. Find J and ω_n in relation to K and a.
   b. Compute K and a so that the rise time is 0.3 s.

Solution
   a. The transfer function of the closed-loop system is

$$G_{cl}(s) = \frac{8K/((s+a)(s+4))}{1+(8K/((s+a)(s+4)))} = \frac{8K}{(s+a)(s+4)+8K} = \frac{8K}{s^2+(a+4)s+8K+4a} \qquad \text{(P9.11.1)}$$

   Since this is a second-order system, it holds that

$$\omega_n = \sqrt{8K + 4a} \qquad \text{(P9.11.2)}$$

$$2J\omega_n = 4 + a \qquad \text{(P9.11.3)}$$

$$(P9.11.2), (P9.11.3) \Rightarrow J = \frac{4+a}{2\omega_n} = \frac{4+a}{4\sqrt{a+2K}} \qquad (P9.11.4)$$

b. The rise time equals approximately to

$$t_r \simeq \frac{2}{\omega_d} \Rightarrow \omega_d = 6.67 \, \text{rad/s} \qquad (P9.11.5)$$

For $J = 0.7$, we have $\omega_n = \omega_d/\sqrt{1-J^2} = 9.34 \, \text{rad/s}$; thus,

$$a = 2J\omega_n - 4 \simeq 9 \quad \text{and} \quad K = \frac{\omega_n^2 - 4a}{8} \simeq 6.4 \qquad (P9.11.6)$$

9.12 Consider the unity feedback closed-loop system with loop transfer function $G(s) = 400/(s(s^2 + 30s + 200))$.

Design a *PID* controller so that the new system has an acceleration steady-state error of 10%, a maximum overshoot of 10%, and a settling time of 2 s.

Solution

The original system is of type 1; thus, the velocity error constant is

$$k_v = \lim_{s \to 0} sG(s) = \frac{400}{200} = 2 \qquad (P9.12.1)$$

Therefore,

$$e_{ss,v} = 1/2 = 0.5 \qquad (P9.12.2)$$

The transfer function of the controller is

$$G_c(s) = K_p \left(1 + T_d s + \frac{1}{T_i s} \right) \qquad (P9.12.3)$$

The compensated system with the controller is of type 2, thus there is a steady-state error

$$e_{ss,v} = 1/k_{ep}$$

where $k_{ep} = \lim_{s \to 0} s^2 G(s)G_c(s) = \frac{400}{s(s^2 + 30s + 200)} K_p \left(1 + T_d s + \frac{1}{T_i s} \right) = 2\frac{k_p}{T_i} \qquad (P9.12.4)$

It must hold that

$$2\frac{K_p}{T_i} = \frac{1}{0.1} \qquad (P9.12.5)$$

We need a 10% maximum overshoot; hence, we know that

$$\exp\left(\frac{-J\pi}{\sqrt{1-J^2}}\right) = 0.1 \Rightarrow J = 0.591 \tag{P9.12.6}$$

From the settling time specification it yields that $t_s = 4/J\omega_n = 2$. Thus,

$$\omega_n = 3.384 \, \text{rad/s} \tag{P9.12.7}$$

The dominant poles are

$$s_{1,2} = -J\omega_n + j\omega_n\sqrt{1-J^2} = 1.997 \pm j2.73 \simeq -2 \pm j2.73 \tag{P9.12.8}$$

The poles of the new closed-loop system result from the following equation:

$$P(s) = 1 + G(s)G_C(s) = 0 \Rightarrow$$
$$(s^3 + 30s^2 + 200s)T_is + 400K_p(T_is + T_iT_ds^2 + 1) = 0 \tag{P9.12.9}$$

or

$$P(s) = s^4 + 30s^3 + (200 + 400K_pT_d)s^2 + 400K_ps + \frac{400K_p}{T_i} = 0 \tag{P9.12.10}$$

But

$$K_p \overset{(P9.12.5)}{=} 5T_i \tag{P9.12.11}$$

thus,

$$P(s) = s^4 + 30s^3 + (200 + 400K_pT_d)s^2 + 400K_ps + 2000 = 0 \tag{P9.12.12}$$

The relationship (P9.12.12) must be fulfilled by the dominant poles $s_{1,2} - 2 \pm j2.73$. Finally, we get two equations: $\text{Re}\{P(s_1)\} = 0$ and $\text{Im}\{P(s_1)\} = 0$ with two unknowns, K_p and T_d.

9.13 Consider a control system with transfer function $G_p(s) = 10/(s(s + 4))$ and a lag controller connected in series. Compute the parameters of the controller for a velocity error constant of $K_v = 50 \, \text{s}^{-1}$ without causing large change to the dominant pole positions, which are $s = -2 \pm j\sqrt{6}$.

Solution

Suppose that the transfer function of the lag controller is

$$G_c(s) = K_c \frac{s + (1/T)}{s + (1/bT)} \quad (b > 1) \tag{P9.13.1}$$

We have

$$K_v = \lim_{s \to 0} sG_c(s)\frac{10}{s(s+4)} = K_c b2.5 = 50 \qquad \text{(P9.13.2)}$$

Thus,

$$K_c b = 20 \qquad \text{(P9.13.3)}$$

We choose

$$K_c = 1, \quad \text{hence } b = 20 \qquad \text{(P9.13.4)}$$

For $T = 10$, the transfer function of the lag controller is given by the relationship (P9.13.5):

$$G_c(s) = \frac{s+0.1}{s+0.005} \qquad \text{(P9.13.5)}$$

For the dominant pole $s = -2 + j\sqrt{6}$, the argument of G_c is

$$Arg(G_c \text{ for } s = -2 + j\sqrt{6}) = \tan^{-1}\frac{\sqrt{6}}{-1.9} - \tan^{-1}\frac{\sqrt{6}}{-1.995} = -1.3616° \qquad \text{(P9.13.6)}$$

The computed value from (P9.13.6) is small, hence the change in the positions of the dominant poles is very small.

The loop transfer function of the system is

$$G_c(s)G(s) = \frac{s+0.1}{s+0.005}\frac{10}{s(s+4)} \qquad \text{(P9.13.7)}$$

The closed-loop transfer function of the system is

$$G_{cl}(s) = \frac{10s+1}{s^3 + 4.005s^2 + 10.02s + 1} \qquad \text{(P9.13.8)}$$

With this compensation it is evident that the steady-state error of the compensated system is 0.02 from 0.4 that it was for the original system.

9.14 A control system has transfer function $G_p(s) = 10/(s(s + 1)(s + 5))$ and a *PID* controller connected in cascade. Compute the parameters of the controller (K_p, T_i, and T_d) by using one of the Ziegler–Nichols methods.

Solution

Since the system to be controlled has an integrator, we will use the second Ziegler–Nichols method. We suppose that $T_i = \infty$ and $T_d = 0$. The transfer function of the closed-loop system is

$$G_{cl}(s) = \frac{K_p}{s(s+1)(s+5) + K_p} \qquad \text{(P9.14.1)}$$

The value of K_p that renders the system critically stable is found by using Routh's tabulation. The characteristic equation of the system is

$$s(s+1)(s+5)+K_p = 0 \Rightarrow s^3 +6s^2 +5s+K_p = 0 \qquad \text{(P9.14.2)}$$

Routh's tabulation is

$$
\begin{array}{c|cc}
s^3 & 1 & 5 \\
s^2 & 6 & K_p \\
s^1 & \dfrac{30-K_p}{6} & \\
s^0 & K_p &
\end{array}
$$

From row s^1, we get the value for K_{pcr}:

$$K_{pcr} = 30 \qquad \text{(P9.14.3)}$$

By setting $K_p = K_{pcr}$, the characteristic equation becomes

$$s^3 +6s^2 +5s+30 = 0 \qquad \text{(P9.14.4)}$$

For finding the period of the sustained oscillations, we substitute $s = j\omega$ at the characteristic equation of the relationship (P9.14.4):

$$(j\omega)^3 +6(j\omega)^2 +5j\omega+30 = 0 \Rightarrow$$

$$6(5-\omega^2)+ j\omega(5-\omega^2) = 0 \Rightarrow \omega = \sqrt{5} \text{ rad/sec} \qquad \text{(P9.14.5)}$$

The period of the sustained oscillations is

$$T_{cr} = \frac{2\pi}{\omega} = 2.8099\,\text{s} \qquad \text{(P9.14.6)}$$

From the Table 9.4 of the second Ziegler–Nichols method, we get the desired parameters of the controller:

$$K_p = 0.6K_{cr} = 18$$

$$T_i = 0.5T_{cr} = 1.405 \qquad \text{(P9.14.7)}$$

$$T_d = 0.125T_{cr} = 0.35124$$

The transfer function of the controller is

$$G_c(s) = K_p\left(1+\frac{1}{T_i s}+T_d s\right) = 18\left(1+\frac{1}{1.405s}+0.35124s\right) = \frac{6.3223(s+1.4235)^2}{s} \qquad \text{(P9.14.8)}$$

The controller has a pole at $s = 0$ and a double zero at $s = -1.4235$.
  The transfer function of the closed-loop system is provided from the relationship (P9.14.9):

$$G_{cl}(s) = \frac{6.3223s^2 +18s+12.811}{s^4 +6s^3 +11.3223s^2 +18s+12.811} \qquad \text{(P9.14.9)}$$

We observe that the maximum overshoot is approximately 62%. If we tune all the parameters of the controller, then we can reduce it. We maintain the proportional gain $K_p = 18$, and by displacing the double zero at $s = -0.65$, the transfer function of the new controller becomes

$$G_c(s) = 18\left(1 + \frac{1}{3.077s} + 0.7692s\right) = \frac{13.846(s+0.65)^2}{s} \qquad \text{(P9.14.10)}$$

This makes the maximum overshoot of the system decrease to 18%.

If we increase the proportional gain to $K_p = 39.42$ and we maintain the double zero position at $s = -0.65$, the transfer function of the controller is

$$G_c(s) = \frac{30.322(s+0.65)^2}{s} \qquad \text{(P9.14.11)}$$

In this case the response speed is increased, but the maximum overshoot is also increased to approximately 28%. This maximum overshoot is very close to the desired one at 25%, and the system responds faster. Hence, the controller of the relationship (P9.14.11) is satisfactory. The new parameters of the controller are

$$K_p = 39.42$$

$$T_i = 3.077 \qquad \text{(P9.14.12)}$$

$$T_d = 0.7692$$

9.15 A control system has the loop transfer function $G(s) = K/s(s/5 + 1)(s/200 + 1)$. Design a lead controller so that the obtained closed-loop system has

  i. A velocity error <0.01

  ii. A damping ratio $J \geq 0.4$ for the dominant poles

Solution

The lead controller has the following form:

$$G_c(s) = \frac{(1/\omega_1)s + 1}{(1/\omega_2)s + 1} \qquad \text{(P9.15.1)}$$

The specification (i) demands the use of proportional control with gain K.

The transfer function of the closed-loop system is

$$W(s) = \frac{K}{s(s/5+1)(s/200+1) + K} \qquad \text{(P9.15.2)}$$

The transfer function of the system is

$$Q(s) = 1 - W(s) = \frac{s(s/5+1)(s/200+1)}{s(s/5+1)(s/200+1) + K} \qquad \text{(P9.15.3)}$$

We apply the final value theorem and for a ramp input signal we get

$$e_{ss} = \lim_{s \to 0} sE(s) = \lim_{s \to 0} sQ(s)\left(\frac{1}{s^2}\right) = \lim_{s \to 0} s\frac{(s/5+1)(s/200+1)}{s(s/5+1)(s/200+1)+K} = \frac{1}{K} \leq 0.1 \Rightarrow K \geq 100 \quad (P9.15.4)$$

In this case the open-loop system is

$$G(s) = \frac{100}{s(s/5+1)(s/200+1)} \quad (P9.15.5)$$

The Bode diagram of the closed-loop system is

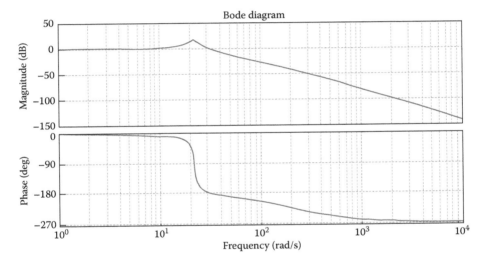

We observe that the closed-loop system is tuned at a frequency of $\simeq 20\,\text{rad/s}$. In order to reduce it, we choose ω_{max}:

$$\omega_{max} = \sqrt{\omega_1 \omega_2} = 20\,\text{rad/s} \quad (P9.15.6)$$

In order to compute the phase of the open-loop system for a dominant pole of the closed-loop system at its natural frequency, we proceed as follows:

$$W(s) = \frac{\omega_n^2}{s^2 + 2J\omega_n s + \omega_n^2} = \frac{G(s)}{1+G(s)} \quad (P9.15.7)$$

where

$$G(s) = \frac{\omega_n^2}{s^2 + 2J\omega_n s} \quad (P9.15.8)$$

But

$$G(i\omega) = \frac{\omega_n^2}{-\omega^2 + i2\xi\omega_n\omega} \tag{P9.15.9}$$

From the relationship (P9.15.9), we get the phase of the open-loop system:

$$\theta(\omega) = 0 - \left[\pi - \tan^{-1}\left(\frac{2J\omega_n}{\omega}\right)\right] = -\pi + \tan^{-1}\left(\frac{2J\omega_n}{\omega}\right)\right] \tag{P9.15.10}$$

For

$$\omega = \omega_n \Rightarrow \theta(\omega_n) = -\pi + \tan^{-1}(2J) \tag{P9.15.11}$$

The Bode diagram of the open-loop system is

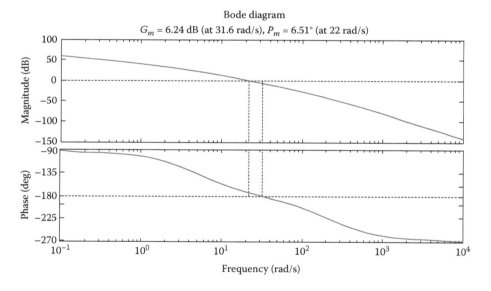

We observe from the diagram that the frequency for which the magnitude of the transfer function is one, that is, the crossover frequency of the open-loop control system, is $\simeq 22\,\text{rad/s}$. For $J = 0.4$ the phase is

$$\theta(\omega_n) \overset{(\text{P5.9.11})}{\cong} -180° + 40° = -140° \tag{P9.15.12}$$

We will try out a phase-lead controller with a high-frequency gain:

$$\alpha = \frac{\omega_2}{\omega_1} = \frac{[1+\sin 45°]}{[1-\sin 45°]} \simeq 5.8 = 15.3\,\text{dB} \tag{P9.15.13}$$

Hence,

$$\omega_2 = 5.8\omega_1 \tag{P9.15.14}$$

By assuming that $\omega_{max} \cong 20\,\text{rad/s}$, we have

$$(\text{P9.15.6}) \overset{(\text{P9.15.14})}{\Rightarrow} \omega_{max} \cong 20 = \sqrt{\omega_1\omega_2} = \omega_1\sqrt{5.8} \Rightarrow$$
$$\omega_1 = 20/\sqrt{5.8} = 8.3\,\text{rad/s} \tag{P9.15.15}$$

$$(\text{P9.15.14}) \overset{(\text{P9.15.15})}{\Rightarrow} \omega_2 = 5.8\omega_1 = 48.1\,\text{rad/s} \tag{P9.15.16}$$

Thus, the transfer function of the phase-lead controller is

$$G_c(s) = \frac{(1/8.3)s+1}{(1/48.1)s+1} \tag{P9.15.17}$$

The loop transfer function of the compensated system is

$$G_{new}(s) = G_c(s)G(s) = \frac{(1/8.3)s+1}{(1/48.1)s+1} \cdot \frac{100}{s(s/5+1)(s/200+1)} \tag{P9.15.18}$$

The relevant Bode diagram is

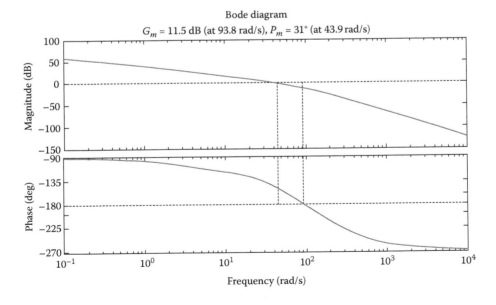

Bode diagram
$G_m = 11.5\,\text{dB (at 93.8 rad/s)}, P_m = 31°\,\text{(at 43.9 rad/s)}$

The phase margin is approximately $30°$.

The Bode plot of the closed-loop system is

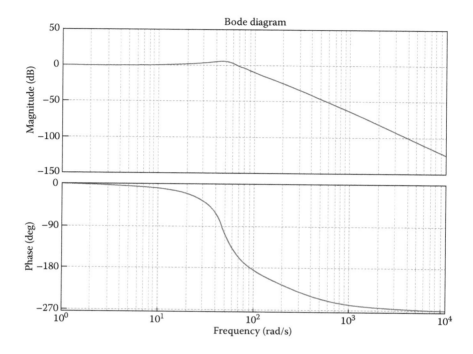

The gain for the transfer function at the resonance frequency is substantially reduced, and it is approximately 15 db.

The dominant poles of the closed-loop system are

$$s_{d_{1,2}} = -14 \pm j48 \tag{P9.15.19}$$

Hence, the natural frequency of the system is

$$\omega_n = \sqrt{14^2 + 48^2} = 50\,\text{rad/s} \tag{P9.15.20}$$

and the damping ratio is

$$J = \frac{14}{50} \simeq 0.3 \tag{P9.15.21}$$

which is relatively close to the second specification.

The design of the lead controller is satisfactory.

10

Simulation Tools

10.1 Introduction

In this chapter, we introduce various computer-based simulation tools that can help in the design of a control system.

A **computer simulation** is a computer program that attempts to simulate an abstract model of a particular system.

A model is a schematic description of a system and can be used for further study of its characteristics.

System modeling is the procedure of expressing a (usually) simplified version of the system in terms of differential equations.

The simulation procedure is implemented in the following steps:

1. The problem and the study procedure are specified.
2. The mathematical model of the system is derived.
3. A suitable simulation software is selected.
4. Simulation tests are run and are compared with the real system (if available).
5. Extensive simulations are run and the results are analyzed.

In this chapter, we introduce the following simulation tools:

- **MATLAB**® is a Mathworks product. It is a very popular software tool designed for scientific and engineering computing. It comes with a control systems toolbox, a toolbox that provides a way for systematically analyzing, designing, and tuning linear control systems.

- **Simulink**® is also a Mathworks product. It provides an interactive graphical environment and a customizable set of block libraries that let you design, simulate, implement, and test a variety of time-varying systems.

- The **Program CC** is a Systems Technology product. It provides a control system design package containing many tools and algorithms of current control system theory and practice.

- **SIMAPP** is another computer simulation software for modeling systems in the time and frequency domains. The model is built visually through block diagrams.

- **SCILAB** is an open-source software similar to MATLAB. It comes with toolboxes appropriate for the simulation, design, and optimization of control systems. Moreover, as an alternative to Simulink, the Scilab team has developed an interactive graphical environment called **XCOS**.

10.2 MATLAB®

We start our introduction to the use of MATLAB in the design and analysis of control systems by discussing some useful commands.

10.2.1 Laplace Transform

The computation of the one-sided ($t \geq 0$) Laplace transform of a function $f(t)$ in MATLAB is performed by the command `laplace(f,s)`, while the inverse Laplace transform of a function $F(s)$ is performed by the command `ilaplace(F,t)`. The variables t and s have to be previously declared as symbolic variables by the command `syms`. For the transformation of a rational function into a sum of partial fractions, the command `[r,p,k] = residue(B,A)` is used. A system of differential equations is solved by the command `dsolve('eqn1','eqn2',...)`. A complex number is declared as $z = a + b*i$, where a and b are constant coefficients. Finally, the commands `real(z)`, `imag(z)`, `abs(z)`, `angle(z)`, `conj(z)` compute the real and the imaginary part, the modulus, the phase, and the complex conjugate of z, respectively.

For example, the Laplace transform of the function $f(t) = te^{-t}u(t)$ is computed as

```
syms t s
f=t*exp(-t);
F=laplace(f,s)
```

The result is

```
F =
1/(1 + s)^2
```

The inverse Laplace transform of $F(s) = 1/(s + 1)^2$ is computed as

```
F=1/(1 + s)^2;
f=ilaplace(F,t)
```

10.2.2 Construction of LTI Models

A linear time-invariant (LTI) system can be defined in MATLAB either in the complex frequency domain as a transfer function, or in the time domain as a state-space model.

Consider a single-input single-output (SISO) system described in the following state-space form:

$$\dot{x} = Ax + Bu, \quad x(0) = 0$$
$$y = Cx + Du \tag{10.1}$$

The system transfer function is given by

$$G(s) = \frac{Y(s)}{U(s)} = \frac{b_m s^m + \cdots + b_2 s^2 + b_1 s + b_0}{a_n s^n + \cdots + a_2 s^2 + a_1 s + a_0} \tag{10.2}$$

Or equivalently, in zero/pole/gain form:

$$G(s) = \frac{Y(s)}{U(s)} = K\frac{(s-z_1)(s-z_2)\cdots(s-z_m)}{(s-p_1)(s-p_2)\cdots(s-p_n)} \tag{10.3}$$

A system is defined in state-space form (10.1) by the MATLAB command ss, in transfer function form (10.2) by the command tf, and in zero/pole/gain form (10.3) by the command zpk. These three commands create a special type of MATLAB variable called object.

The command sys = ss(A,B,C,D) creates an *ss* object *SYS* that represents the continuous-time state-space model given in (10.1). For example, a system described by the state-space matrices $A = \begin{bmatrix} -2 & -1 \\ 1 & -2 \end{bmatrix}$, $B = \begin{bmatrix} 0 \\ 1 \end{bmatrix}$, $C = [1 \quad 1]$, $D = 1$ is created by the command

```
sys = ss([-2 -1;1 -2],[0; 1],[1 1],[1])
```

The variable sys is a state-space object. The system transfer function is obtained by

```
G = tf(sys)
```

Another syntax of command tf is sys = tf(num,den), where num and den are the coefficients of the numerator and denominator of (10.2), respectively. In this case, sys is called a tf object. For example, the transfer function $G(s) = 2s/s + 2$ can be obtained as follows:

```
num=[2 0];
den=[1 2];
g=tf(num,den)
```

or directly by typing

```
g=tf([2 0],[1 2])
```

The transfer function $G(s) = \dfrac{2s}{s+2}e^{-5s}$ is obtained by the command

```
g=tf([2 0], [1 2], 'inputdelay', 5)
```

or

```
g=tf([2 0], [1 2], 'outputdelay', 5)
```

The presence of the exponential function can create some computational problems. In order to overcome these problems, one can use a Padé approximation. **Padé approximant** is the approximation of a function (in this case of the exponential function) by a rational function of given order. The higher the order of the polynomials, the better the approximation gets. Table 10.1 introduces Padé approximations of first-, second-, and third-order polynomials.

TABLE 10.1

Padé Approximations of an Exponential Function

First order $n = 1$	$e^{-t_d s} = \dfrac{-s + \dfrac{2}{t_d}}{s + \dfrac{2}{t_d}}$
Second order $n = 2$	$e^{-t_d s} = \dfrac{s^2 - \dfrac{6}{t_d}s + \dfrac{12}{t_d^2}}{s^2 + \dfrac{6}{t_d}s + \dfrac{12}{t_d^2}}$
Third order $n = 3$	$e^{-t_d s} = \dfrac{-s^3 + \dfrac{12}{t_d}s^2 - \dfrac{60}{t_d^2}s + \dfrac{120}{t_d^3}}{s^3 + \dfrac{12}{t_d}s^2 + \dfrac{60}{t_d^2}s + \dfrac{120}{t_d^3}}$

The command [num,den] = pade(T,n) returns the *n*-th order Padé approximation of the continuous time delay exp(–*T**s) in transfer function form. For example, a second-order Padé approximation of the function

e^{-3s} is obtained as follows:

```
td = 3;
n = 2;
s = tf('s');
sys_tf = exp(-td*s);
sys_pade_tf = pade(sys_tf,n)
```

The command sys_zpk = zpk(z,p,k) returns a *zpk* object sys_zpk, where *z*, *p*, and *k* are the zeros, poles, and gain of relationship (10.3).

The command sys_zpk = zpk(sys) converts the (*ss* or *tf*) object *sys* to a *zpk* object sys_zpk.

If an *ss*, *tf*, or *zpk* object is already created, we can derive information about the system by using the commands ssdata, tfdata, and zpkdata. More specifically, the command [num,den] = tfdata(sys) returns the numerator and denominator of the transfer function *sys*. The command [A,B,C,D] = ssdata(sys) returns the *A*, *B*, *C*, *D* matrices of the state-space model *sys*. Finally, the command [z,p,k] = zpkdata(sys) returns the zeros, poles, and gain of the LTI model *sys*. One other useful command is the command get. The command get(sys) displays all property names and their current values for object *sys*. The commands pole(sys) and zero(sys) compute the poles and the zeros of the system *sys*.

In the following example, a *zpk* object G is created by the zeros z = [–1 –3] and the poles p = [0 –2 –4]. Next, the command [num,den]=tfdata(G,'v') returns the numerator and denominator as row vectors and by the command Gs = tf(num,den) we create the transfer function $G(s) = (s + 1)(s + 3)/s(s + 2)(s + 4)$. Finally we use the command [z,p,k] = zpkdata(sys,'v') in order to get the zeros *z* and poles *p* as column vectors:

```
z = [-1 -3];
p = [0 -2 -4];
K = 1;
G = zpk(z,p,K)
[num,den]=tfdata(G,'v')
Gs = tf(num,den)
[zeros,poles,gain]=zpkdata(Gs,'v')
```

Another useful command is the command `frd`. The command `sys_frd=frd(sys,w)` creates a frequency-response data model `sys_frd`, where `sys` represents the system transfer function and `sys_frd` contains the values of the system's frequency response evaluated at the frequency points w.

For example, the frequency response of a system with transfer function $G(s) = \dfrac{0.9}{0.1s+1}$ is $G(j\omega) = \dfrac{0.9}{0.1j\omega+1} = \dfrac{0.9}{\sqrt{(0.1\omega)^2+1}} e^{-j\tan^{-1}(0.1\omega)}$. The frequency response of the system is evaluated as follows:

```
omega = logspace(-2,2,40);
sys1 = tf(0.9,[.1 1]);
sys1g = frd(sys1,omega)
```

The command `[r,p,k] = residue(b,a)` finds the residues, poles, and direct term of a partial fraction expansion of the ratio of two polynomials $B(s)/A(s)$.

In the case of distinct poles, a transfer function of the form $G(s) = A(s)/B(s)$ can be written in partial fraction form as $G(s) = \dfrac{A(s)}{B(s)} = \dfrac{r_1}{s-p_1} + \dfrac{r_2}{s-p_2} + \cdots + \dfrac{r_n}{s-p_n} + K(s)$. In the case of a multiple pole (e.g., p_1) with multiplicity m, the transfer function in partial fraction form is written as $G(s) = \dfrac{A(s)}{B(s)} = \dfrac{r_{11}}{s-p_1} + \dfrac{r_{12}}{(s-p_1)^2} + \cdots \dfrac{r_{1m}}{(s-p_1)^m} + \dfrac{r_n}{s-p_n} + K(s)$.

For example, the transfer function $G(s) = 4s^2 + 4s + 4/s^2(s^2 + 3s + 2)$ can be written in partial fraction form as $G(s) = \dfrac{-3}{s+2} + \dfrac{4}{s+1} - \dfrac{1}{s} + \dfrac{2}{s^2}$. The MATLAB code is

```
n=[0 0 4 4 4];
d=[1 3 2 0 0];
[r,p,k]=residue(n,d)
```

The command `size(sys)` returns the number of inputs and outputs of the model `sys`, and also the number of states if `sys` is an *ss* object.

The command `hasdelay(sys)` returns true (1) if the LTI model *sys* has input, output, or internal delays, and false (0) otherwise.

The command `k = dcgain(sys)` computes the steady-state gain of the LTI model *sys*.

The command `pzmap(sys)` plots the poles and zeros of the system *sys* in the complex plane, while the syntax `[p,z]=pzmap(sys)` returns the poles and zeros without drawing a plot.

We use the system with transfer function $G(s) = 4s^2 + 4s + 4/s^2(s^2 + 3s + 2)$ to introduce the aforementioned commands.

```
n=[0 0 4 4 4];
d=[1 3 2 0 0];
sys=tf(n,d)
size_sys=size(sys)
delay_sys=hasdelay(sys)
dcgain_sys=dcgain(sys)
pzmap(sys)
```

10.2.3 Systems Interconnections

In this section, we introduce commands for the various types of interconnections between (sub)systems.

The command sys = series(sys1,sys2) connects two LTI models *sys1* and *sys2* in series.

Similarly, the command sys = parallel(sys1,sys2) connects two LTI models *sys1* and *sys2* in parallel.

The command sys = feedback(sys1,sys2,sign) creates a closed-loop feedback system SYS. The system *sys1* is placed at the direct branch while the system *sys2* is placed at the feedback branch. By default sign = -1, that is, negative feedback is assumed. To apply positive feedback, we set sign = 1.

Consider two systems with transfer functions $G_1(s) = (s + 1)/(s + 3)$ and $G_2(s) = 10/(s(s + 2))$. If the two systems are connected in series, the total transfer function is $G_s(s) = G_1(s) \cdot G_2(s)$, if they are connected in parallel, the transfer function is $G_p(s) = G_1(s) + G_2(s)$, while a feedback connection results in the transfer function $G_{fd}(s) = G_1(s)/1 + G_1(s)G_2(s)$. Indeed,

```
G1=tf([1 1],[1 3]);
G2=tf(10,conv([1 0],[1 2]));
Gs=series(G1,G2)
Gs1=G1*G2% Gs = Gs1
Gp=parallel(G1,G2)
Gp1 = G1 + G2% Gp = Gp1
Gcl=feedback(G1,G2)
```

Another useful command is the command append. Its syntax is sys=append(sys1, sys2,…,sysN). It gives the transfer function of *N* systems appended, as shown in the following figure.

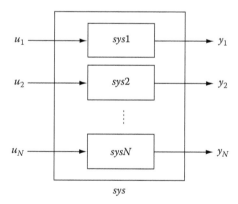

sys

Suppose that $G_1(s)$ and $G_2(s)$ are the transfer functions of the systems *sys1* and *sys2*. The command G=append(G1,G2) creates a system with transfer function $G(s) = \begin{bmatrix} G_1(s) & 0 \\ 0 & G_2(s) \end{bmatrix}$. For example, suppose that $G_1(s) = 1/(2s + 3)$, $G_2(s) = 4/(5s + 6)$.

```
num1 = [1]; den1 = [2 3];
num2 = [4]; den2 = [5 6];
sys_tf1 = tf(num1,den1)
sys_tf2 = tf(num2,den2)
sys=append(sys_tf1, sys_tf2)
```

The command sys = connect(sys1,sys2,…,inputs,outputs) constructs the aggregate model for a given block-diagram interconnection of LTI models.
    The block diagram connectivity can be specified in two ways:

1. First, we name the input and output signals of all LTI blocks in the block diagram, including the summation blocks. A summation block is created with the command sumblk. The aggregate model *sys* is then built by the command connect.
2. In this approach, first combine all LTI blocks into an aggregate, unconnected model blksys using append. Then construct a matrix *q*, where each row specifies one of the connections or summing junctions in terms of the input vector *u* and output vector *y* of blksys.

Consider the following block diagram. Suppose that $G(s) = 3/(s + 4)$ and $H(s) = 2/(s + 5)$. The MATLAB code that computes the transfer function of the total system is given in the following figure:

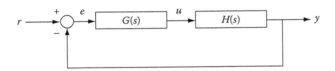

First we define all the transfer functions of the partial models.

```
G=tf(3,[1 4]);
H=tf(2,[1 5]);
```

First way: name-based interconnection

```
G.InputName = 'e'; G.OutputName = 'u';
H.InputName = 'u'; H.OutputName = 'y';
Sum = sumblk('e','r','y','+-');
TF1 = connect(H,G,Sum,'r','y')
```

Second way: index-based interconnection

```
BLKSYS = append(G,H);
```

%U = inputs to *C,G*. Y = outputs of *C,G*
%Here $Y(1)$ feeds into $U(2)$ and $-Y(2)$ feeds into $U(1)$
$Q = [2\ 1; 1\ -2]$;
%External I/Os: *r* drives $U(1)$ and *y* is $Y(2)$

```
TF2 = connect(BLKSYS,Q,1,2)
```

10.2.4 Conversions between Various Forms of LTI Objects

A system expressed in state-space form can be converted to a transfer function form with the command ss2tf and to a zero/pole/gain form with the command ss2zpk. A system in transfer function is converted to state-space form with the command tf2ss. The syntaxes are [num, den]=ss2tf(A,B,C,D), [z,p,k] ss2zpk(A,B,C,D), and [A,B,C,D] = tf2ss(num,den), respectively.

Remarks:

- The three forms *ss*, *zpk*, and *tf* are equivalent, but it is more precise to work with state-space models.
- When converting from one form to another, it is possible to experience pole displacement.

Consider a system with transfer function $G(s) = 1/(100s^2 + 5s + 0.05)^2$. The equivalent state-space model is obtained by the following MATLAB code:

```
num = [1];
den = [100 5 0.05];
den = conv(den,den);
sys_tf = tf(num,den);
[A,B,C,D] = tf2ss(num,den)
```

Next, we derive the *zpk* model in two ways: The first way is directly from the system transfer function. The second way is to declare the variable *s* as a *zpk* object.

% First way

```
z = [];
p1 = roots([100 1 0.01]);
p = [p1.', p1.'];
k = 1/100^2;
sys_zpk1 = zpk(z,p,k)
```

% Second way

```
s = zpk('s');
sys_zpk3 = 1/(100*s^2 + s + 0.01)^2
```

10.2.5 System Analysis in the Time Domain

The command impulse(sys) computes the impulse response of the system *sys*. The following syntaxes are available:

- impulse(sys,tfinal)
- impulse(sys,t)
- impulse(sys1,sys2,...,sysn,t)
- [y,t] = impulse(sys)
- [y,t,x] = impulse(sys)

The command step(sys) computes the step response of the system *sys*. The following syntaxes are available:

- step (sys,tfinal)
- step (sys,t)
- step (sys1,sys2,...,sysn,t)
- [y,t] = step (sys)
- [y,t,x] = step (sys)

The command lsim(sys,u,t) computes the response of the system *sys* to the input signal *u* for the time interval *t*. The following syntaxes are available:

- lsim (sys,u,t,x0)
- lsim (sys1,sys2,…,sysn,t,x0)
- [y] = lsim (sys,u,t)

The command initial(sys,x0) plots the undriven response of the state-space model *sys* (created with ss) with initial condition *x0* on the states.

Consider a unity feedback with loop transfer function $G(s) = 1/(s + 1)$. The following code computes the impulse response, the step response, and the ramp response of the system:

% System definition

```
a=[1];
b=[1 1];
sys1=tf(a,b);
sys2=1;
sys=feedback(sys1,sys2)
```

% Impulse response

```
[y,t]=impulse(sys);
plot(t,y)
```

% Step response

```
[y,t]=step(sys);
```

% Ramp response

```
t=[0:0.01:20];
u=t;
[y,T]=lsim(sys,u,t);
plot(T,y,t,u,'o')
```

10.2.6 System Analysis in the Frequency Domain

Suppose that the input to an LTI system is a sinusoidal function of the form $u(t) = A \sin(\omega t)$. Suppose also that the poles of the system transfer function are in the left half of the complex plane. Then, after the transient period, the system output is given by $y(t) = A_1 \sin(\omega t + \varphi)$. The amplitude A_1 and the phase φ are related to the angular frequency ω of the input signal according to the following equations:

$$\frac{A_1}{A} = |G(j\omega)| \quad \text{and} \quad \varphi = Arg(G(j\omega)) = \tan^{-1}\frac{Im(G(j\omega))}{Re(G(j\omega))} \tag{10.4}$$

Bode diagrams are based on these two relations, that is, are plots of the magnitude and the phase angle versus the angular frequency ω. A Bode diagram is created in MATLAB with the command bode(sys), where *sys* is an *ss*, *tf*, or *zpk* model.

The following syntaxes are available:

- `bode(sys,[wmin, wmax])`, where *wmin* and *wmax* are the minimum and maximum values of the angular frequency.
- `bode(sys,w)`, where *w* is a vector of angular frequencies. The syntax `bode(sys1,sys2,…, sysn,w)` graphs the Bode response of multiple LTI models on a single plot.
- `[mag,phase] = bode(sys,w)` returns the response magnitudes and phases in degrees. No plot is drawn on the screen.

The command `logspace(X1, X2, N)` generates a row vector of *N* logarithmically equally spaced points between the decades 10^{X1} and 10^{X2}.

Finally, the command `[gm,pm,wcg,wcp] = margin(sys)` computes the gain margin *gm*, the phase margin *pm*, and the associated frequencies *wcg* and *wcp*, for the SISO open-loop model *sys*.

Consider a system with transfer function $G(s) = 5/(4s^2 + 4s + 1)$. The following code plots the system's Bode diagram and computes the gain and phase margins as well as the resonant frequency and the bandwidth of the system.

% Bode diagram

```
num = [5];
den = [4 4 1];
sys = tf(num,den);
bode(sys)
```

% Gain margin and phase margin bandwidth

```
[Gm,Pm,Wcg,Wcp] = margin(sys)
```

% Resonant frequency and bandwidth

```
[mag, phase, w] = bode (sys, w)
[Mp, k] = max (mag)
resonant_ peak = 20 * log (10*(Mp))
resonant _frequency = w(k)
n=1
while 20*log(10*(mag(n))) > -3
n=n + 1
end
bandwidth = w(n)
```

The command `nyquist(sys)` draws the nyquist plot of the LTI model *sys*. Other possible syntaxes are

- `nyquist (num, den, w)`
- `nyquist (A, B, C, D)`
- `nyquist (A, B, C, D, w)`

where *w* is the frequencies vector.

Consider the loop transfer function $G(s) = 1/(s^3 + 2s^2 + 5s + 1)$. The MATLAB code that plots the Nyquist diagram of the open-loop system is

```
num = [0 0 0 1];
den = [1 2 5 1];
nyquist(num,den)
```

10.2.7 Root Locus

The command `rlocus(sys)` computes and plots the root locus of the SISO LTI model *sys*. The root-locus plot shows the trajectories of the closed-loop poles when the feedback gain K varies from 0 to ∞. The syntax `[r,k] = rlocus(sys)` returns the matrix R of complex root locations for the gains K. Other possible syntaxes are

- `[r, k] = rlocus (num, den)`
- `[r, k] = rlocus (A, B, C, D)`
- `[r, k] = rlocus (A, B, C, D, K)`
- `[r, k] = rlocus (sys)`

The command `rlocfind(num,den)` computes root-locus gains for a given set of roots. It can be used for interactive gain selection from the root-locus plot.

Consider a system with transfer function $G(s) = 2(s + 1)/(2s^2 + s + 8)$. The root locus is easily found by typing

```
sys=tf ([2 1], [2 4 8]);
rlocus (sys)
```

10.2.8 Pole Placement

With MATLAB, it is very easy to choose the desired poles of a system according to the feedback law $u = -Kx$. The command `acker` provides pole placement gain selection using Ackermann's formula. More specifically, the command `k= acker(a,b,p)` calculates the feedback gain matrix K such that the single-input system $\dot{x} = Ax + Bu$ with a feedback law of $u = -Kx$ has closed-loop poles at the values specified in vector P, that is, $P = \text{eig}(A - B^*K)$.

The command `Co=ctrb(sys)` returns the controllability matrix of the state-space model *sys* with realization (A, B, C, D), while the rank of the controllability matrix Co is found with the command `r=rank(Co)`. Similarly, the command `Co = obsv(sys)` returns the observability matrix of the state-space model *sys* with realization (A, B, C, D).

10.2.9 Two Useful Tools

Finally we should mention two nice tools with a graphical user interface, available in MATLAB called **LTI Viewer** and **SISO Design Tool**. To enable the LTI Viewer, simply type `ltiview` in the command prompt. Then, from the menu of the LTI Viewer choose to import a system from the workspace. We can immediately see the system's step response, impulse response, etc. Moreover, we can see various characteristics of each response such as the settling time or the rise time.

The command `sisotool` opens the SISO Design Tool. This graphical user interface lets you design SISO compensators by graphically interacting with the root locus, Bode, and Nichols plots of the open-loop system.

Problems

10.1 Consider a second-order system with closed-loop transfer function

$$G(s) = \frac{3}{s^2 + 2s + 10}$$

Compute
a. The peak time
b. The percent overshoot
c. The rise time
d. The settling time

Solution

The final value of the step response is computed according to the relationship

$$Final\_value = \lim_{s \to 0} sG(s)\frac{1}{s} = G(0) = 0.3$$

```
final_value=0.3;
num=3; den=[1 2 10];
[y,x,t]=step(num,den);
[Y,k]=max(y);
time_to_peak=t(k)
percent_overshoot=100*(Y-final_value)/final_value
```

%Computation of rise time

```
n=1;
    while y(n)<0.1*final_value,n=n + 1; end
m=1;
    while y(m)<0.9*final_value, m=m + 1; end
risetime=t(m)-t(n)
```

%Computation of settling time

```
l=length(t);
while(y(l)>0.98*final_value)&(y(l)<1.02*final_value)
l=l-1;
end
step(num,den)
```

10.2 Consider a second-order system with closed-loop transfer function

$$G(s) = \frac{16}{s^2 + 4s + 16}$$

Plot the step response of the system and study the effect in the time response of the system, when adding additional poles and zeros.

Solution

To start, we compute the step response of the system and then we study the effect of adding additional poles and zeros.

```
zeta = 0.5;
wn = 4;
num0 = wn^2; den0 = [1 2*zeta*wn 0]; % transfer function
[ncl0,dcl0] = feedback(num0,den0,1,1,-1);
t = linspace(0,4,1001); % time vector
ys0 = step(ncl0,dcl0,t); % step response
plot(t,ys0)
title('Step Response: zeta = 0.5, wn = 4r/s')
```

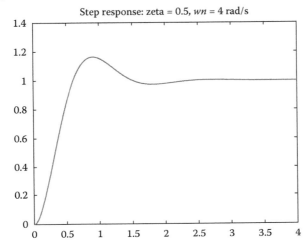

% Adding a zero

```
z = [0.2 0.5 1 2 5 10]; % various values for the zero
for i = 1:length(z)
numz(i,:) = (wn^2/z(i)) * [1 z(i)];
[nclz(i,:),dclz(i,:)] = feedback(numz(i,:),den0,1,1,-1);
clpz(:,i) = roots(dclz(i,:));
ysz(:,i) = step(nclz(i,:),dclz(i,:),t); % new step responses
end
plot(t,ysz),
ylabel('Amplitude'),title('Step Response with extra zero'),
```

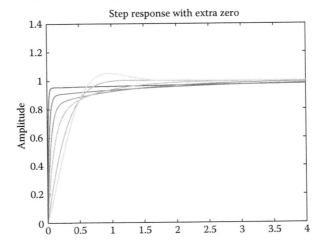

% Adding a pole

```
p = [1 2 5 10 20]; % various values for the pole
t = 5*t; % new time vector
for i = 1:length(p)
nump(i) = p(i) * num0;
denp(i,:) = conv(den0,[1 p(i)]);
[nclp(i,:),dclp(i,:)] = feedback(nump(i),denp(i,:),1,1,-1);
clpp(:,i) = roots(dclp(i,:));
ysp(:,i) = step(nclp(i,:),dclp(i,:),t); % new step responses
end
plot(t,ysp),
title('Step Response with extra pole')
```

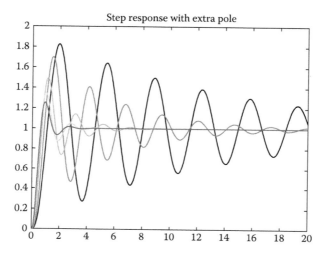

10.3 Consider a system described by following state-space equations:

$$\frac{dx_1(t)}{dt} + 2x_1(t) - x_2(t) = u_1(t)$$

$$\frac{dx_2(t)}{dt} - x_1(t) + x_2(t) = 2u_2(t)$$

$$x_1(0) = x_2(0) = 0$$

a. Create an *ss* object that describes the system.

b. Find the equivalent *tf* and *zpk* objects.

c. Create again an *ss* object from the *tf* of the *zpk* objects. Is this *ss* object identical to the one created in query (a)?

Solution

The state-space matrices are

$$A = \begin{bmatrix} -2 & 1 \\ 1 & -1 \end{bmatrix}, \quad B = \begin{bmatrix} 1 & 0 \\ 0 & 2 \end{bmatrix}, \quad C = \begin{bmatrix} 0 & 5 \end{bmatrix}, \quad D = \begin{bmatrix} 0 & 0 \end{bmatrix} \qquad \text{(P10.3.1)}$$

We provide two different solutions:

% First solution

```
A = [-2,1;1,-1]; B = [1,0;0,2]; C = [0,5]; D = [0,0];
sys_ss1 = ss(A,B,C,D)
sys_tf1 = tf(sys_ss1)
sys_zpk1 = zpk(sys_ss1)
```

% Second solution

```
A = [-2,1;1,-1]; B = [1,0;0,2]; C = [0,5]; D = [0,0];
[num11,den11] = ss2tf(A,B,C,D,1);
[num12,den12] = ss2tf(A,B,C,D,2);
sys_tf11 = tf(num11,den11);
sys_tf12 = tf(num12,den12);
sys_tf2 = [sys_tf11, sys_tf12]
[z11,p11,k11] = ss2zp(A,B,C,D,1);
[z12,p12,k12] = ss2zp(A,B,C,D,2);
sys_zpk11 = zpk(z11,p11,k11);
sys_zpk12 = zpk(z12,p12,k12);
sys_zpk2 = [sys_zpk11, sys_zpk12]
sys_ss2 = ss(sys_tf1)
sys_tf3 = tf(sys_ss2)
```

The model has two inputs and one output. It is different but equivalent to the first *ss* model as it yields the same transfer function matrix.

10.4 Consider a system with transfer function $G(s) = 10/(s^2 + 3s + 10)$. Compute for $t = 5\,s$ the time response of the system for $u_1(t) = 2t$ and $u_2(t) = 0.5t^2$.

Solution

```
t=[0:0.1:5]'; u1=2*t; u2=0.5*t.^2;
G=tf(10,[1 3 10]) %System
y=lsim(G,u1,t);plot(t,y,t,u1); %time response and input for the first query
y=lsim(G,u2,t);plot(t,y,t,u2); %time response and input for the second query
```

10.5 Consider the spring–mass system depicted in the figure below

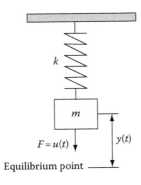

Given that $M = 10$, $k = 5$, $b = 2$,

a. Compute and plot the time response of the system to the input $u(t) = 7\sin(3t)$, if at time $t = 0$, the mass is at position $y(0) = 1\,\text{m}$ and its velocity is $y'(0) = 2\,\text{m/s}$.

b. Give a state-space representation of the system and compute the system's free response.

c. Plot the step response, the impulse response, and compute from your curves various characteristics of the systems (rise time, settling time, etc.)

Solution

a. The differential equation that describes the system is

$$M\frac{d^2y(t)}{dt^2} + b\frac{dy(t)}{dx} + ky(t) = u(t) \qquad (\text{P10.5.1})$$

The solution of the differential equation is easily computed by the command dsolve:

```
y=dsolve('10*D2y+2*Dy+5*y=7*sin(3*t)','y(0)=1,Dy(0)=2')
ezplot(y,[0,50])
ylim([-3 4])
```

The mass position over time is depicted in the figure below

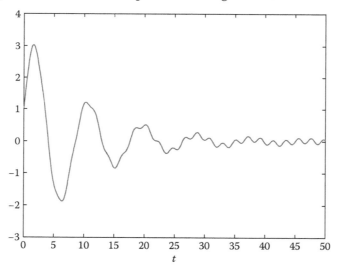

b. The spring–mass state-space representation is

$$\begin{bmatrix} \dot{x}_1 \\ \dot{x}_2 \end{bmatrix} = \begin{bmatrix} 0 & 1 \\ -k/M & -b/M \end{bmatrix} \begin{bmatrix} x_1 \\ x_2 \end{bmatrix} + \begin{bmatrix} 0 \\ 1/M \end{bmatrix} u = \begin{bmatrix} 0 & 1 \\ -0.5 & -0.2 \end{bmatrix} \begin{bmatrix} x_1 \\ x_2 \end{bmatrix} + \begin{bmatrix} 0 \\ 0.1 \end{bmatrix} u \qquad (\text{P10.5.2})$$

$$y = \begin{bmatrix} 1 & 0 \end{bmatrix} \begin{bmatrix} x_1 \\ x_2 \end{bmatrix} \qquad (\text{P10.5.3})$$

The MATLAB implementation is

```
a=[0,1;-0.5,-0.2];
b=[0;0.1];
c=[1,0];
sys=ss(a,b,c,0)
```

The system simulation is provided by the command `lsim`. First we define the initial conditions, the input signal, and simulation time:

```
t=[0:0.1:100];
y0=[1;1.5];
u=zeros(1,1001);
y=lsim(sys,u,t,y0);
plot(t,y)
```

As the input signal is zero the graph that yields from the `lsim` command is in fact the initial condition ($y(0) = 1$ and $y'(0) = 2$) response of the system.

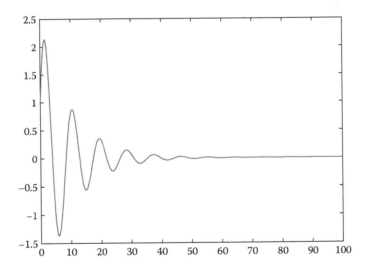

The maximum displacement from the equilibrium point is computed as

```
mx=max(abs(y))
```

Note that the initial condition response could have been computed more easily as

```
initial(sys,y0,100)
```

where *y0* are the initial conditions and 100 is the simulation time.

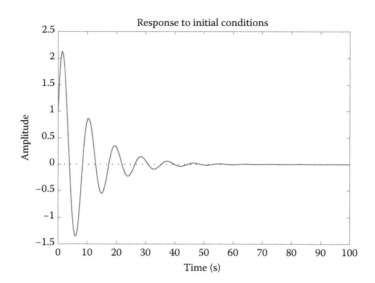

c. The impulse response of the system is computed as

```
impulse(sys,100)
```

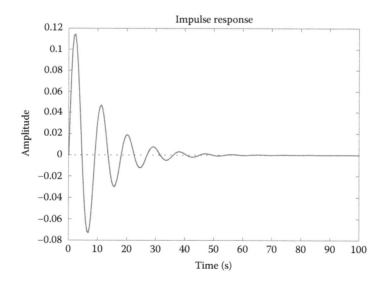

while the step response is computed as

```
step(sys)
```

We right click on the figure and choose the characteristics of the response that we want to appear in the figure.

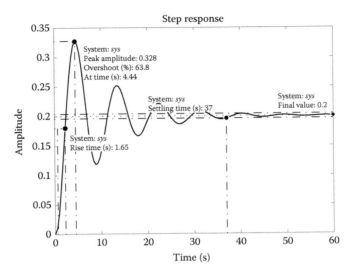

10.6 Consider a system with transfer function $G_p(s) = 1/(s^2 + s + 1)$. Study the effect in the system performance if we connect in series (a) a *P* controller with $K_p = 10$; (b) a *PD* controller with $K_p = 10$, $K_d = 1$; and (c) a *PID* controller with $K_p = 100$, $K_d = 100$, and $K_i = 1$.

Solution

a. First we study the effect of the *P* controller

```
Gp=tf(1,[1 1 1]); % Transfer function of the system
Kp=10;
P=tf(Kp,1); % Transfer function of the controller
Gcl=feedback(series(Gp,P),1); % Transfer function of the closed-loop system
step(Gp,Gcl,':'); % Step response of the system without and with a P % controller
legend('Gp', 'Gcl')
```

From the diagram, we see that the use of a *P* controller reduces the rise time but produces significant oscillations and increases the percent overshoot and the steady-state error.

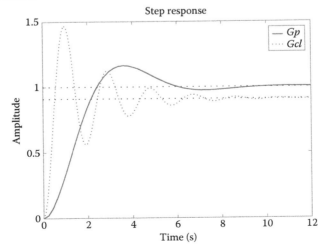

b. The system with a *PD* controller is implemented as follows:

```
Kp=10; Kd=1;
PD=tf([Kd Kp],1); % Transfer function of the PD controller
Gcl=feedback(series(Gp,PD),1); % Transfer function of the closed-loop system
step(Gp,Gcl,':');
legend('Gp', 'Gcl')
```

The *PD* controller reduces the rise time and, compared to the *P* controller, provides a lower percent overshoot and fewer oscillations.

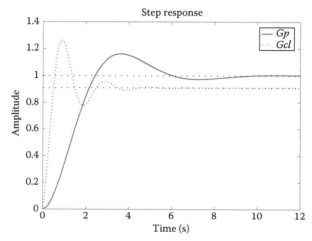

The system with a *PID* controller is implemented as follows:

```
Kp=100; Kd=100; Ki=10;
PID=tf([Kd Kp Ki],[1 0]); % PID controller
Gcl=feedback(series(Gp,PID),1); % Closed-loop system
step(Gp,Gcl,':');
legend('Gp', 'Gcl')
```

In this case the response of the system is almost perfect. The rise time is 0.0223 s, the settling time is 0.00403 s, and the steady-state error is zero.

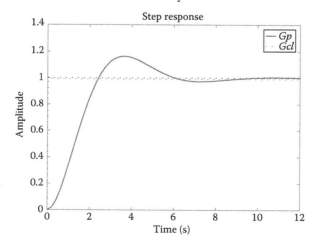

10.7 A velocity control system is depicted in the following block diagram. The transfer
function of the process is $G_p(s) = 10/(24s + 1)$.

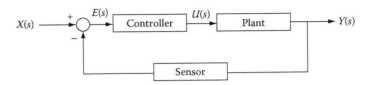

The sensor input is the angular velocity signal and the output is a voltage signal. The
system is controlled by a voltage ranging from 0 to 50 V and the maximum angular
velocity is 1500 rads. Thus, the ratio of the sensor is 1/30. We want to improve the system performance by using a P controller with proportional gain (a) $K_p = 10$, (b) $K_p =$
100, and (c) $K_p = 1000$.

Solution
From the transfer function of the process, we conclude that the gain constant is 10 and the
time constant is 24 s. The step response of the system is computed as follows:

```
Gp=tf([10],[24 1]); % Transfer function of the process
step(Gp)
```

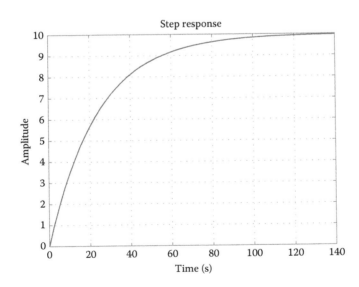

The required gain for the velocity in order to reach 1500 rad is 1500/10 = 150. The step
response in this case is given by

```
step(150*Gp);grid on;
```

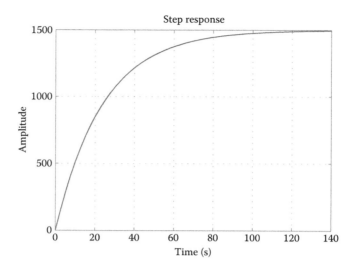

a. We begin our study with the *P* controller with gain $K_p = 10$. After defining the closed-loop system, we compute the steady-state gain and the time constant of the system and we compare the step responses of the two systems (uncompensated and compensated). The response of the closed-loop system is multiplied by 50, which is the value of the input voltage.

```
Gp=tf([10],[24 1]); % Process transfer function
Kp=10;
P=tf(Kp,1) % Controller transfer function
Gcl=feedback(series(Gp,P),1/30); % Closed-loop system transfer function
Ks=dcgain(Gcl) % Steady-state gain
Ts=-1/pole(Gcl) % Time constant
step(150*Gp,50*Gcl);
```

The rise time is reduced (i.e., the system becomes faster) but a significant steady-state error occurs.

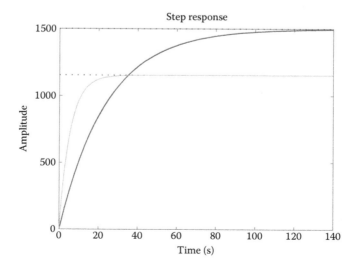

b. In the case of a P controller with $K_p = 100$ we have

```
Kp=100;
P=tf(Kp,1)
Gcl=feedback(series(Gp,P),1/30);
Ks=dcgain(Gcl)
Ts=-1/pole(Gcl)
step(150*Gp,50*Gcl);
```

The gain is now 29.1262 and the time constant is 0.699 s.

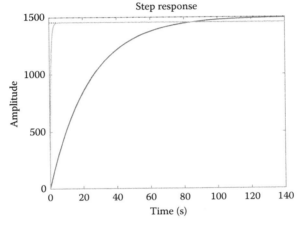

By increasing the proportional gain of the controller we have managed to reduce the rise time. Moreover, the steady-state error is quite small.

c. Finally, in the case of a P controller with $K_p = 1000$ we have

```
Kp=1000;
P=tf(Kp,1)
Gcl=feedback(series(Gp,P),1/30);
Ks=dcgain(Gcl)
Ts=-1/pole(Gcl)
step(150*Gp,50*Gcl)
```

The gain constant is 29.9103 and the time constant is 0.0718 s.

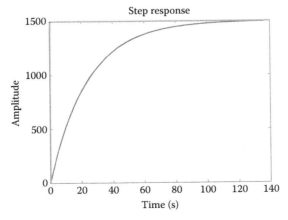

The performance of the system is now very satisfactory. The rise time is very low and the steady-state error tends to zero.

10.8 Consider a system with transfer function $G_p(s) = \dfrac{4}{4s^5 + 92s^4 + 1045s^3 + 5550s^2 + 10625s}$

    a. Find the time required for the step response of the system to reach 90% of its final value.

    b. Repeat the first query after connecting a P controller with $K_p = 1$ to the system.

    c. Compute the parameters of a PD controller so that the time required for the step response of the system to reach 90% of its final value is reduced by 6000 times and the maximum overshoot is less than 10%.

Solution

    a. The step response of the system is computes as follows:

```
Gp=tf([4],[4 92 1045 5550 10625 0]); % Process transfer function
step(Gp);
```

The system does not reach a steady state, thus it is impossible to answer the first query.

    b. By connecting a P controller to the system we observe that the time required for the step response to reach 90% of its final value is 6120 s.

```
Gp=tf([4],[4 92 1045 5550 10625 0]); % Process transfer function
Kp=1;
P=tf(Kp,1)  % Controller transfer function
Gcl=feedback(series(Gp,P),1); % Closed-loop system transfer function
step(Gcl)
```

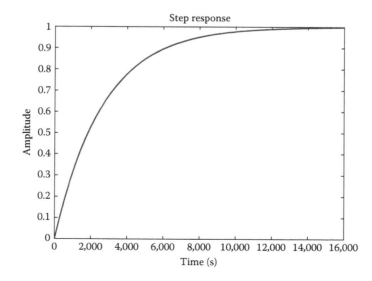

c. The values of K_p and K_d that result in rise time (6120/6000) = 1.02 s and percent overshoot less than 10% are computed experimentally.

```
Kp=100;Kd=100; % First attempt is for Kp = 100 and Kd = 100;
PD=tf([Kd Kp],1); % Controller transfer function
Gcl=feedback(series(Gp,PD),1);
step(Gcl)
```

For K_p = 100 and K_d = 100, the system does not present overshoot but still the rise time is quite large.

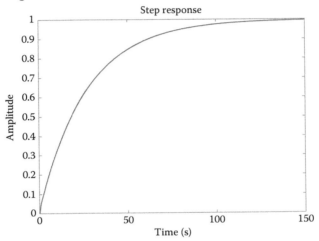

After various experiments we conclude that for K_p = 4700 and K_d = 850 the time needed for the step response to reach 90% of its final value is 0.977 s, while the percent overshoot is 9.13%:

```
Kp=4700;Kd=850;
PD=tf([Kd Kp],1); % Controller transfer function
Gcl=feedback(series(Gp,PD),1);
step(Gcl)
```

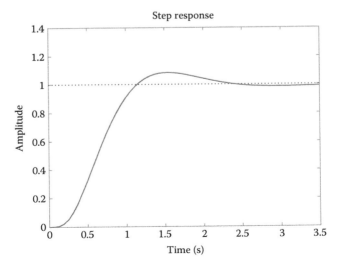

10.9 Consider a process with transfer function $G_p(s) = 1/(s + 1)^3$. Study the effect in the system's step response when connecting in series (a) a *P* controller, (b) a *PI* controller, and (c) a *PID* controller.

Solution

a. In order to examine the system performance when we apply proportional control (*P* controller), we plot the step response of the compensated system for various values of the gain K_p:

```
Gp=tf(1,[1 3 3 1]);
for Kp=[0.1:0.1:1],
Gcl=feedback(Kp*Gp,1);
step(Gcl);hold on;
end
```

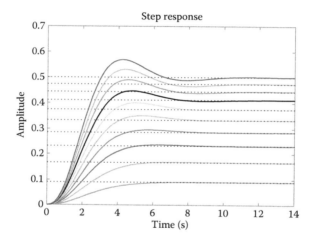

We also plot the root locus of the system's characteristic equation:

```
rlocus(Gp,[0,15]) % Root locus of the system
```

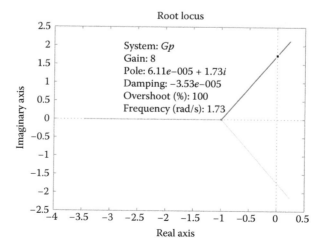

From the root-locus plot, we conclude that the system is stable for $0 < K_p < 8$.

b. We select $K_p = 1$, and connect the *PI* controller. We plot the step response of the system for various values of the parameter T_i:

```
Gp=tf(1,[1 3 3 1]);
Kp=1;
s=tf('s')
for Ti=[0.7:0.1:1.5]
Gpi=Kp*(1 + 1/Ti/s); % Transfer function of the PI controller
Gcl=feedback(Gp*Gpi,1); % Transfer function of the system
step(Gcl), hold on
end
xlim([0 30])
hold off
```

For $T_i < 0.6$ the system turns unstable. When T_i increases, the maximum overshoot is getting smaller but the rise time increases.

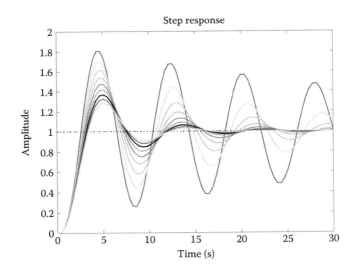

c. We select $K_p = 1$ and $T_i = 1$ and we connect the *PID* controller. We plot the step response of the system for various values of the parameter T_d:

```
Kp=1;
Ti=1;
for Td=0.1:0.2:2
Gpid=Kp*(1 + 1/Ti/s + Td*s); % PID controller transfer function
Gcl=feedback(Gp*Gpid,1);
step(Gcl), hold on
end
xlim([0 30])
```

When T_d increases, the overshoot is reduced but the settling time remains almost the same.

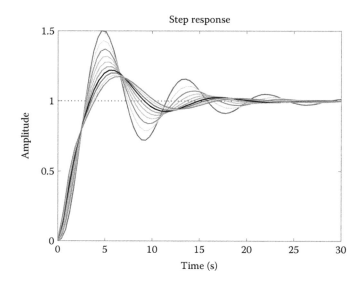

Step response

10.10 Consider the process $G_p(s) = (1/0.5s^2 + 6s + 10.01)$. Compensate the system by using (i) a *PID* controller connected in cascade; (ii) a controller designed with the use of the root-locus method; (iii) a controller designed with the frequency response method such that the step response of the closed-loop system has percent overshoot less 5%, settling time less than 2 s, and steady-state error less than 1%.

Solution

i. The transfer function of a *PID* controller is $G_c(s) = G_{PID}(s) = \dfrac{K_d s^2 + K_p s + K_i}{s}$

The values of the parameters K_d, K_p, K_i are computed experimentally. To start, we apply proportional control with $K_p = 100$ and plot the step response of the closed-loop system:

```
num=[1];
den=[0.5 6 10.01];
Gp=tf(num,den);
Kp=100;
Gcl=feedback(Kp*Gp,1)
step(Gcl);
title('Step response with a P controller')
```

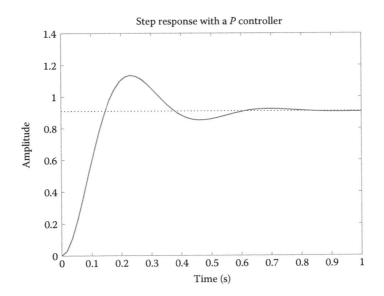

Step response with a *P* controller

From the step-response graph we observe that the steady-state error is less than 1% but the percent overshoot is more than 5%. In order to reduce the overshoot we have to try different values for the parameters K_d and K_i. The parameter $K_p = 100$ remains constant. We start our experiment with $K_d = K_i = 1$:

```
Kp=100;
Ki=1;
Kd=1;
Gc=tf([Kd Kp Ki],[0 1 0]);
Gcl=feedback(Gc*Gp,1)
step(Gcl)
title('Step response with a PID controller, Kp=100,Ki=1,Kd=1')
```

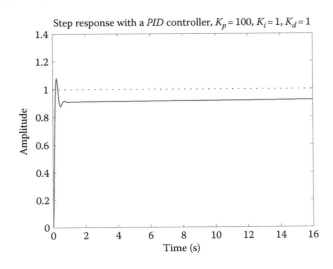

Step response with a *PID* controller, $K_p = 100$, $K_i = 1$, $K_d = 1$

In this case the settling time is very large. We increase K_i to 200 and K_d to 10 and rerun our program:

```
Kp=100;
Ki=200;
Kd=10;
Gc=tf([Kd Kp Ki],[0 1 0]);
Gcl=feedback(Gc*Gp,1)
step(Gcl);axis([0 1 0 1.2])
title('Step response with PID controller, Kp=100,Ki=200,Kd=10')
```

From the graph we observe that all the specifications are fulfilled.

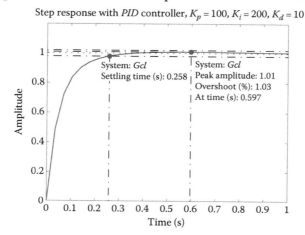

Step response with *PID* controller, $K_p = 100$, $K_i = 200$, $K_d = 10$

ii. For the requirement about the overshoot we derive that the damping ratio is approximately $J = 0.8$. The root locus of the uncompensated system is plotted as follows:

```
num=[1];
den=[0.5 6 10.01];
Gp=tf(num,den);
rlocus(Gp);
title('Root locus without compensation')
grid
```

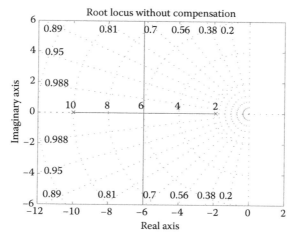

The rise time and the maximum overshoot must be as small as possible. A large damping ratio corresponds to a point on the root locus close to the real axis. A fast system response corresponds to a point on the root locus left of the imaginary axis. The gain that corresponds to a point of the root locus is computed by the command rlocfind.

```
[K,poles]=rlocfind(Gp)
```

We choose the point $-6 + j2.3$ on the root locus. This choice is made due to the relationship $J = \cos \theta \Rightarrow 0.8 = \cos \theta \Rightarrow \theta \simeq 37°$.
   We get the following result:

```
selected_point =
-5.9627 + 2.3939i
K =
10.8550
poles =
-6.0000 + 2.3937i
-6.0000 - 2.3937i
```

The step response of the closed-loop system with gain $K = 10.8550$ is obtained with the following code:

```
[numcl, dencl]=feedback(K*num,den,1,1);
step(numcl, dencl);axis([0 3 0 0.6])
title("Step response with gain")
```

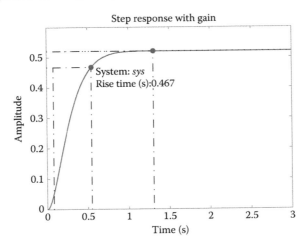

We observe that the system is overdumped and the rise time is satisfactorily small. The only problem is the introduced steady-state error. If we increase the gain, in order to reduce the error we will also increase the overshoot. The solution is to add a lag controller to the system.
   A lag controller with transfer function $G_{lag}(s) = s + 1/s + 0.01$ is selected:

```
num=[1];
den=[0.5 6 10.01];
numlag = [1 1];
denlag = [1 0.01];
nums = conv (num, numlag); % numerator of the loop transfer function
```

```
dens = conv (den, denlag); % denominator of the loop transfer function
rlocus (nums, dens);
title ('Root locus with lag controller')
sgrid
```

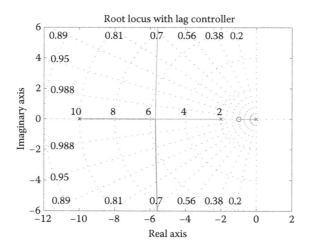

As previously, we type the command

```
[k, poles] = rlocfind (nums, dens)
```

and we choose a point that corresponds to the desired damping ratio. The obtained value for the gain is approximately 20. The step response of the system is computed as follows

```
[numcl, dencl] = feedback (k*nums, dens,1,1);
step (numcl, dencl)
title ('Step response with lag controller')
```

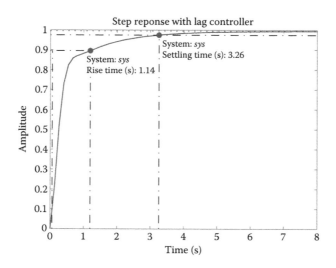

The obtained step response is not satisfactory as the settling time is quite large. We can also see that although the gain was selected according to the damping ratio, the percent overshoot is not close to 5%. This is caused by the use of the lag controller.

We rerun our program choosing a gain close to 53. From the step-response graph we observe that all the specifications are fulfilled.

iii. Compensation using the Bode diagram.

The Bode diagram of the uncompensated system is plotted with the following MATLAB code:

```
num=[1];
den=[0.5 6 10.01];
bode(num,den);
title('Bode diagram of the uncompensated system')
```

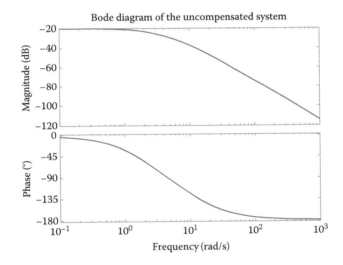

The phase margin is larger than 60° when the frequency is less than 10 rad/s. If we add gain to the system, the bandwidth will become 10 rad/s, which is sufficient for a phase margin of 60°. The gain at 10 rad/s is about −40 dB or 0.01.

This is easily checked from the command

```
[mag, phase, w] = bode(num,den,10)
```

which returns

```
mag =
0.0139
phase =
-123.6835
w =
10
```

In order to increase the gain to 1, we multiply the numerator by $1/0.0139 \simeq 72$. The new Bode diagram is shown in the following figure.

```
bode(num*72,den);grid on
title('Bode diagram with additional gain but without controller')
```

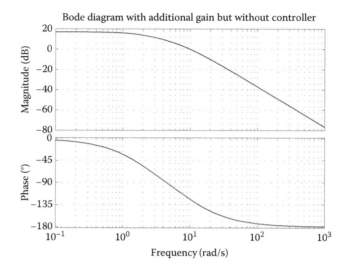

Bode diagram with additional gain but without controller

The phase margin is increased significantly. The step response of the closed-loop system is

```
num=[72];
den=[0.5 6 10.01];
[numcl,dencl]=feedback(num,den,1,1);
step(numcl,dencl);
title('Step response with additional gain but without controller')
```

The settling time is small but the overshoot and the error are quite large. The overshoot can be reduced by reducing the gain, but in this case the error can become very large.

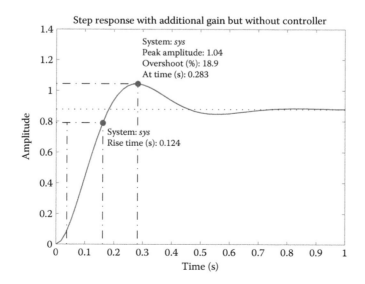

Step response with additional gain but without controller

In order to reduce the steady-state error by $1/0.01 = 100$ times, we choose to connect in cascade a lag controller with transfer function $G_{lag}(s) = s + 1/s + 0.1$. We also reduce the gain to 50. The Bode diagram of the open-loop compensated system is shown below.

```
num=[50];
den=[0.5 6 10.01];
numlag=[1 1];
denlag=[1 0.1];
nums=conv(num,numlag);
dens=conv(den,denlag);
bode(nums,dens)
title('Bode diagram with a lag controller'); grid
```

The phase margin is satisfactory. The steady-state error must be 1/40 or 1%.

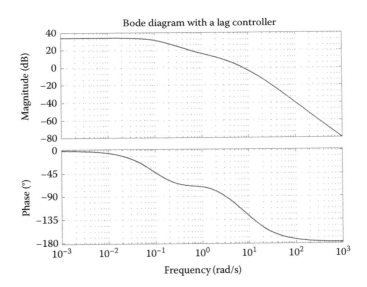

The step response of the closed-loop compensated system is obtained as follows:

```
num=[50];
den=[0.5 6 10.01];
numlag=[1 1];
denlag=[1 0.1];
nums=conv(num,numlag);
dens=conv(den,denlag);
[numcl,dencl]=feedback(nums,dens,1,1);
step(numcl,dencl);axis([0 10 0 1.2])
title('Step response of the closed loop compensated system')
```

We observe that the steady-state error is less than 1%, the percent overshoot is 5%, and the settling time is approximately 2 s.

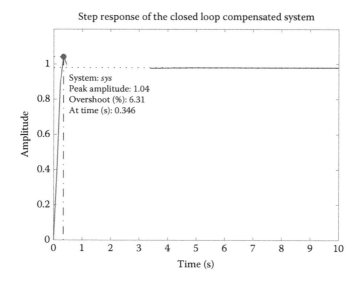

Step response of the closed loop compensated system

10.11 The block diagram of a control system is depicted in the figure below. Find the transfer function of the system.

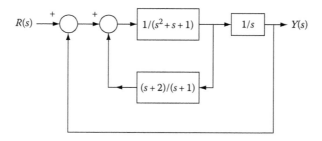

Solution

The MATLAB code is
% Transfer function of the inner feedback system

```
num1= [1];
den1= [1 1 1];
num2= [1 2];
den2= [1 1];
[num, den] =feedback (num1, den1, num2, den2)
```

% Transfer function of the total system

```
num3=[1];
den3=[1 0];
[num4,den4]=series(num,den,num3,den3);
[numcl,dencl]=feedback(num4,den4,1,1);
printsys(numcl,dencl)
num/den =
            s + 1
    ---------------------------
    s^4 + 2 s^3 + 3 s^2 + 4 s + 1
```

10.12 Consider a system with transfer function $G_p(s) = 1/(s(s + 1)(0.2s + 1))$. Design an observer and a controller so that the damping ratio of the system is $J = 0.707$ and the natural frequency is $\omega_n = 3$ rad/s.

Solution

The system must be expressed in state space form. In a state-space representation we can design a controller and an observer such that the poles are placed in the desired positions.

If we choose to have a second-order system, a very helpful command is the command ord2. The command [A,B,C,D] = ord2(wn,z) returns the state-space representation of the continuous second-order system with natural frequency *wn* and damping ratio *z*.

For a third-order system, we select the two poles that were computed from the ord2 command and choose an additional pole that is 10 times faster from the previous two.

We will use the command acker to compute the feedback gain for the system described in the state space. Note that we could have used equivalently the command place instead of acker.

Next we confirm that the poles are the desired ones by using the command roots.

If the states are not accessible to the controller we have to design an observer that estimates the states from the output. The poles of the observer must be 10 times faster from the poles of the controller. The gain is again computed with the command acker.

Finally, we use the command damp to explore the damping characteristics that correspond to each pole. The damping ratio and the natural frequency of the two first eigenvalues (that correspond to the poles created with ord2) are 0.707 and 3 rad/s, respectively. The other poles are significantly faster.

```
numGp=1;
denGp=conv(conv([1 0], [1 1]),[0.2 1]);
[Ag,Bg,Cg,Dg]=tf2ss(numGp,denGp);%State-space representation
damping=0.707;
wn=3;
[num2,den2]=ord2(wn,damping);% Creation of a second-order system with
%J = 0.707 and ωn = 3 rad/s
dominant=roots(den2); % Computation of the dominant poles
desired_poles=[dominant' 10*real(dominant(1))] % Computation of the %desired poles
k=acker(Ag,Bg,desired _ poles) % Pole placement gain selection
Asf=Ag-Bg*k; Bsf=Bg;Csf=Cg;Dsf=0;
[numsf,densf]=ss2tf(Asf,Bsf,Csf,Dsf); % Closed-loop transfer function
Printsys(numsf,densf)
```

```
roots(densf) % Confirmation of the poles
observed_poles=10*desired_poles; % Pole of the observator
L=acker(Ag',Cg',observed_poles)
Areg=[(Ag-Bg*k) Bg*k; zeros(size(Ag)) (Ag-L'*Cg)];
Breg=[Bg; zeros(size(Bg))];
Creg=[Cg zeros(size(Cg))];
Dreg=0;
[numreg,denreg]=ss2tf(Areg,Breg,Creg,Dreg);
printsys(numreg,denreg) % Closed-loop transfer function
damp(denreg);
```

10.13 Plot the Nyquist diagram of the system with transfer function $G(s) = 40/(s + 0.01)$ $(s + 4)(s + 1 + 6j)(s + 1 - 6j)$. Examine the stability of the system.

Solution

A Nyquist diagram is plotted in MATLAB as follows:

```
[num,den]=zp2tf([],[-0.01 -4 -1 + 6*i -1-6*i],40);
nyquist(num,den);
```

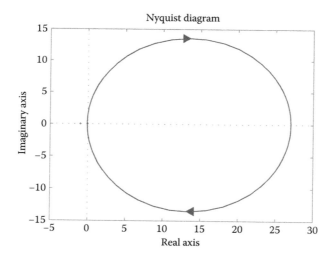

At first sight it appears that the system is stable as the whole Nyquist diagram seems to be at the left half of the complex plane. However, if we zoom close to the axis origin, we observe that the system becomes unstable when the gain is 120. The limits of the two axes change as follows:

```
axis([-0.5 0.5 -0.5 0.5])
```

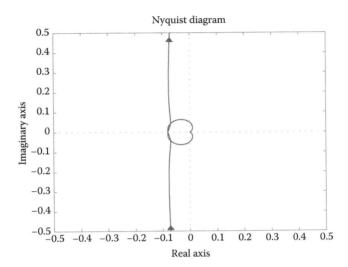

The conclusion is that when the poles of the system are close to the imaginary axes the command nyquist is not appropriate for the study of the system.

10.14 Consider a system with transfer function $G(s) = 1/(s(s + 1)(0.2s + 1))$. Plot the root locus and design a controller such that the damping ratio is as large as possible and the natural frequency is 0.707.

Solution
```
num1=1;
den1=conv(conv([1 0], [1 1]), [0.2 1]);
rlocus(num1,den1);
axis([-4 0.1 -0.1 2.5]);
damping=0.707;
wn=1:1:4;
sgrid(damping,wn)
[k,poles]=rlocfind(num1,den1)
[numcl,dencl]=feedback(k*num1,den1,1,1);
roots(dencl)
hold on
num2=[1 1];
den2=[0.1 1];
num=conv(num1,num2);
den=conv(den1,den2);
rlocus(num,den);axis([-4 0 0 2.5]); sgrid(damping,wn)
hold off
```

Initially we plot the root locus of the system's characteristic equation. We only care for the second quadrant of the complex plane as the system is stable (thus we do not need the first quadrant) and due to symmetry about the real axis we do not need the third and fourth quadrants. The pole at −5 will move to infinity. The command sgrid(J,wn) plots constant damping and frequency lines for the damping ratios in the vector *J* and the natural frequencies in the vector *wn*. Here, $J = 0.707 = \cos \theta \Rightarrow \theta = 45$ and $\omega_n = 1, 2, 3, 4 \, \text{rad/s}$. We conclude that to ensure that $J = 0.707$, the natural frequency of a *P* controller must be close to 0.5 rad/s and the gain is $k = 0.4$. The gain is computed by typing the command `rlocfind` and selecting the point where the locus intersects with the straight line with slope 45°. We then confirm our results by finding the poles of the closed-loop system. The closed-loop system is constructed with the command `feedback` while its poles are computed by the command `roots`. The damping ratio of the computed poles is approximately $J = 0.707$ and the natural frequency is 0.64.

To improve the natural frequency of the closed-loop system we connect a phase-lead controller, that is, we place a pole to the left of the root locus. We select a controller with transfer function $G_c(s) = K\dfrac{s+1}{0.1s+1}$, for which it holds that $\lim_{s \to 0} G_c(s)G(s) = k$.

The controller effect is depicted in the figure as a line that leaves the real axis from the point –2. The damping ratio is 0.707. The gain is 2.7.

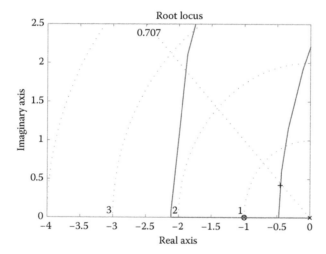

10.15 Compute the response of the system with transfer function $G(s) = 100(s + 2)/(s + 1)$ $(s + 0.2 - 10i)(s + 0.2 -10i)$ to the input signals (i) $r(t) = e^{-t}$ and (ii) $r(t) = e^{-0.2t} \sin(10t)$.

Solution

```
[num,den]=zp2tf([-2],[-1 -0.2 + 10*i -0.2-10*i],100);
t=[0:0.1:20];
u=exp(-t);
subplot(211), lsim(num,den,u,t)
hold on
plot(t,u)
title('Reponse to input exp(-t)')
hold off
unew=exp(-0.2*t).*sin(10*t);
subplot(212), lsim(num,den,unew,t)
hold on
plot(t,unew)
title('Response to input exp(-0.2*t)sin(10t)')
hold off
```

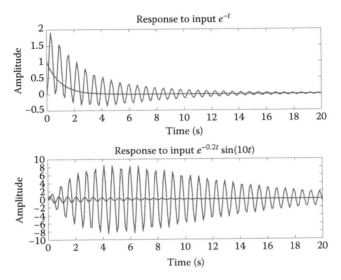

Response to input e^{-t}

Response to input $e^{-0.2t} \sin(10t)$

10.16 A unity feedback control system includes a process with transfer function $GH(s) = 3/(s(s + 1)(0.5s + 1))$. Find a phase-lag controller that provides the system with a phase margin of 45°.

Solution

First we plot the open-loop Bode plot with the gain and phase margins marked with a vertical line by using the command `margin`:

```
num=[3];
den=conv(conv([1 0],[1 1]),[0.5 1]);
margin(num,den)
```

The frequency that satisfies our requirements is $\omega_c = 1.41$ rad/s.

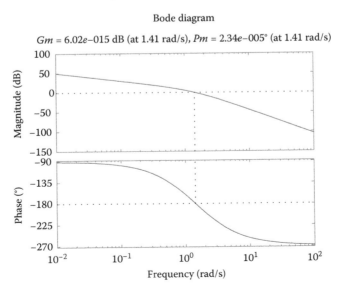

Bode diagram

$Gm = 6.02e{-}015$ dB (at 1.41 rad/s), $Pm = 2.34e{-}005°$ (at 1.41 rad/s)

We place the zero of the phase-lag controller $G_{lag}(s) = (s + z/s + p)$ a decade below the critical frequency, that is, $z = 0.14$. The pole of the compensator is computed as follows:

The phase margin must be $45°$. From the Bode plot we get $-20 \log K = -10.3 \Rightarrow K \simeq 3.16$. Thus, $p = z/K = 0.044$

The system is

$$G_{lag}(s)GH(s) = \frac{3(s+z)}{(s+p)s(s+1)(0.5s+1)} \Rightarrow$$

$$G_{lag}(s)GH(s) = \frac{3(s+0.14)}{(s+0.044)s(s+1)(0.5s+1)}$$

The Bode diagram with system compensation is shown in the following figure.

```
num=[3];
den=conv(conv([1 0],[1 1]),[0.5 1]);
nums=[3 3*0.14];
dens=conv(den,[1 0.044]);
margin(nums,dens)
```

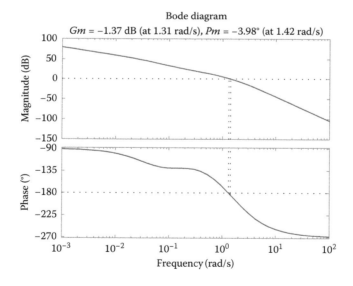

10.17 The impulse response of a second-order system is $h(t) = \dfrac{h(0)}{\sqrt{1-J^2}}e^{-J\omega_n t}$ $\sin\left(\omega_n\sqrt{1-J^2}t+\theta\right), \theta = \cos^{-1} J$. Plot the impulse response for $\omega_n = \sqrt{2}$ rad/s and $J_1 = 3/2\sqrt{2}, J_2 = 1/2\sqrt{2}$, and $h(0) = 0.15$.

Solution

```
h0=0.15;wn=sqrt(2);
zeta1=3/(2*sqrt(2));zeta2=1/(2*sqrt(2));
t=0:.1:10;
th1=acos(zeta1); th2=acos(zeta2)
c1=(h0/sqrt(1-zeta1^2)); c2=(h0/sqrt(1-zeta2^2));
h1=c1*exp(-zeta1*wn*t).*sin(wn*sqrt(1-zeta1^2)*t + th1);
h2=c2*exp(-zeta2*wn*t).*sin(wn*sqrt(1-zeta2^2)*t + th2);
env1=c2*exp(-zeta2*wn*t); % upper envelope
env2=-env1; % lower envelope
plot(t,h1,t,h2,'--',t,env1,':',t,env2,':'),grid
legend('Overdamped', 'Underdamped')
```

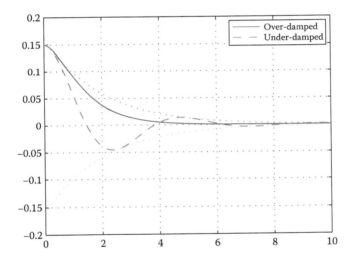

10.18 Consider a system with closed-loop transfer function $G(s) = 1/s^2 + 2Js + 1$. Plot the step responses of the system for $J = 0, 0.2, 0.4, 0.6, 0.8$, and 1.

Solution

```
t=0:0.1:10;
hold on
for J=0:0.2:1
G=tf(1,[1 2*J 1]);
step(G,t)
end
hold off
title('Step response');
gtext('J=0'); gtext('J=0.2'); gtext('J=0.4'); gtext('J=0.6');
  gtext('J=0.8'); gtext('J= 1');
```

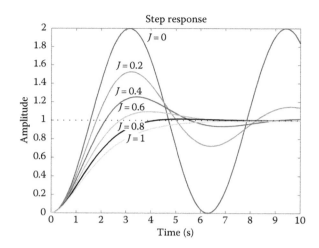

10.19 Plot in one figure for $K = 1$, 10, and 20 the Bode diagrams of the closed-loop transfer
function $G_{cl}(s) = \dfrac{K}{s(s+1)(s+5)+K}$.

Solution

```
K=1; G1= tf(K,conv(conv([1 0],[1,1]),[1 5]) + K)
K=10; G2= tf(K,conv(conv([1 0],[1,1]),[1 5]) + K)
K=20; G3= tf(K,conv(conv([1 0],[1,1]),[1 5]) + K)
bode(G1,G2,':',G3,'.'); legend('K=1','K=10','K=20')
```

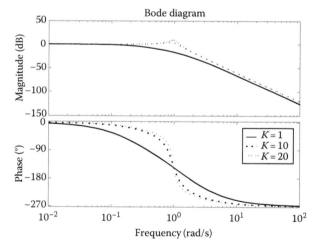

10.20 Consider a unity feedback system with open-loop transfer function $G(s) = (2s + 3)/(s^2 + 3s + 6)$. Compute and plot the error $e(t) = y(t) - u(t)$, where $u(t)$ is the unit-step function.

Solution

```
numg=[2 3];
deng=[1 3 6];
sysg=tf(numg,deng); % Open-loop transfer function
sys=feedback(sysg,1); % Closed-loop transfer function
```

```
[y,t]=step(sys); % Step response
ref = ones(1,length(t));
err = ref - y';% Error computation
plot(t,err,t,y,':' ); legend('error','output')
```

At the steady state, the value of the step response is approximately 0.33, hence the error is close to 0.67.

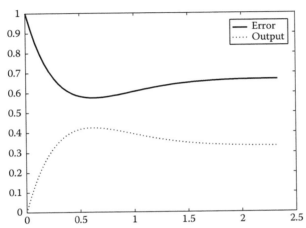

10.21 Consider a process with transfer function $G_p(s) = 4500K/s(s + 361.2)$.
Design a *PD* controller such that the steady-state error for a unit ramp input signal is less than 0.000443, the settling time is less than 0.005 s, and the rise time is less than 0.005 s.
What would be the differences if you have used a *PI* controller?

Solution

The open-loop system transfer function with a *PD* controller is

$$G_{ol}(s) = GcGp(s) = (k_p + k_d s)\left(\frac{4500K}{s(s + 361.2)}\right)$$

$$= (K \cdot k_p + K \cdot k_d \cdot s)\left(\frac{4500}{s^2 + 361.2 \cdot s}\right) \Rightarrow \qquad \text{(P10.21.1)}$$

$$G_{ol}(s) = GcGp(s) = (k'_p + k'_d \cdot s)\left(\frac{4500}{s^2 + 361.2 \cdot s}\right)$$

The closed-loop transfer function is

$$G_{cl}(s) = \frac{G_{ol}(s)}{1 + G_{ol}(s)} = \frac{(k'_p + k'_d \cdot s) \cdot 4500}{s^2 + (361.2 + k'_d \cdot 4500) \cdot s + k'_p \cdot 4500} \qquad \text{(P10.21.2)}$$

The steady-state error is computed by

$$e_v = \frac{1}{K_v} \qquad \text{(P10.21.3)}$$

where

$$K_v = \lim_{s \to 0} s \cdot G_{ol}(s) = \lim_{s \to 0} s \cdot (K \cdot k_p + K \cdot k_d \cdot s)\left(\frac{4500}{s^2 + 361.2 \cdot s}\right) = \frac{4500 \cdot K \cdot k_p}{361.2}$$

From the requirement: error <0.000443, we get

$$\frac{361.2}{4500 \cdot K \cdot k_p} \leq 0.000443 \Rightarrow K \cdot k_p \geq \frac{361.2}{4500 \cdot 0.000443}$$

$$\Rightarrow K \cdot k_p \geq 181.189 \qquad\qquad\qquad\qquad\qquad\qquad \text{(P10.21.4)}$$

The MATLAB code that computes the steady-state response of the system is

```
Kd=1;
K=1;
Kp=250;
t=[0:0.00001:0.01];
Gp=tf([4500],[1 361.2 0]); % Process transfer function
Gc=tf([Kd*K Kp*K],[1]); % PD controller transfer function
Go=series(Gc,Gp); % Open-loop transfer function
G=feedback(Go,1); % Closed-loop transfer function
new_sys=series(tf([1],[1 0]),G); % System that gives a ramp response to a unit-step input %
    signal
step(new _ sys,t); % Ramp response
hold on;
step(tf([1],[1 0])); % Input signal
hold off;
```

From the figure we see that the error specification is fulfilled.

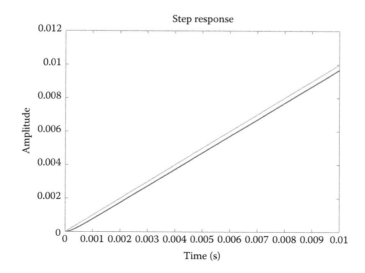

We want the settling time to be less than 0.005 s. For a second-order system with characteristic polynomial $s^2 + 2J\omega_n s + \omega_n^2 = 0$, the settling time is computed approximately as $T_s = \dfrac{4}{J\omega_n}$. We ignore the effect of the zeros and we compute the settling time as

$$T_s = \frac{8}{361.2 + k_d' \cdot 4500} \tag{P10.21.5}$$

Thus,

$$\frac{8}{361.2 + k_d' \cdot 4500} \leq 0.005 \Rightarrow k_d' \geq 0.27529 \tag{P10.21.6}$$

For confirmation we plot the step response of the closed-loop system for $k_p' = 190$ and $k_d' = 0.4$:

```
Kd=0.4;
K=1;
Kp=190;
t=[0:0.00001:0.01];
Gp=tf([4500],[1 361.2 0]);
Gc=tf([Kd*K Kp*K],[1]);
Go=series(Gc,Gp);
G=feedback(Go,1);
step(G,t)
hold on;
```

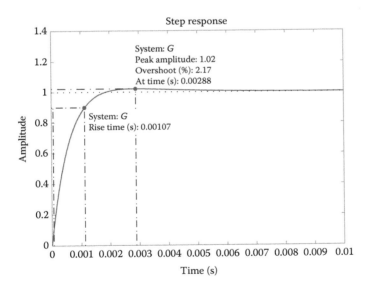

By repeating the procedure for other values of k_d' we observe that for $k_d = 0.2753$, the settling time is 0.00501 s. This satisfies the specification. For values greater than 0.2753 the settling time is further reduced.

Moreover, recall that in systems with overshoot, the settling time is always larger than the rise time. Hence, we have also fulfilled the rise-time specification. But the system does not have overshoot for all values of k_p' and k_d'. A system has overshoot if the roots of its characteristic equation are complex. Thus, its determinant must be negative:

$$(361.2 + k_d' 4500)^2 - 4k_p' 45000 < 0 \tag{P10.21.7}$$

For $k_d' > 0$ and for $k_p' = 190$, Equation P10.21.7 is true for $k_d' < 0.330694$. For example, for $k_d' = 0.32$, the step response of the system is

```
Kd=0.32;
K=1;
Kp=190;
Gp=tf([4500],[1 361.2 0]);
Gc=tf([Kd*K Kp*K],[1]);
Go=series(Gc,Gp);
G=feedback(Go,1);
Step(G),axis([0 0.01 0 1.4])
```

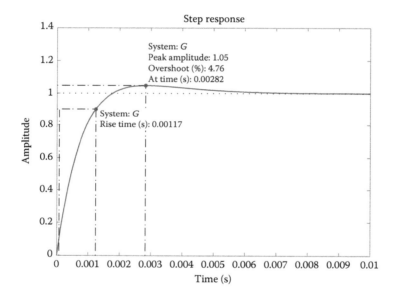

The rise time is 0.00117 s and the peak time is 0.00282 s, that is, both are less than 0.005 s.

For $k_d' = 0.5$ the step response is

```
Kd=0.5;
K=1;
Kp=190;
Gp=tf([4500],[1 361.2 0]);
Gc=tf([Kd*K Kp*K],[1]);
Go=series(Gc,Gp);
G=feedback(Go,1);
Step(G),axis([0 0.01 0 1.4])
```

Again the specifications are met.

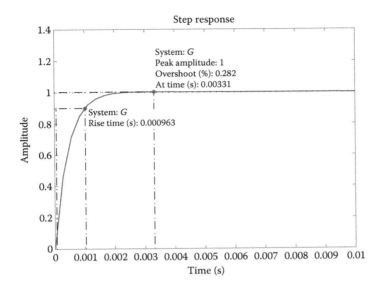

In general, by increasing k_p' we reduce the steady-state error, the rise time, and the settling time. However, the overshoot is increased. Overshoot can be decreased with the proper choice of k_d'.

For instance, for $k_d' = 0.32$ and $k_p = 190$, we observe that the overshoot is increased, while the steady-state error, the rise time, and the settling time are reduced:

```
Kd=0.32;
K=1;
Kp=190;
Gp=tf([4500],[1 361.2 0]);
Gc=tf([Kd*K Kp*K],[1]);
Go=series(Gc,Gp);
G=feedback(Go,1);
step(G);
```

Keeping k_p' equal to 400 we can give various values to k_d' and observe the step responses of the system.

```
Kp=400;
Kd=0.3;
K=1;
Gp=tf([4500],[1 361.2 0]);
Gc=tf([Kd*K Kp*K],[1]);
Go=series(Gc,Gp);
G=feedback(Go,1);
step(G),axis([0 0.01 0 1.4])
```

The increase of k'_p reduces the overshoot while the rise time is not significantly affected. Thus, a good combination is to select a large value for k'_p (e.g., $k'_p = 400$). Then for reducing the overshoot, we select a proper k'_d (e.g., $k'_d = 1$). The response for these two values is shown in the figure below.

```
Kp=400;
Kd=1;
K=1;
Gp=tf([4500],[1 361.2 0]);
Gc=tf([Kd*K Kp*K],[1]);
Go=series(Gc,Gp);
G=feedback(Go,1);
step(G),axis([0 0.01 0 1.4])
```

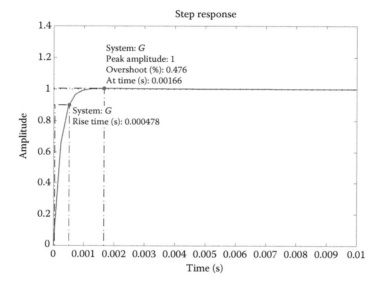

From the graph, we obtain:

   Steady-state error: 0.000207
   Percent overshoot: 4.8%
   Rise time: 0.00115 s
   Settling time: 0.000818 s

These results are satisfactory, thus a suitable *PD* controller must have $k'_p = 400$ and $k'_d = 1$. Therefore,

$$K \cdot k_d = 400 \Rightarrow k_d = \frac{400}{K} \tag{P10.21.8}$$

$$K \cdot k_p = 1 \Rightarrow k_p = \frac{1}{K} \tag{P10.21.9}$$

A *PI* controller compared to a *PD* controller is less powerful in controlling the transient state. On the contrary, a *PI* controller has better performance in the control of the steady-state as it can reduce the steady-state error without the need to use large gains.

Moreover, with the use of a *PI* controller the velocity error of the system is eliminated and the acceleration error becomes finite.

To conclude, these two types of controllers improve the system performance in different ways. Thus, in order to achieve a total good performance, it is sometimes necessary to use a *PID* controller.

10.22 Consider a unity feedback system with loop transfer function $G(s) = \dfrac{K}{s(s+5)(s+12)}e^{-0.5s}$.

Find the overshoot of the closed-loop system when the input signal is the unit-step function. Use a Padé approximation for the exponential term

Solution

```
numg1=1;
deng1=poly ([0 -5 -12]);
G1 = tf(numg1,deng1) % Transfer function without delay
[numg2, deng2] = pade(0.5,5); % Padé approximation
G2 = tf (numg2,deng2)
G =G1*G2
K = input('Type gain, K'); % The user inserts the gain
T = feedback (K*G,1);
step(T)
title (['Step response for K =',num2str(K)])
```

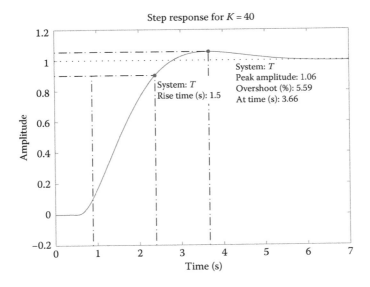

10.3 Simulink®

Simulink is a program developed by Mathworks, suitable for modeling, simulating, and analyzing multi-domain dynamic systems. Its primary interface is a graphical block diagramming tool and a customizable set of block libraries. It offers tight integration with the rest of the MATLAB environment and can either drive MATLAB or be scripted from it.

Problems

10.23 In this simple example, we will introduce how to create a signal $x(t) = A\cos(\omega t + \varphi)$ with $\omega = 5\,\text{rad/s}$, $\varphi = \pi/2$ and $A = 2$.

Solution

From the Simulink libraries we drag and drop to the empty model the following blocks:

Blocks	Library
Ramp	Sources
Constant	Sources
Gain	Math operation
Sum	Math operation
Product	Math operation
Trigonometry function	Math operation
Scope	Sinks
Mux	Signal routing

And we connect them as follows.

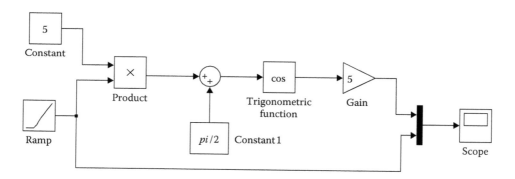

By double clicking a block, a dialogue window appears, where we can change the block's default values. The use of the multiplexer results in the plot of the input t together with the output $x(t)$. Then at the upper-right part of the model window we set the simulation time (here is 10 s) and press the "start simulation" button to begin our simulation. To display the diagram, we must double click on the scope block.

10.24 Plot the step response of the system depicted in the following block diagram.

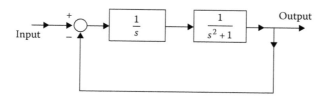

Solution

First we have to insert the following blocks to the model:

a. A *unit-step function* block where we set step time = 1 s, initial value = 0, final value = 1, and sample time = 0

b. A *sum* block, where at the "List of signs" tab we set |+− instead of |++

c. An *integrator* block

d. A *transfer function* block where we set numerator = [0 0 1], denominator = [1 0 1], and absolute tolerance = 0

We connect the blocks as shown in the diagram and start our simulation. If needed, we can adjust simulation parameters such as the sampling time from the menu "Simulation-> Configuration Parameters" or simply by typing "Ctrl + E." The result appears by double clicking the scope block. At the figure, we might need to autoscale our diagram. This can be done by right clicking the figure and choose "autoscale."

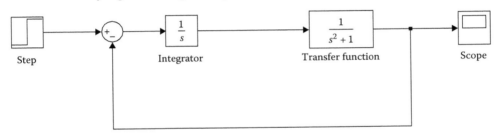

The closed-loop transfer function is

$$G(s) = \frac{1}{(s+0.6823)(s^2 - 0.6823s + 1.466)} = \frac{Y(s)}{X(s)} \tag{P10.24.1}$$

The system output is an increasing oscillation as the system is not stable. The system is not stable as it has a real pole $s = -0.6823$ at the left-half complex plane and two complex poles $s = 0.34 \pm j1.1617$ at the right-half complex plane.

10.25 Consider a spring–mass system, where the spring is connected in parallel with the mass. The mathematical model is given by the equation: $m\ddot{x} + b\dot{x} + kx = f(t)$. Compute and plot the step response of the system.

Solution

From the Simulink libraries, we choose the following blocks:

Block	Library
Step	Sources
Gain	Math operation
Sum	Math operation
Integrator	Continuous
Scope	Sinks
To workspace	Sinks

We solve the equation for the acceleration and get $\ddot{x} = \frac{1}{m}(f(t) - b\dot{x} - kx)$.

We connect the blocks as shown in the figure below, based on the previous equation. Note that we can rotate any block by typing `ctrl + R`. We change the values of each block appropriately. Here we assume $m = 2.0$, $b = 0.7$, $k = 1$.

The block `To Workspace` saves its input to the MATLAB workspace.

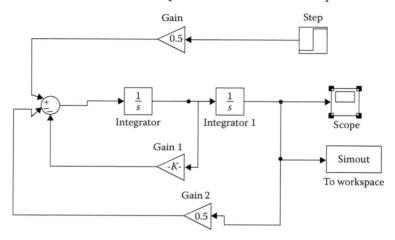

The result of the simulation as it appears in the `scope block` is depicted in the next figure. Alternatively we can use the block `simout`. Before running the simulation we have to change its format from structure to array. This is done by double clicking the block and changing its last attribute. In addition, if we go to "`Configurations Parameters->Data Import/Export`" we see that the simulation time is saved in a variable `tout`. Thus, by typing in the MATLAB command prompt the command `plot(time,simout)`, we obtain the same result.

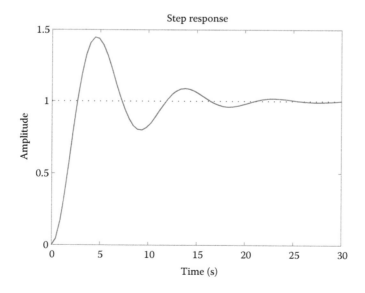

An easier way to simulate the spring–mass system is shown in the following figure. The transfer function of the system is defined and connected directly with the unit-step input signal. The obtained result is the same.

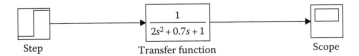

10.4 Program CC (Comprehensive Control)

Program CC is a computer-aided control system analysis and design program quite similar to MATLAB. Before trying to solve problems with Program CC, we introduce some of its basic commands.

10.4.1 Transfer Function

Suppose, for example, that we want to define the transfer function $G(s) = \dfrac{10(s+1)}{s(s^2 + 2s + 100)}$.

At the command prompt, we type g=10*(s + 1)/(s*(s^2 + 2*s + 100)).

To display the transfer function, simply type g or display(g) and the transfer function is depicted:

$$g(s) = \frac{10(s+1)}{a(s^2 + 2s + 100)}$$

We can convert a given transfer function in other forms. Some useful commands are introduced in the following table:

Command	Result	Comments
display(g)	$g(s) = \dfrac{10(s+1)}{s(s^2 + 2s + 100)}$	Transfer Function as ratio of polynomials
single(g)	$g(s) = \dfrac{10s + 10}{s^3 + 2s^2 + 100s}$	Executes the operations in numerator and denominator
pzf(g)	$g(s) = \dfrac{10(s+1)}{s[(s+1)^2 + 9,95^2]}$	Zero pole gain form
shorthand(g)	$g(s) = \dfrac{10(1)}{(0)[0,1,\ 10]}$	Writes the transfer function in a short format
unitary(g)	$g(s) = \dfrac{10(s+1)}{s(s^2 + 2s + 100)}$	The coefficient of the higher order term becomes unity
tcf(g)	$g(s) = \dfrac{0,1(s+1)}{s(0,01s^2 + 0,02s + 1)}$	The constant terms of the polynomials become unity
poles(g)	$0 + 0j$ $-1 + 9.9498744j$ $-1 - 9.9498744j$	Computes the poles of the transfer function
pfe(g)	$\dfrac{0,1}{s} - \dfrac{0,1s + 9,8}{[(s+1)^2 + 9,95^2]}$	Converts the transfer function in partial fraction form

The inverse Laplace transform of a function $g(s)$ is computed by the command `ilt(g)`. For instance, by typing `ilt(g)`, we get

```
g(t) = 0,1 + sin(9,95t-0,1002)*exp(-t) for t >= 0.
```

10.4.2 System Interconnections

Consider two transfer functions $g_1(s)$ and $g_2(s)$:

To connect them in series simply type `g=g1*g2`

To connect them in parallel simply type `g=g1+g2`

If we want to connect them with negative feedback the command is `g=g1/(1+g1*g2)` while if we want positive feedback the command is `g=g1/(1-g1*g2)`. Alternatively, we can use the *feedback operator* (|), that is, by typing `g1|g2` the result is `g1/(1+g2*g1)`. For unity feedback, simply type `g1|1`.

10.4.3 Time Response

The commands `time(g)` and `sim(g)` plots the unit-step response of system g. The command `(ty,y)=sim(g,ut,u)` returns the response to a time-series input.

To see other possible syntaxes of these commands, simply type `help` and the command name, for example, `help sim`.

Example

Compute the step response of a system with transfer function $G(s) = s - 5/s^2 + 3s + 2$ and the ramp response of the same system. The program for the step response is

```
g=(s-5)/(s^2 + 3*s + 2)
time(g)
title('Step response')
```

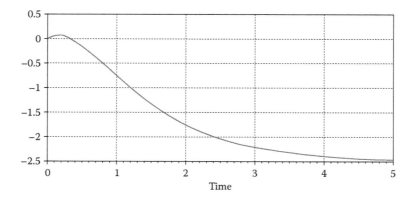

To plot the ramp response the program is

```
t=0:.2:10
x=t;
y=sim(g,x,t)
plot(t,y)
title('Ramp response')
```

10.4.4 Bode Diagram: Gain and Phase Margins

A Bode diagram is plotted by typing the command bode(g). The command bode(w,g) plots the bode diagram over the frequencies stored in vector w. The command margin(g) returns text containing phase, delay, gain, and mp margins. The phase margin is 180° plus the phase where the magnitude is unity. The delay margin is the delay with phase lag equal to phase margin. The gain margin equals to 1/gain where the phase is −180°. Last, the mp margin is the local maximum of $g/(1 + g)$.

Example

Plot the Bode diagram of the system with transfer function $G(s) = \dfrac{2000}{(s+2)(s+7)(s+16)}$

The program is

```
g=2000/((s + 2)*(s + 7)*(s + 16))
bode(g)
```

The solid curve depicts the magnitude, while the dotted curve depicts the phase.

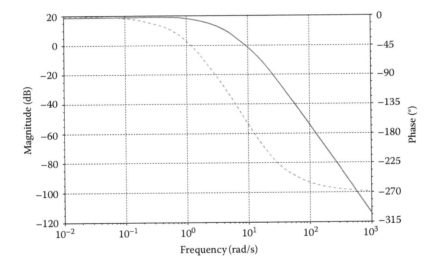

By typing

```
margin(g)
```

we get the following information:

```
At w= 9,18r/s, Phase margin= 19,79 deg, Delay margin= 0,0376s
At w= 9,76r/s, Mp= 3,16 (10,00dB)
At w= 12,6r/s, Gain margin= 1,86 (5,40dB)
```

While by typing

```
point(g,1)
```

we get information of the Bode diagram for the specific frequency we have selected (here $\omega = 1\,\text{rad/s}$):

```
At s = -0 + 1j
g(s) = 6,195 - 4,887j
Magnitude = 7,890 (17,94 dB)
Phase = -38,27 deg
```

The command `y=freq(g,w)` returns the frequency response $g(j\omega)$ for the vector of frequencies w. A good way to construct the frequency vector w is by using the command `freqvec`. In the following code we define with the command `freqvec` a frequency vector from 10^{-2} to $10^2\,\text{rad/s}$ with 100 points:

```
w=freqvec(.01,100,100)
y=freq(g,w)
bode(w,y)
```

10.4.5 Root-Locus Design

The command `rootlocus(g)` plots the root locus of the characteristic equation of a system with transfer function $g(s)$. The command `rootlocus(k,g)` plots the root locus for gains in the vector k.

Example

Plot the root locus of a system with loop transfer function $GH(s) = \dfrac{4}{s(s^2 + 2.828s + 4)}$

The code is

```
g=4/(s*(s^2 + 2.828*s + 4))
rootlocus(g)
```

The root locus is depicted in the figure.

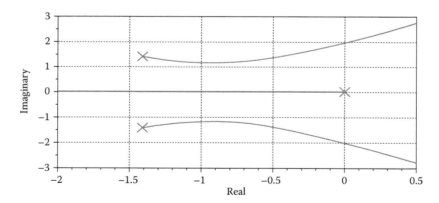

10.4.6 State-Space Representation

A system is represented in the state space as $\begin{aligned}\dot{x} &= Ax + Bu \\ y &= Cx + Du\end{aligned}$. To define a state-space representation in Program CC, we first define the matrices a,b,c, and d and then we use the command `p=pack(a,b,c,d)`. The dimensions of the matrices are returned by the command `what(p)`.

A system in state-space form can be converted to transfer function form by the commands `fadeeva(p)` and `gep(p)`.

Example

Consider a system with state-space representation

$$\begin{bmatrix} \dot{x}_1 \\ \dot{x}_2 \end{bmatrix} = \begin{bmatrix} 0 & 1 \\ 0 & 0 \end{bmatrix} \begin{bmatrix} x_1 \\ x_2 \end{bmatrix} + \begin{bmatrix} 0 \\ 1 \end{bmatrix} u$$

$$y = \begin{bmatrix} 1 & 0 \end{bmatrix} \begin{bmatrix} x_1 \\ x_2 \end{bmatrix} + [0]u$$

To define the system, we type

```
a=(0,1; 0,0)
b=(0; 1)
c=(1,0)
d=0
p=pack(a,b,c,d)
```

The transfer function of the aforementioned system is $G(s) = 1/s^2$. $G(s)$ is computed by

```
g=fadeeva(p)
g
```

A system expressed in transfer function form can be converted in state-space form with the commands **ccf(g)**, **ocf(g)**, and **dcf(g)**. The first command returns a controllable canonical form realization, the second one returns an observable canonical form realization, and the last one returns a diagonal canonical form realization.

Example

Consider a system with transfer function $G(s) = \dfrac{27}{(s+3)(s^2+3s+9)}$.

By using the command `ccf(g)`, we get the following controllable canonical form realization:

$$A = \begin{bmatrix} 0 & 1 & 0 \\ 0 & 0 & 1 \\ -27 & -18 & -6 \end{bmatrix}, \quad b = \begin{bmatrix} 0 \\ 0 \\ 1 \end{bmatrix}, \quad c^T = [27 \quad 0 \quad 0], \quad d = [0]$$

By using the command `ocf(g)`, we get the following observable canonical form realization:

$$A = \begin{bmatrix} -6 & 1 & 0 \\ -18 & 0 & 1 \\ -27 & 0 & 0 \end{bmatrix}, \quad b = \begin{bmatrix} 0 \\ 0 \\ 27 \end{bmatrix}, \quad c^T = \begin{bmatrix} 1 & 0 & 0 \end{bmatrix}, \quad d = [0]$$

while by using the command `dcf(g)`, we get the diagonal canonical form realization:

$$A = \begin{bmatrix} -1.5 & 2.6 & 0 \\ -2.6 & -1.5 & 0 \\ 0 & 0 & -3 \end{bmatrix}, \quad b = \begin{bmatrix} 2 \\ 0 \\ 1 \end{bmatrix}, \quad c^T = \begin{bmatrix} -1.5 & -0.87 & 3 \end{bmatrix}, \quad d = [0]$$

10.4.7 Systems with Time Delay

The command g=pade(tau,order) sets g(s) equal to the Padé approximation of e^{-s*tau}, where tau is the delay.

Example

Consider the system with transfer function $G(s) = \dfrac{27}{(s+3)(s^2+3s+9)}e^{-0.5s}$. To apply a third-order Padé approximation, the code is

```
g=27/((s + 3)*(s^2 + 3*s + 9))
g=g*pade(-0.5,3)
sho(g)
```

The result is

$$g(s) = \frac{27(-s^3 + 24s^2 - 240s + 960)}{(s+3)(s^2 + 3s + 9)(s^3 + 24s^2 + 240s + 960)}$$

Suppose now that for the given process we want to design a unity feedback system with the following specifications: a steady-state error lower than 2%, settling time less than 3 s, and overshoot less than 10%.

The step response of the system is

time(g)

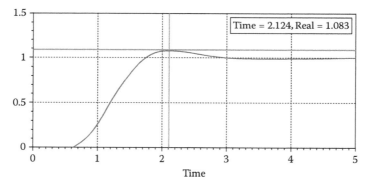

From the step response diagram, we observe that the overshoot is 8.3% and the settling time is 2.7 s.

We will first try integral control. The bandwidth of the closed-loop system will be about 1/3 of the desired rise time, that is, 1 rad/s. We set the controller equal to 1/s and adjust the gain such that the crossover frequency of *gk(s)* is 1 rad/s. The code is

```
gk=g/s
gk(j*1)
ans = -0.9241708 - 0.3801822j
gk=gk/abs(gk(j*1))
k=gk/g
k
```

$$k(s) = \frac{1.001}{s}$$

The closed-loop transfer function is computed by

```
cl=gk|1
```

We plot the Bode diagram of the loop transfer function with the controller *gk(s)* and the step response of the closed-loop system

```
subplot(121); bode(gk); subplot(122); time(cl)
```

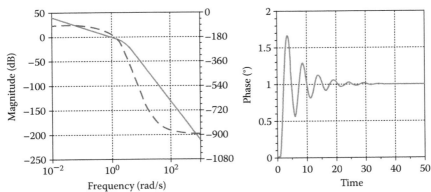

From the step-response graph we see that the settling time is significantly larger than 3 s. The gain and phase margins are

```
margin(gk)
At w= 1r/s, Phase margin= 22.36 deg, Delay margin= 0.39s
At w= 1.2r/s, Mp= 3.89 (11.79dB)
At w= 1.32r/s, Gain margin= 1.32 (2.42dB)
```

Another approach to the problem is to select a desired response for the closed-loop system and design a compensator by which we approximate the desired response while retaining the system stable. The code is

```
desired=15/(s^3 + 6*s^2 + 15*s + 15)
k=imc(g,desired)
```

The command k=imc(g,f) returns the controller k so that the closed-loop system is stable and is optimally close to f. The transfer function of the controller (shorthanded written) is

```
sho(k)
```

$$k(s) = \frac{0.556(3)[0.5, 3](9.289)[0.724, 10.17]}{(0)[0.437, 3.996][0.693, 10.86](11.46)}$$

The compensator is of sixth order but all the specifications are met. The transfer function of the closed-loop system is obtained as follows:

```
gk=g*k
cl=gk|1
sho(cl)
```

$$cl(s) = \frac{-15(-9.289)[-0.724, 10.17]}{(2.322)[0.724, 2.542] (9.289)[0.724, 10.17]}$$

```
subplot(121); bode(gk,'redo'); subplot(122); time(cl)
```

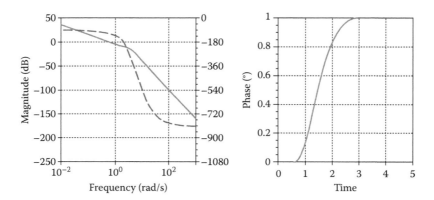

Finally note that the command y=pid(a,b,c) returns the proportional integral derivative controller y(s) = a + b/s + c*s.

10.4.8 State Feedback Pole Placement

The command k = poleplace(p,vec) places the desired poles in the complex plane. The argument p can be a quadruple (system in state-space form) or transfer function, the argument vec is the vector of the desired poles, and k is a full state feedback or dynamic compensator.

Consider, for example, the process with transfer function $G(s) = \dfrac{s}{s^2 + s + 1}$. We want to design a compensator such that the poles of the closed-loop system are $p = [-1\ -2\ -3]$. The code is

```
g=s/(s^2 + s + 1) % process transfer function
k=poleplace(g,[-1 -2 -3]) % compensator design
cl=(g*k)|1% closed loop system
pole_g=poles(g) % poles of the G(s)
pole_cl=poles(cl) % poles of the closed loop system
```

The result is

```
pole_g =
-0,5000000 + 0,8660254j
-0,5000000 - 0,8660254j
pole_cl =
-1,0000000
-2,0000000
-3,0000000
```

10.5 Simapp

Simapp is computer simulation software for modeling systems in the time and frequency domains. The model is built visually through block diagrams. We introduce Simapp through a set of examples.

Problems

10.26 Consider the system depicted in the following Simapp block diagram. The transfer function of the plant is $G_p(s) = 1/((2s + 1)(s + 1))$. The transfer function of the PI controller is $G_c(s) = G_{PI}(s) = \dfrac{s+1}{s}$. The measuring system in the feedback loop is given by $G(s) = e^{-0.1s}$, that is, it has a dead time of 0.1 s. The gain of the amplifier is equal to 10 and the input signal is $x(t) = \sin(2\pi f t)$, where $f = 0.25$ Hz. Compute and plot the time response of the system as well as the Bode and Nyquist diagrams.

Solution

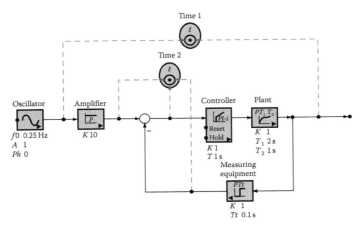

We connect two *time probes* from the library *plots* that are depicted by the dotted lines. The first one (*Time1*) shows the input signal and the system output, and the second one depicts the output from the amplifier, the output from the measuring equipment and the error. We run the simulation and we get the following graphs:

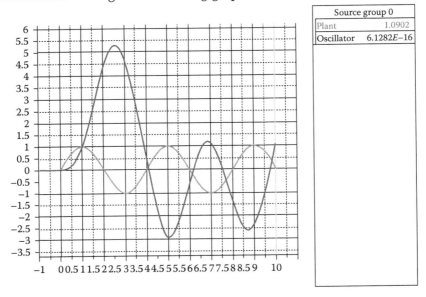

Source group 0	
Plant	1.0902
Oscillator	6.1282E−16

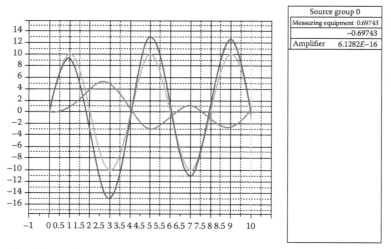

Source group 0	
Measuring equipment	0.69743
	−0.69743
Amplifier	6.1282E−16

In order to plot the Bode and Nyquist diagrams we connect (shown in dotted lines) two *frequency probes* again taken from the library *plots* and run the simulation. The new Simapp block diagram and the graphs stemming from the *frequency probes* are shown below.

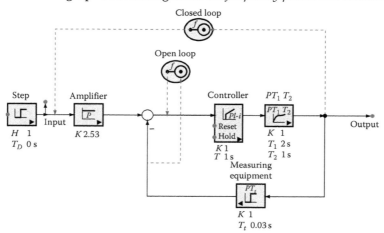

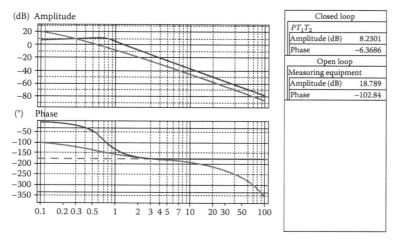

Closed loop	
$PT_1 T_2$	
Amplitude (dB)	8.2301
Phase	−6.3686
Open loop	
Measuring equipment	
Amplitude (dB)	18.789
Phase	−102.84

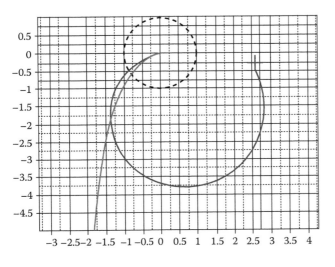

Closed loop		
$PT_1 T_2$		
		$2.5634 - 0.28611j$
Open loop		
Measuring equipment		
		$-1.9327 - 8.4815j$

10.27 Consider the RC bandpass filter of the figure given below

   a. Implement and simulate the system in Simapp.

   b. Compute the time response and the frequency response of the system for a unit-step input signal. The values of the system parameters are $R_1 = 5\,k\Omega$, $C_1 = 100\,\mu F$, $R_2 = 10\,k\Omega$, $C_2 = 200\,\mu F$.

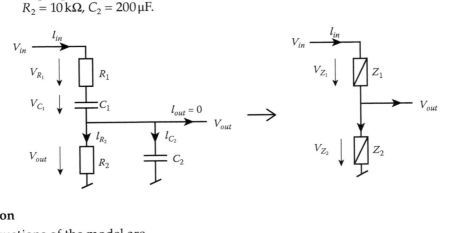

Solution

The equations of the model are

$$V_{in} - V_{out} = I_{in}Z_1 \Rightarrow I_{in} = \frac{V_{in} - V_{out}}{Z_1} \quad \text{and} \quad V_{out} = I_{in}Z_2 \qquad (P10.27.1)$$

where

$$Z_1 = R_1 + \frac{1}{C_1 s} = \frac{sR_1C_1 + 1}{C_1 s} \quad \text{and} \quad Z_2 = \frac{R_2 \cdot \dfrac{1}{sC_2}}{R_2 + \dfrac{1}{sC_2}} = \frac{R_2}{sR_2C_2 + 1} \qquad (P10.27.2)$$

The output voltage V_{out} is

$$V_{out} = \frac{Z_2}{Z_1 + Z_2} = \frac{\dfrac{1}{Z_1} \cdot Z_2}{1 + \dfrac{1}{Z_1} \cdot Z_2}$$

(P10.27.3)

The Simapp block diagram is

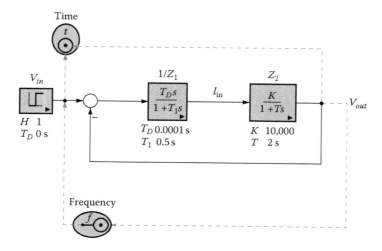

The time and gain constant are computed as

$$T_D = C_1, \quad T_1 = R_1 C_1$$
$$K = R_2, \quad T = R_2 C_2$$

(P10.27.4)

We run the simulation and obtain the graph of the input signal (unit-step function) and the step response of the circuit.

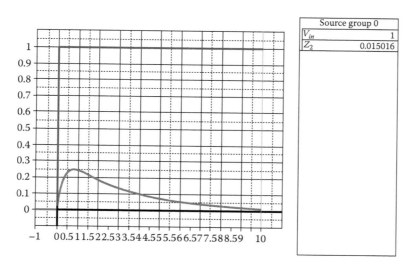

The frequency response graph is the following:

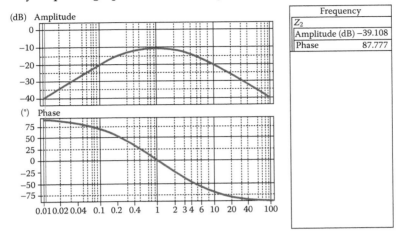

Frequency	
Z_2	
Amplitude (dB)	−39.108
Phase	87.777

10.28 Simulate the depicted mass–spring damper system.

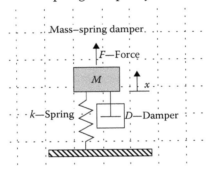

Mass–spring damper.

Solution

The relevant equations are

$$F - M\ddot{x} - D\dot{x} - Kx = 0 \quad \text{or} \quad M\ddot{x} = F - D\dot{x} - Kx \tag{P10.28.1}$$

The system is implemented in Simapp as follows.

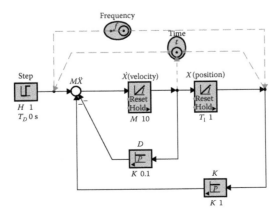

The simulation results depict the unit-step input, the position, and the velocity of the mass.

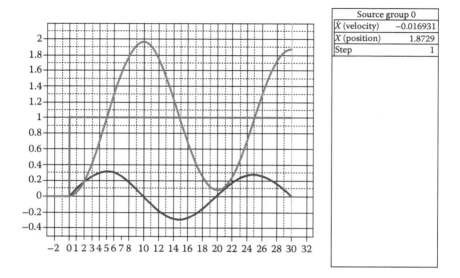

Source group 0	
$\dot{X}$ (velocity)	−0.016931
X (position)	1.8729
Step	1

The frequency response graph depicts the amplitude and the phase of the position control system.

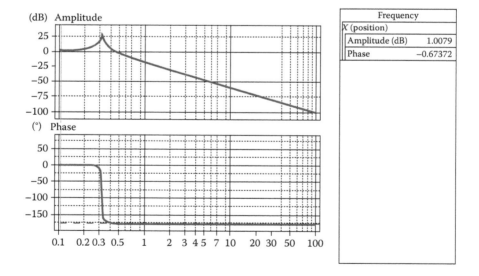

Frequency	
X (position)	
Amplitude (dB)	1.0079
Phase	−0.67372

10.29 Consider a system with loop transfer function $G(s) = \dfrac{K}{1+Ts}e^{-T_d s}$, where $K = 2$, time constant $T = 0.0159\,\text{s}$, and dead time $0.00159\,\text{s}$. Design a controller such that the closed-loop system has percent overshoot less than 8%, the settling time is less than $0.03\,\text{s}$, and the steady-state error is zero.

Solution

The Simapp implementation is shown as follows. We apply a unit-step input and execute the simulation.

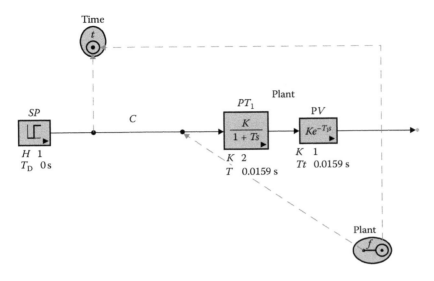

The step response of the system is plotted together with the input signal.

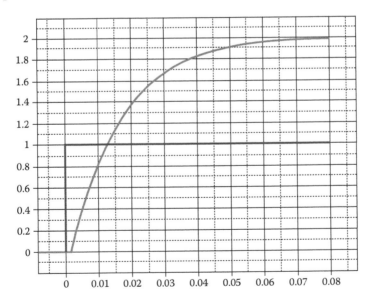

We observe a delay of 0.00159 s in the response of the system. Furthermore, we notice that the output reaches the value 2 (which is the value of the gain) after 0.08 s and that the output reaches a 63% of its final value after 0.0159 s. The steady-state error is 100%.

In order to improve the system performance we will apply the two Ziegler–Nichols methods to design a suitable *PID* controller.

The closed-loop system with a *PID* controller is implemented as follows.

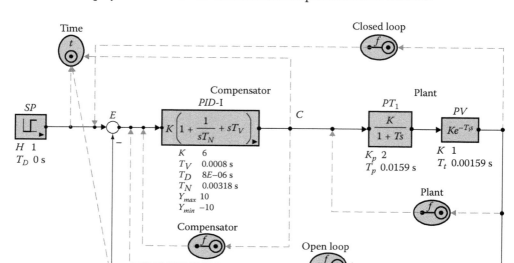

The parameters *Kp*, *Ti*, and *Td* are computed according to the first Ziegler–Nichols method for the following specifications: $T = 0.0159$, $L = 0.00159$, and $K = 2$.

	Kp	Ti	Td
P	$T/L/K = 5$		
PI	$0.9T/L/K = 4.5$	$L/0.3 = 0.0053$	
PID	$1.2T/L/K = 6$	$2L = 0.00318$	$L/2 = 0.0008$

In Simapp there are available various types of *PID* controllers (see library `Controllers`). We select a controller *PID – I*. The integration term of this controller provides an additional pole (T_D) that reduces the high value of the controller's frequency response. The transfer function of the *PID – I* controller is

$$G_{PID-I}(s) = \frac{C(s)}{E(s)} = K_p\left(1 + sT_v + \frac{1}{sT_n}\right) = \frac{K_p(s^2 T_v T_n + sT_n + 1)}{sT_n} \tag{P10.29.1}$$

or

$$G_{PID-I}(s) = \frac{C(s)}{E(s)} = \frac{K(sT_{r1}+1)(sT_{r2}+1)}{s} \tag{P10.29.2}$$

We do not take into account the additional pole assuming that $T_D \ll T_V$. In our case $T_D = 1\% \times T_V$. The designed controller has two identical zeros (($T_{r1} = T_{r2}$) and an integrator. We run the simulation and obtain the following figure, where the input, the output from the controller, and the system output are depicted.

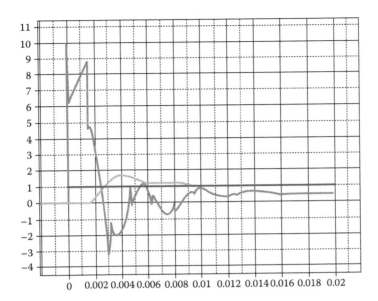

We observe that at time 0.0039 s the system has a maximum overshoot of 1.7127, but after 0.01 s, it converges to the input. The overshoot is quite large hence the designed controller does not provide a satisfactory compensation.

We will try to tune the controller according to the second Ziegler–Nichols method. We insert in the direct loop a proportional controller and increase its gain until the system becomes critically stable. For $K_{cr} = 8.1$ the system's time response is depicted in the next figure. The oscillations' period is $T_{cr} = 0.0061$ s.

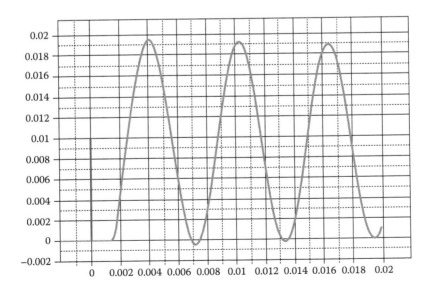

The parameters of the controller are computed for these two values ($T_{cr} = 0.0061$ s και $K_{cr} = 8.1$):

	Kp	Ti	Td
P	$0.5\,K_{cr} = 4.05$		
PI	$0.45\,K_{cr} = 3.645$	$T_{cr}/1.2 = 0.0051$	
PID	$0.6\,K_{cr} = 4.86$	$T_{cr}/2 = 0.00305$	$T_{cr}/30 = 0.00076$

The step response of the system is depicted in the figure below.

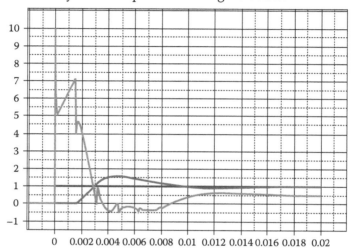

The system performance is now somehow better but still not satisfactory. The maximum overshoot is 1.5848 (at time 0.0046 s) and the settling time is still less than 0.01 s. The system is more stable than before as the gain margin is 5 dB and the phase margin is 36°.

To achieve better compensation, we design our system using other compensation combinations like the one depicted in the following figure. In this case, we do not use differential control.

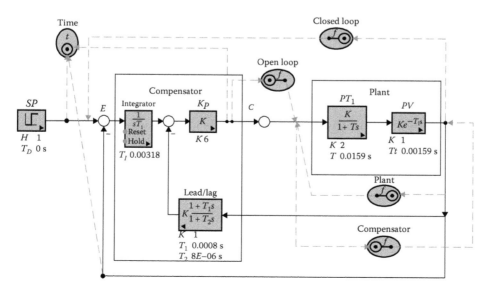

The results in this case are better. The percent overshoot is 9% and the controller is not very noisy. The derived graph is depicted below.

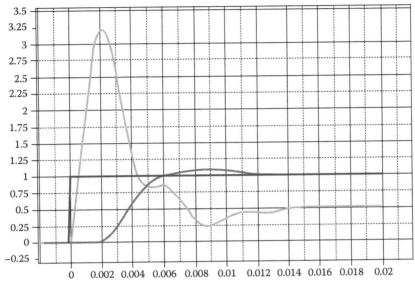

Now, we want to design a controller such that the closed-loop system has a percent overshoot less than 8%, the settling time is less than 0.03 s for ±2% divergence, and the steady-state error is zero.

The transfer function of the closed-loop system is of the form

$$G(s) = \frac{PV(s)}{SP(s)} = \frac{\omega_n^2}{s^2 + 2J\omega_n s + \omega_n^2}, \tag{P10.29.3}$$

where
   ω_n is the natural frequency
   J is the damping ratio

The following relationships hold:

$\omega_d = \omega_n \sqrt{1 - J^2}$ is the frequency with damping.

$t_r \simeq 1.8/\omega_n$ is the rise time.

$M_p = e^{-J\pi/\sqrt{1-J^2}}$ is the maximum overshoot.

$t_s = 4/J\omega_n$ is the settling time for ±2% divergence.

Such a design involves the nonexistence of zeros in the closed loop, the overshoot is subject to the damping ratio and the rise time depends on ω_d.

From the specifications, the damping ratio is $J = 0.62$ and the phase margin is $100J = 60°$.

We add an integrator to the system, to ensure zero steady-state error. Moreover, we need a lead/lag compensator to add phase margin to the system. Lead/lag compensators have a pole and a zero. The zero is placed at the lower frequency, while the pole is placed at the higher frequency.

The transfer function of the lead/lag compensator is of the form $G_{lead/lag}(s) = \dfrac{1+sT}{1+asT}$.

The maximum phase of the compensator is at frequency $\omega_{max} = \dfrac{1}{T\sqrt{a}}$. The relationship is $a = \dfrac{1-\sin\varphi_{max}}{1+\sin\varphi_{max}}$, where φ_{max} is the maximum phase.

We run the simulation solely with integral control for the open-loop system and we derive that for $\omega_n \simeq 230\,\text{rad/s}$ the gain is $-52.81\,\text{dB}$ and the phase is $-186°$. Therefore, the controller must be tuned in such a way that the gain at $\omega = 230\,\text{rad/s}$ is $52.81\,\text{dB}$. For this reason, we set the gain of the lead/lag compensator equal to 92. The Simapp implementation is depicted below.

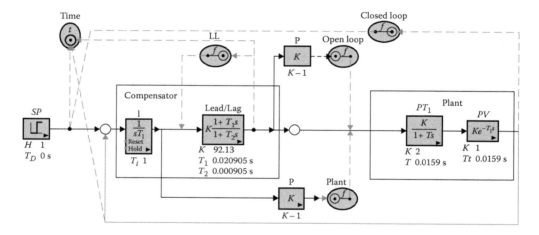

We run the simulation and from the time response figure we observe that the percent overshoot is 5% and the settling time is 0.03 s. The system performance is now satisfactory.

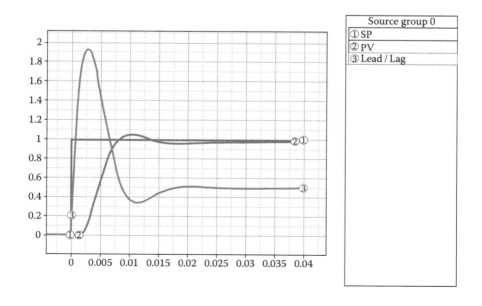

10.6 Scilab

Scilab is quite similar to MATLAB. Suppose, for example, that we wish to solve the system of the following equations:

$$3x + y - z = 0$$
$$2x - 6y + 3z = 2$$
$$4x - 2y + 8z = 1$$

We simply have to type

```
A = [3 1 -1; -2 -6 3; 4 -2 8];
b = [0; 2; 1];
x=A\b
```

We now introduce the basic commands of Scilab that are applicable to control system analysis.

a. Definition of a symbolic variable

```
s=%s
```

b. Numerator and denominator construction

```
num=2*s + 1
den=2*s.^2 + 48*s + 8
```

c. Roots of a polynomial

```
r=roots(den)
```

d. Construction of a rational polynomial

```
tf=num/den
```

e. Polynomial from matrix elements

```
p=poly([1 2 3],'z','coeff')
```

f. Polynomial from its roots

```
pol=poly([-1 -2],'s')
```

g. Transfer function creation

```
s=%s;
num=2*s + 3;
den=2*s^2 + 3*s + 1;
p=syslin('c',num/den)
```

or alternatively

```
s = poly(0, "s");
L = syslin('c', 3e4 * (0.05*s + 1)^2/((s + 1)^3 * (0.01*s + 1)));
```

h. Graph of poles and zeros

```
plzr(p)
```

i. Root-locus plot

```
s = poly(0, "s");
L = syslin('c', 3e4 * (0.05*s + 1)^2/((s + 1)^3 * (0.01*s + 1)));
evans(L, 2.6);
```

// Root-locus plot (second way)

```
n=2 + s;
d=7 + 5*s + 3*s^2;
TF=syslin('c',n,d)
evans(TF2,20)
xgrid
```

// Evaluation of *K* for a specific point of the root locus

```
k=-1/real(horner(TF,[1,%i]*locate(1)))
```

The command `locate` returns the coordinates of a point of the root locus selected with the mouse, while the command `horner` evaluates the polynomial for that point.

j. Bode diagram

```
s = poly(0, "s");
L = syslin('c', 3e4 * (0.05*s + 1)^2/((s + 1)^3 * (0.01*s + 1)));
bode(L);
```

// another example for Bode diagram

```
s=%s;
num=1;
den=s + 1;
tf=syslin('c',num/den)
x=0:0.1:2*%pi;
m=sqrt(1 + x^2);
mag=m^(-1);
bode(tf,0.01,100);
```

k. Gain and phase margins

// Gain margin

```
h=syslin('c',-1 + %s,3 + 2*%s + %s^2)
[g,fr]=g_margin(h)
[g,fr]=g_margin(h-10)
nyquist(h-10)
```

// Phase margin

```
h=syslin('c',-1 + %s,3 + 2*%s + %s^2)
[p,fr]=p_margin(h)
[p,fr]=p_margin(h + 0.7)
nyquist(h + 0.7)
t=(0:0.1:2*%pi)';plot2d(sin(t),cos(t),-3,'000')
```

l. Impulse, step, and ramp responses

```
s=%s;
num=36;
den=36 + 3*s + s^2;ç
TF=syslin('c',num,den)
t=linspace(0,5,500);
imp_res=csim('imp',t,TF); // Impulse response
plot(t,imp_res),xgrid(),xtitle('Impulse response','time','response');
step_res=csim('step',t,TF); // Step response
plot(t,step_res),xgrid(),xtitle('Step response','time','response');
ramp_res=csim(t,t,TF); //Ramp response
plot(t,ramp_res),xgrid(),xtitle('Ramp response','time','response');
```

m. Plot of ramp input signal and of ramp response in the same figure

```
s=poly(0,'s')
H=syslin('c',1/(1 + s))
t=0:0.1:5;
deff('u=ramp(t)','u=2*t');
y=csim(ramp,t,H); plot(t,y,t,2*t)deff('u=ramp(t)','u=2*t')
```

n. State-space representation

```
A = [-5 -1;6 0];
B = [-1; 1];
C = [-1 0];
D =0;
Sss = syslin('c',A,B,C,D)
```

// State-space matrices extraction

```
[A,B,C,D]=abcd(Sss)
```

// State-space equation

```
ssprint(Sss)
```

// Pole zero plot

```
plzr(Sss);sgrid
```

// controllable canonical form

```
[Ac,Bc,U,ind]=canon(A,B)
```

// observable canonical form

```
[Ac,Bc,U,ind]=contr(A,B)
```

// controllability matrix

```
Cc=cont_mat(A,B)
```

o. Conversion from state-space form to transfer function form and vice versa

```
SS1=tf2ss(TF) // Transfer Function to state space
TF1=ss2tf(SS1) // State space to Transfer Function
```

p. Determinant of matrix *A* and inverse of matrix *A*

```
s=poly(0,'s');
a=[0s + 1; s + 2s];
det(a)
inv(a)
```

q. System interconnection with negative feedback

// TF1 is the transfer function of the process and TF2 is the transfer function of the feedback loop.

```
TFcl=TF1/.TF2
```

// Second way

```
TFcl=TF1/(1 + TF1*TF2)
```

// Unity feedback

```
TFcl=TF*(1 + TF)^(-1)
```

r. Transfer Function with dead time

We will make a Padé approximation of the exponential term. A first-order Padé approximation is $e^{-Ds} = \dfrac{2-Ds}{2+Ds}$ while a fifth-order Padé approximation is

$$e^{-Ds} = \frac{30240 - 15120Ds + 3360D^2s^2 - 420D^3s^3 + 30D^4s^4 - D^5s^5}{30240 + 15120Ds + 3360D^2s^2 + 420D^3s^3 + 30D^4s^4 + D^5s^5}$$

We create the function `delay.sci`.

```
function[y]= delay(s,D)
y =(30240-15120*D*s+3360*D^2*s^2-420*D^3*s^3

    +30*D^4*s^4-D^5*s^5)/(30240+15120*D*s

    +3360*D^2*s^2+420*D^3*s^3+30*D^4*s^4+D^5*s^5)
endfunction
```

And then we execute it as follows:

```
s=%s; TF=syslin('c',delay(s,2)/(1+s));
t=0:0.1:10; y=csim('step',t,TF);plot(t,y);
```

s. Pole placement

```
A=rand(3,3);B=rand(3,2);
F=ppol(A,B,[-1,-2,-3]);
spec(A-B*F)
```

XCOS (ex **Scicos**) is quite similar to Simulink. The models are created in an interactive graphical environment using various blocks.

Example

Consider a spring–mass system described by the differential equation $m\ddot{x} + c\dot{x} + kx = f(t)$ or equivalently by $\ddot{x} = \dfrac{1}{m}(f(t) - c\dot{x} - kx)$, where $m = 0.5\,\text{kg}$, $c = 0.35\,\text{N s/m}$, and $k = 0.5\,\text{N/m}$. Simulate the position $x(t)$ for a unit-step input.

Solution

The XCOS implementation is shown in the following block diagram.

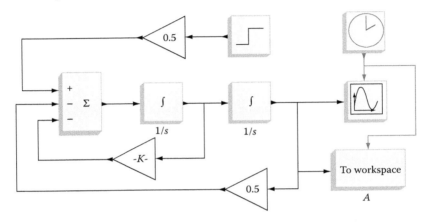

From the menu Simulate -> Setup we set the simulation time to 30 s and execute the simulation from the menu Simulate -> Run. The result is position $x(t)$ versus the time t.

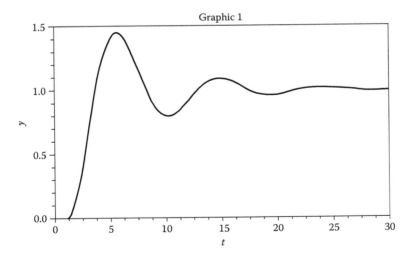

Bibliography

1. A. Palamides, A. Veloni, *Signals and Systems Laboratory with MATLAB*, CRC Press, Boca Raton, FL, 2011.
2. F. Golnaraghi, B. Kuo, *Automatic Control Systems*, John Wiley & Sons, Hoboken, NJ, 2009.
3. N. Nise, *Control Systems Engineering*, 5th Edn., John Wiley & Sons, Hoboken, NJ, 2007.
4. M. Fogiel, *Automatic Control Systems/Robotics Problem Solver*, Research & Education Association, Piscataway, NJ, 1982.
5. R. Dorf, R. Bishop, *Modern Control Systems*, 10th Edn., Prentice Hall, Upper Saddle River, NJ, 2004.
6. R. Bateson, *Introduction to Control System Technology*, 7th Edn., Prentice Hall, Upper Saddle River, NJ, 2001.
7. B. Friedland, *Control System Design: An Introduction to State—Space Methods* (Dover Books on Engineering), Dover Publications, Mineola, NY, 2005.
8. P. Lewis, C. Yang, *Basic Control Systems Engineering*, Prentice Hall, Upper Saddle River, NJ, 1997.
9. L. Fenical, *Control Systems Technology*, Thomson/Delmar Learning, Clifton Park, NY, 2006.
10. T. Mcavinew, R. Mulley, *Control System Documentation: Applying Symbols and Identification*, ISA, Research Triangle Park, NC, 2004.
11. R. Dorf, R. Bishop, *Modern Control Systems*, Addison-Wesley, Menlo Park, CA, 1994.
12. D. Sante, *Automatic Control System Technology*, Prentice Hall, Upper Saddle River, NJ, 1980.
13. W. Wolovich, *Automatic Control Systems: Basic Analysis and Design* (Oxford Series in Electrical and Computer Engineering), Saunders, Philadelphia, PA, 1993.
14. C. Johnson, H. Malki, *Control Systems Technology*, Prentice Hall, Upper Saddle River, NJ, 2001.
15. G. Franklin, J. Powell, A. Emami-Naeini, *Feedback Control of Dynamic Systems*, 6th Edn., Pearson Prentice Hall, Upper Saddle River, NJ, 2009.
16. S. Tripathi, *Modern Control Systems: An Introduction* (Engineering Series), Infinity Science Press LLC, Hingham, MA, 2008.
17. B. Friedland, *Advanced Control System Design*, Prentice Hall, Upper Saddle River, NJ, 1996.
18. W. Levine, *The Control Systems Handbook*, 2nd Edn.: *Control System Advanced Methods*, 2nd Edn. (*Electrical Engineering Handbook*), CRC Press, Boca Raton, FL, 2010.
19. E. Rozenwasser, R. Yusupov, *Sensitivity of Automatic Control Systems* (Control Series), CRC Press, Boca Raton, FL, 1999.
20. R. Stefani, B. Shahian, C. Savant, G. Hostetter, *Design of Feedback Control Systems* (Oxford series in Electrical and Computer Engineering), Oxford University Press, Inc., New York, 2001.
21. W. Palm, *Control Systems Engineering*, John Wiley & Sons, Inc., New York, 1986.
22. C. Smith, A. Corripio, *Principles and Practices of Automatic Process Control*, Wiley-VCH Verlag GmbH & Co. KGaA, Weinheim, Germany, 2005.
23. R. Jacquot, *Modern Digital Control Systems* (Electrical and Computer Engineering), Marcel Dekker, Inc., Monticello, New York, 1994.
24. V. Zakian, *Control systems Design: A New Framework*, Springer-Verlag, London, U.K., 2005.
25. R. Williams, D. Lawrence, *Linear State—Space Control Systems*, John Wiley & Sons, Inc., Hoboken, NJ, 2007.
26. F. Colunius, W. Kliemann, L. Grüne, *The Dynamics of Control (Systems & Control: Foundations & Applications)*, Birkhäuser, Boston, MA, 2000.
27. A. Langill, *Automatic Control Systems Engineering*, Vol. 1, Prentice Hall, Englewood Cliffs, NJ, 1965.
28. S. Shinners, *Modern Control System Theory and Design*, 2nd Edn., John Wiley & Sons, Inc., Hoboken, NJ, 1998.

29. R. Dukkipati, *Analysis & Design of Control Systems Using MATLAB*, New Age International Publishers, Daryaganj, New Delhi, 2009.

30. D. Frederick, J. Chow, *Feedback Control Problems Using MATLAB and the Control System Toolbox* (Bookware Companion Series), Brooks/Cole Thomson Learning, Belmont, CA, 1999.

31. R. Bishop, *Modern Control Systems Analysis and Design Using MATLAB*, Addison-Wesley Longman Publishing Co. Inc., Boston, MA, 1993.

32. N. Leonard, W. Levine, *Using MATLAB to Analyze and Design Control Systems*, 2nd Edn., Benjamin/Cummings Publishing Company, Inc., Redwood City, CA, 1995.

33. B. Kisacanin, G. Agarwal, *Linear Control Systems: With Solved Problems and MATLAB Examples* (University Series in Mathematics), Kluwer Academic/Plenum Publishers, New York, 2001.

34. A. Cavello, R. Setola, F. Vasca, *Using MATLAB, Simulink and Control System Tool Box: A Practical Approach*, Prentice Hall PTR, Upper Saddle River, NJ, 1996.

35. B. Lurie, P. Enright, *Classical Feedback Control: With MATLAB and Simulink*, 2nd Edn. (Automation and Control Engineering), CRC Press, Boca Raton, FL, 2011.

36. R. Bishop, *Modern Control Systems Analysis and Design Using MATLAB and Simulink*, Addison-Wesley Longman Publishing Co. Inc., Boston, MA, 1997.

37. D. Hanselman, B. Kuo, *MATLAB Tools for Control System Analysis and Design/Book and Disk* (The MATLAB Curriculum), Prentice Hall, Upper Saddle River, NJ, 1995.

38. B. Shahian, M. Hassul, *Control System Design Using MATLAB*, Prentice Hall, Englewood Cliffs, NJ, 1993.

39. K. Singh, G. Agnihotri, *System Design through MATLAB, Control Toolbox and SIMULINK*, Springer-Verlag, London, U.K., 2000.

40. N. Nise, *Control Systems Engineering with MATLAB*, Tutorial Version, Addison-Wesley Publishing Company, Reading, MA, 1995.

41. B. Lurie, *Classical Feedback Control: With MATLAB* (Automation and Control Engineering), Marcel Dekker, New York, 2000.

42. P. Chau, *Process Control: A First Course with Matlab* (Cambridge Series in Chemical Engineering), Cambridge University Press, Cambridge, U.K., 2002.

43. N. Nise, *MATLAB Tutorial Update to Version 6 to Accompany Control Systems Engineering*, John Wiley & Sons, New York, 2002.

44. C. Houpis, J. Dazzo, S. Sheldon, *Linear Control System Analysis and Design with Matlab*, 5th Edn., Marcel Dekker, Inc., New York, 2009.

45. L. Wanhammar, *Analog Filters Using MATLAB*, Springer Science+Business Media, New York, 2009.

Index

Milton Keynes UK
Ingram Content Group UK Ltd.
UKHW052025071024
449327UK00027B/2436

9 780367 382056